APPLICATION OF
ARTIFICIAL INTELLIGENCE
IN PROCESS CONTROL

APPLICATION OF ARTIFICIAL INTELLIGENCE IN PROCESS CONTROL

Lecture notes ERASMUS intensive course

Edited by

L. BOULLART

A. KRIJGSMAN

and

R. A. VINGERHOEDS

PERGAMON PRESS

OXFORD · NEW YORK · SEOUL · TOKYO

UK   Pergamon Press Ltd, Headington Hill Hall,
            Oxford OX3 0BW, England
USA   Pergamon Press Inc., 660 White Plains Road,
            Tarrytown, New York 10591-5153, USA
KOREA   Pergamon Press Korea, KPO Box 315, Seoul 110-603, Korea
JAPAN   Pergamon Press Japan, Tsunashima Building Annex,
            3-20-12 Yushima, Bunkyo-ku, Tokyo 113, Japan

Copyright © 1992 Pergamon Press Ltd

First edition 1992

Library of Congress Cataloging in Publication Data
A catalogue record for this book is available from the
Library of Congress.

British Library Cataloguing in Publication Data
A catalogue record for this book is available from the
British Library.

ISBN 0 08 042016 8 Hardcover
     0 08 042017 6 Flexicover

Transferred to digital printing 2005

Contents.

Preface.

This book is the result of a united effort of six European universities to create an overall course on the application of artificial intelligence (AI) in process control. Starting from an initiative of the Automatic Control Laboratory of the University of Ghent, Belgium, an intensive course was set up, aiming at university engineering students in their final study year.

The proposal was followed by five other universities in the European Community (University College of Swansea, Delft University of Technology, Queen Mary and Westfield College, Universitat Politècnica de Catalunya and Universidad Politècnica de Valencia), bringing the total to 6 universities in 4 different countries. In October, 1989, the application was submitted to the European Community, to be carried out in the framework of the ERASMUS program. This program stimulates education in Europe in important areas. The idea of a joint course for the application of AI in process control fits perfectly in this program.

AI technology is a relatively young discipline. Many research groups are working in the domain, but especially on the application in process control where special problems are encountered, no single university can offer a complete course on the subject. AI is, however, very important for the European industry and its position in competition with the U.S.A. and Japan. Therefore a combined effort is necessary to stimulate students to step into the field.

The project was accepted and in September 1991 the first edition of the intensive course was held in Ghent. More than 100 students from the participating universities came together for two weeks, to get lectures from the specialists from those universities. The results were excellent and, with some minor changes, new editions of the course are held in September 1992 in Swansea and in September 1993 in Barcelona and Valencia.

The final goal is that the course should be held every year at another university in Europe and the lecture notes should spread out over other European universities. In this way not only can a good educational program be offered to the students, but also joint research initiatives between the participating universities are stimulated.

The texts in this book serve as lecture notes for the course. They give a good overview of the domain, the problems that can be encountered, the possible solutions using AI techniques and how to successfully apply these techniques.

In part I, a general introduction to the field of A.I. is given. Thereby not only logic and expert system techniques are discussed, but an in-depth overview of neural network techniques is presented as well.

Part II focuses on control engineering and shows how control systems have evolved over the years. From there an insight is given in where and how artificial intelligence techniques can be used to improve control systems.

In part III, real-time issues are discussed. The importance of a good and solid communication between the components of control systems is discussed and it is shown which criteria are important to take into account. Furthermore, real-time expert systems are introduced and their working is explained.

Part IV illustrates the use of artificial intelligence techniques for designing control systems. It shows how the software environments for design tasks have evolved over the years and where expert system techniques can give new impulses.

In part V, intelligent control itself is presented. In four chapters fuzzy (adaptive) control and the use of neural networks for control are discussed.

Part VI finally, focuses on supervisory control, monitoring and fault diagnosis and on the use of expert system techniques in numerical optimization.

The book combines the ideas in research and education of six prominent universities. It is our hope that it will bring a start to what can be seen as a standard for education in this fascinating domain.

Acknowledgment. The editors wish to thank the authors of the chapters in this book for their enthusiastic cooperation. In particular, we wish to thank A.J. Krijgsman and M.G. Rodd for their help in editing the book.

R.A. Vingerhoeds and L. Boullart
University of Ghent, Automatic Control Laboratory [1]
editors
June 1992.

---

[1]The editors can be contacted at:
Grotesteenweg Noord, 6, B-9052 Zwijnaarde, Belgium

Contributing authors.

University of Ghent
L. Boullart
R.A. Vingerhoeds

University College of Swansea
M.G. Rodd
C.P. Jobling

Queen Mary and Westfield College
J. Efstathiou
R. Vepa

Delft University of Technology
P.M. Bruijn
R. Jager
A.J. Krijgsman
H.B. Verbruggen

Universitat Politècnica de Catalunya
G. Cembrano
M. Tomás Saborido
G. Wells

Universidad Politècnica de Valencia
P. Albertos
A. Crespo Lorente
M. Martínez
F. Morant
J. Picó

Lawrence Livermore National Laboratories
G.J. Suski

PART I

ARTIFICIAL INTELLIGENCE

Road Map.

The first part of the course is a *Basic Course in Artificial Intelligence* for the graduate student, who doesn't need a fundamental study in the field as such, but who nevertheless needs a study which is thorough enough to be used as a fundament to other disciplines. The target field primarily is process control, because of some applications chosen, but it has been kept sufficiently loose from this in order to be more widely useful.

There are eight chapters:

A Gentle Introduction to Artificial Intelligence, with Appendix: 'A Rule Based System called "CLIPS"'. This chapter gives a general introduction in the field of A.I. which covers also briefly topics from other chapters in order to keep consistency. One part is more descriptive, while another deals with some topics more technically in depth. The appendix gives a short overview of the production rule system 'CLIPS', which is used throughout many examples in several chapters.

Knowledge Representation by Logic. This chapter presents the fundamental underlying elements when logic is used for inference. It is useful for understanding Prolog and production rules, for the basic resolving algorithms are being discussed.

Object-Orientation and Object-Oriented Programming. The fundamental underlying elements of the Object Oriented (O.O.) formalism are presented. Some O.O.-languages are compared in order to help the reader making the right decision if necessary. Therefore this chapter is far more general than the A.I.-field alone.

Expert System Case Study: The Chocolate Biscuit Factory. This chapter presents an educational case study of a rule based expert system. Its purpose is not to present a typical application but merely to offer an educational aid to the reader to understand the basic principles of both production rules, reasoning methodologies and expert systems.

Using A.I.-Formalisms in Programmable Logic Controllers. Here a typical application in process control is presented, where both Prolog and production rules are demonstrated in programmable logic controllers.

An introduction to Expert System Development. A complete survey of the activities involved in setting up an expert system project will be given. This involves selecting appropriate knowledge acquisition techniques, knowledge representations, the appropriate expert system shells, etc.

Introduction to Fuzzy Logic and Fuzzy Sets. Here fuzzy logic is introduced. It allows the programmer to use fuzzy data, like the terrain is high (as opposed to the terrain is 1345 m).

An Introduction to Neural Networks. A complete survey on neural networks is presented. Starting from the biological neural networks, the link is made to artificial neural networks, as they are used in AI.

This road map should enable the reader to make up his own choice in this basic course, depending on his skills and future use.

A Gentle Introduction to Artificial Intelligence.

Prof.Dr.ir Luc Boullart

Automatic Control Laboratory, University of Ghent.

1 Introduction.

'Intelligence' is commonly considered as the ability to collect knowledge and to reason with this knowledge in order to solve certain class of problems. There are several reasons to try to catch this intelligence in computers ('**Artificial Intelligence**': A.I.):

- to acquire a better insight in the human reasoning process by designing a computer-model 'in his own image';

- to ease the use of computers, by giving them a more 'human' face;

- to create the possibility to solve very complex problems, which could not be solved by standard programming, or could only by the expense of huge efforts;

In the early stage of computing (early 50's), it was completely unthinkable computers would ever do something else than computing ballistic curves. The problem undoubtedly lies in the terminology itself, where 'computer' always associates with 'counting' and 'calculating', although the machines in principle could manage all kind of symbols. Even John von Neumann has argued in his last publications that computers never would reach any stage of intelligence.

The first applications of A.I. existed already in the 50's and the early 60's especially in the form of chess and checkers programs. The purpose of the designers thereby was to get an insight in the essential reasoning process of famous chess players: there was a general presumption that some kind of general principle lied on the base of each intelligent behavior. Therefore, chess programs were for a long time considered as a kind of ultimate benchmark for A.I.. To-day, such machines play at a sufficient level to win in 99% of the cases. On the other hand, specialists have realized that the intelligent behavior of such programs only illustrate a few aspects of the reasoning process: a good chess player is only (not less, but also not more) than a good chess player. This change in attitude was very characteristic, and meant a swap from research in general reasoning principles to the recognition of more specific knowledge (facts, experimental knowledge), and its impact in specific domains. This swing in mentality in the scientific research was triggered by experiments in specific projects, which demonstrated that it was possible to use large amounts of knowledge in an 'intelligent' way, leading to results which could never have been reached by human action alone.

2 Definition of Artificial Intelligence.

As a relatively young science, there is no strict definition of A.I.. A definition often referred to, especially in scientific research, is to regard A.I. as a collection of techniques to handle knowledge in such way as to obtain new results and inferences which are not explicitly ('imperatively') programmed. Using A.I., of course involves a strong need for a number of tools and methods. These methods and techniques are in many cases transferable to other application areas. A much broader definition of A.I. therefore contains also all applications which employ A.I.-techniques, whether or not they enable the inference of new knowledge. Some definitions by prominent A.I.-specialists are the following.

- *"Artificial intelligence is the science of making machines do things that require intelligence if done by men"* *(Marvin Minsky, MIT).*

- *"The goals of the fields of Artificial Intelligence can be defined as ... <attempting> to make computers more useful <and> to understand the principles that make intelligence possible"* *(Patrick Winston, MIT).*

- *"The field of artificial intelligence has as its main tenet that there are indeed common processes that underline thinking and perceiving, and furthermore that these processes can be understood and studied scientifically ... In addition, it is completely unimportant to the theory of A.I. who is doing the thinking or perceiving: man or computer. This is an implementation detail"* *(Nils Nilsson, Stanford University).*

- *"A.I. research is that part of computer science that investigates symbolic, non algorithmic reasoning processes and the representation of symbolic knowledge for use in machine intelligence"* *(Edward Feigenbaum, Stanford University).*

A.I. has undoubtedly a growing impact and enjoys an increasing interest from many (especially industrial) users. Computers are performing, all things well considered, rather good, even when processing large amounts of data (knowledge). On the other hand A.I. is not a magic toolbox to solve all problems, but the advance towards more complex ones which could not have been readily solved by 'normal' imperative programming methods, is quite noticeable. The usefulness of the techniques in well defined application areas, has pulled A.I. out of its ivory tower.

One of the key questions which may arise is: when a system will it be intelligent? Although chess programs were considered long time as a benchmarch, there exist a so called *Turing test* (Alan Turing). This test mainly consists in a scene whereby a person by means of a terminal interrogates both a computer and another (invisible) person. Turing states a system is intelligent when the interrogator cannot distinguish between the computer and the physical person. Until now, this experiment has only succeeded in some well defined and strictly limited situations. This is caused by the fact the computer cannot yet interpret subtle nuances, read between lines, nor construct a 'general' knowledge in many different fields. More useful tests, e.g. exist in the execution of extensive realistic test cases in the

6

specific domain of the system under development. In this way, a typical medical diagnosis expert system (PUFF), scored as follows:

- 96% agreement between the system and the diagnosis of the specialist whose knowledge has been transferred;

- 89% agreement of the system with other independent specialists;

- 92% agreement among independent specialists.

3 Application areas of A.I..

Although the application field of A.I. is very wide and certainly not fully explored, research has given noticeable results in the form of useful tools and applications. These applications can be divided in a number of specific domains.

3.1 Expert Systems.

An expert system applies A.I.-techniques to an amount of knowledge concerning a well defined area, and stored in a data base. The ultimate purpose is to mimic and surpass the human expert. The efficiency of expert systems at this time is mainly determinated by the quality and the quantity of the gathered knowledge. A number of special techniques exist of course to deal with this knowledge on a intelligent basis, but there are no real generic reasoning techniques. Nevertheless, results are sometimes quite amazing.

3.2 Systems for natural language expression.

The purpose here is to engage a communication with computer systems in a more 'natural language' approach instead of a traditional procedural language. Thereby both written and spoken text will be used. Written text is processed via keyboards, printers, video systems a.o., while speech is processed with special hardware systems for recognition and synthesis. This application area furthermore splits up into two branches:

- natural language regarded as pure text manipulation (syntactic);

- natural language with recognition and production of a meaningful content (semantic);

Possible applications could be:

- interfacing to data bases, software systems, expert systems, robots, etc.;

- translations of written text from one natural language into another;

- document processing: 'understanding' written documents in order to summarize, indicate important elements, interrogate, etc. ...

3.3 Automatic Process Control.

Industrial process automation is undoubtedly an important application area for A.I. There are a number of important items in this field.

Operator assistance: The problem tackled here is that operators of complex control systems, such as power plants, are confronted with an overwhelming amount of information during alarm situations. Flashing lights, buzzers, printers ejecting page long lists of log values, and many other trigger signals have to be assimilated. For example, in a nuclear environment, some 800 alarm messages can be generated within 2 minutes. Furthermore, the connection between separate alarms may be very complex, if at all present. It is impossible to keep a total view of all events and actions that may influence the situation at hand. The human task to perform here is very difficult. Even people who are trained to do this kind of work make mistakes, purely because of the heavy load with which their decision-making process is faced under the described conditions. Problems of cognitive overload on operators of complex systems have not always been taken seriously. Quite often, people have had to work under immense psychological stress in situations where the available information was not enough to provide even minimal control, and where the consequences of a mistake were phenomenal. It was mainly due to major accidents in chemical and nuclear plants that research was undertaken to explore possible solutions. Artificial intelligence can provide a reasoning support, because they are not influenced by stress, and have no Monday-morning problems. They can follow the same pattern of thought under any condition, and advise humans on which decision to focus.

Process control design: in the situation where a process engineer has to set-up a new complex process scheme, an expert computer program may engage in a conversation with him to finalize into a suitable process control strategy. This will have a growing impact in the future because the requirements in to day's control are much higher and much more algorithmic than they used to.

Process automation: inside the control loop itself, A.I. can have many benefits. In a steadily increasing complexity of control strategies, there may be a claim for a higher, much more abstract level, where a choice for the optimal control strategy is made relying upon real-time information from the process. Such systems can be rather simple, but will surely gain only benefits when research will have delivered much more insight into the symbolic reasonings which form the roots of such strategies.

Controlling Complex Systems. Large scale systems, such as chemical reactors, traffic control systems and real-time shop floor planners are very complex. They are not always well understood, and even if they are, the need may be felt to simplify the model as much as possible Changing system configuration is expensive, making mistakes is even more expensive. Controlling complex processes can be difficult for multiple reasons. Process loops may be tightly coupled, so that eventual control systems must be

8

multidimensional or of a higher order. Parameter values may depend in a nonlinear way upon the current working point of the system. Also, signals may be subjected to heavy noise, which is automatically discarded by human operators, but looks like a transient error to a machine controller. Fortunately, human operators are often experts in keeping the control systems on the right track. If a certain control loop trips, by which we mean that its parameters steadily deviate from their setpoints or from the optimal strategy, thus gradually approaching instability, a skilled operator will limit the effects of such an error on the rest of the process by manually taking over some of the control actions. In doing this the human operator discriminates between normal and erroneous situations.

Controller Jacketing. Although control system theory has delivered new methods for optimal and adaptive control, the algorithms used in industrial environments are very stable. PID controllers are still everywhere in existing industrial process control systems. But even though these algorithms are well understood, they are never used without a number of surrounding heuristics, such as wind-up protection, smoothers for manual to automatic transition, transient filtering, etc... These features have little to do with control theory, but are extremely important in practical controllers. Artificial intelligence techniques could provide a jacket for control algorithms, allowing for decision support under conditions the controller would judge as abnormal.

Measurement techniques: in measurement and monitoring situations it is common knowledge that the amount of information flowing through is enormous; until now we have learned to capture the picture of the process by performing classical signal processing techniques. The time has come now to add another level of abstraction in the form of symbolic manipulation (e.g. pattern matchings) which undoubtedly offer new insights into the real-time capabilities of such systems.

Speech and image processing: this application field is now receiving new incentives because of the possibilities offered by commercially available tools; however the processing power and real-time capabilities because of the high speed data-stream still pose problems.

3.4 Automatic programming and intelligent tools.

The goal in this area is to use A.I. both for automatic design and automatic transformation of programs. The system thereby requires a deep knowledge of:

- general programming principles,

- the programming language itself;

- the hardware and the operating system under which the program will run;

- the specific application domain.

9

The latter is not obvious, but enables the system to take into account the boundary conditions, and so a more efficient implementation. In the domain of automatic programming, several disciplines can be distinguished: automatic program synthesis, automatic program verification, optimizing compilers ...

4   What is special on A.I. techniques?

Although in an A.I. environment a large part of the traditional computer techniques and conventional algorithmic imperative programming styles may be used, in some aspects a significant difference can be observed.

In a traditional programming style, the programmer writes an instruction sequence which follows a clear path from the problem definition towards the solution. Each branch in the sequence has carefully been considered in advance, and every deviation is a fortiori a bug. Advances in 'traditional' computer science are mainly improvements in processing speed and in enhanced programming comfort by using more efficient tools and languages.

A.I. on the other hand, deals more with symbols. These symbols represent quantities from the real problem. But, instead of making calculations with the quantities, the symbols are manipulated and the mutual relations between the symbols are processed. This manipulation in fact is performed as much as possible without any explicit knowledge of the contents. The solution of an A.I.-problem is first of all a search for a functional solution, and actual values are only assigned at the last minute ('instancing'). From the foregoing discussion it follows it is not only necessary to store information (on facts e.g.) into a data base, but especially the relations between those quantities inside a *knowledge base*. This knowledge contains the structure of the information, rather than the information itself.

In A.I.-systems, there is no single way to represent knowledge. A.I.-methods store the knowledge information following certain techniques, and formulate procedures to deduce ('infer') new useful information. The knowledge contains facts on objects, relations, actions which are possible on those objects and also the so called 'meta-knowledge': i.e. knowledge about knowledge. The meta-knowledge determinates how the knowledge is structured and which relations and values are assigned whenever no explicit indication is given at the instancing of the object.

5   Special Languages for A.I..

Because of the typical characteristics of A.I., classical programming languages as they have been used for imperative programming styles, are mostly unsuitable. Therefore a number of specialized languages exist. Important requirements for A.I.-languages are:

- manipulation of objects (especially lists);

- the possibility for general manipulation of a wide area of different data-types;

- late binding of the exact value and typing of the objects;

- pattern recognition in structures;

- interactive operational procedures;

- interactive methods ('inference') for automatic decision making and deducing.

In a first phase, A.I.-languages were derived from enhanced versions of traditional languages (e.g. SAIL from ALGOL). Sometimes they deal only with partial aspects of the required techniques (e.g. XEROX-KRL : pattern recognition) or they may be constrained to specific hardware. Those languages are currently replaced by more general languages from which PROLOG and LISP are the most important. Besides these tools, some manufacturers and/or scientific research-centers have developed their own languages which afterwards spread out to a more general use. Important representatives thereby are SMALLTALK-80 (Xerox-Parc), OPS-5 (Carnegie Mellon, and commercialized by DEC).

Another important issue in A.I.-programming are the *Shells*. A Shell is normally not a programming language, but rather a user environment. Such an environment comprises a programming language built inside interactive development tools: editors, debuggers, tracers ... The debugging of A.I. is at least a magnitude more difficult than 'flat' programming languages, due to its non-procedural behavior. In many cases there is no sequential execution determined by a 'program counter'. Therefore the user needs strong trace back facilities (on program 'break points'), leading him through the dynamic behavior of the program modules. High computer power and large multi-window workstations are thereby no superfluous luxury. The trend in A.I.-environments is more and more to contain more than one single programming language or system: combinations of A.I.-languages (e.g. Prolog), production rule systems and object oriented programming style are often coupled with normal procedural languages (mostly C or C++), to offer the user for each subproblem the most appropriate tool.

In this text the following typical languages will be introduced briefly.

Lisp: probably the most general A.I.-language, and prominent for symbolic processing.

Prolog: a typical example of a logic oriented language, both for general programming and logic statements proving; it is also a language with an intrinsic inference mechanism,

Smalltalk-80: the typical and most prominent object oriented language, built inside a dynamic environment.

The reader should consider this as as a first and very brief introduction, just to feel the typical 'flavor' of the language and the corresponding programming style. In any case he should complement this reading with a more in depth study of *Logic and Logic proving* (to understand thoroughly Prolog behavior), and *Object Orientation* as a more general programming and design formalism.

5.1 Lisp.

LISP ('List Processor') has been developed by John McCarthy at MIT at the end of the 50's. Lisp especially deals with symbols in the form of atoms and lists. The basic element from Lisp is an *Atom* (elementary string), and a list constructed from atoms or other lists.

Example

```
(1 2 3)
(A LIST WITH ME)
(5 MIXED / SINGS)
((NAME FIRSTNAME)(ADDRESS PHONE) CITY)
```

Procedure calls are done by lists too (in prefix notation). *Data- and program structures thereby have identical structures.*

Example

```
(+ 1 2)              → returns the result of 1 + 2;
(+ L 6)              → returns the result of L + 6 (L = variable)
(OPERATION OPERAND)  → performs operation on operand.
```

As basic primitives, LISP contains some procedures to manipulate lists:

CAR: returns the first element of a list (or the null-list);

CDR: returns a list without the first element (or the null-list).

Example

```
(CAR '(THIS IS NOTHING NEW))    → returns THIS
(CDR '(FIRST DAY OF SEMINAR))   → returns (DAY OF SEMINAR)
```

If the list X is bounded to ((a) (B (c) D)), then

```
(CAR (CAR (CDR (CAR ((CDR L))))))    will return c
```

In shortform-notation, this can also be written as

```
(CAADR (CADR L))
```

Indeed

```
L → CDR returns ((B (c) D))
  → CAR returns (B (c) D)
  → CDR returns ((c) D)
  → CAR returns (c)
  → CAR returns c
```

The real power of LISP comes when procedures are defined. A procedure can be seen as a strict functional definition, and this definition may be recursive. Such procedures are in no way different from the basic LISP procedures: self-made procedures, after eventual compilation, may reside into a library, and can be regarded as a true extension of the language. As an example of a recursive definition, the following defines a functional which investigates whether an atom is an element ('member') of a list:

12

```
(defun memb (atm llst)
  (cond
   ((null llst) nil)
   (( eq  (car llst) atm) T)
   (T (memb atm (cdr llst))))))
```

This definition states:

- if we have a null-list: return 'no' (NIL);

- if the first element of the list is an atom: returns 'yes' (T="true")

- in all other cases, apply the same functional to the tail of the list.

Because program structures and data structures are fundamentally identical, LISP allows to *manipulate both program and procedures as symbols*. Towards the user, LISP appears as an interactive language implemented around a highly integrated workstation (mostly single-user). Its interactive nature enables quick and easy developments, which can be tested at once. There are several LISP-implementations, sometimes on specific LISP-oriented architectures. Nevertheless, LISP implementations have converged considerably ('Common Lisp').

Application Example: 'Towers of Hanoi'

Following a very old tradition, *Monkeys in the Underworld of Hanoi* have to move stones from one pile to another, whereby one temporary ('buffer') pile may be used, and never a larger stone can take place on a smaller one. The LISP-solution handy uses the the recursive function-definition mechanism to provide an elegant solution definition for this game.

```
(defun move-series (origin target storage number-of-disks)
  (cond
   ((eq number-of-disks 1) (move-disk origin target 1))
   ( T
      (move-series origin storage target (- number-of-disks 1) )
      (move-disk origin target number-of-disks)
      (move-series storage target origin (- number-of-disks 1) ))))

(defun move-disk (origin target disk)
  (print (list 'Move 'disk disk 'from origin 'to target) ))

(defun tower-of-hanoi (number-of-disks)
  (move-series 'Pile-A 'Pile-B 'Pile-C number-of-disks))
```

When triggering the main-function by a call, such as

13

```
(TOWER-OF-HANOI 4)
```

the following results are print out :

```
(Move disk 1 from Pile-A to Pile-C)
(Move disk 2 from Pile-A to Pile-B)
(Move disk 1 from Pile-C to Pile-B)
(Move disk 3 from Pile-A to Pile-C)
(Move disk 1 from Pile-B to Pile-A)
(Move disk 2 from Pile-B to Pile-C)
(Move disk 1 from Pile-A to Pile-C)
(Move disk 4 from Pile-A to Pile-B)
(Move disk 1 from Pile-C to Pile-B)
(Move disk 2 from Pile-C to Pile-A)
(Move disk 1 from Pile-B to Pile-A)
(Move disk 3 from Pile-C to Pile-B)
(Move disk 1 from Pile-A to Pile-C)
(Move disk 2 from Pile-A to Pile-B)
(Move disk 1 from Pile-C to Pile-B)
```

Although Lisp is intended first of all to be a functional language, the language contains also a number of structures to enable ('flat') imperative programming:

- **prog ( )** instruction to start an imperative sequence;

- a 'label/GO label' and 'DO' mechanism to construct loops;

- a number of directives which in the functional sense return a fixed value, and as a 'side-effect' performs some imperative action: e.g.
  PRINT: print a list or an atom.
  SETQ: set a global variable to a specified value.

- Furthermore, the user himself may define efficient control structures (e.g. WHILE REPEAT / ...).

The example below shows both a functional and an imperative Lisp solution for calculating the sum of a series of numbers.

1. **Functional:**

```
(defun sum(series)
  (cond ((null series) 0)
  (T (+ (car series) (sum (cdr series)))))))
```

2. **Imperative:**

14

```
(defun sum1 (series)
  (prog (accu)
    (setq accu 0)
    begin
    (cond ((null series) (return accu))
          (T (setq accu (+ accu (car series)))))
    (setq series (cdr series))
    (go begin)))
```

The reader should be aware that this is of course not the whole of LISP, which contains much more features, such as function definitions based on the λ-calculus. A great experience is necessary before the user will 'speak fluent LISP'.

5.2   Prolog.

PROLOG ("Programming in Logic") has been developed around 1970 at the Marseille university by Alain Colmerauer. At the department for A.I. at the University of Edinburgh, Prolog has been much elaborated by e.g. Warren and Mellish.

Prolog is based on the so called *Predicate Logic*, and can be regarded as a formal implementation into a language of this type of logic. Predicate Logic differs from simple Propositional Logic by pronouncing more general statements, such as "If someone ... then that individual ...", i.e. the concept of a *variable* which may be instantiated to any individual element plays an essential role. In order to understand profoundly the internal Prolog mechanisms, the reader should study more in detail logic systems in general, and *Predicate Logic* in particular.

The language starts from objects and the relation between objects. It proceeds then towards a consultation of this information by questioning. We can distinguish 3 phases:

1. formulating *facts*;

2. formulating *rules*, to establish the relation between those facts;

3. formulating questions.

Facts and rules form the so called *data base*.

Example.

A. 1. Jan is adult
   2. Piet is adult
   3. Marc is a child

B. If X is adult, he drives a car

15

C. 1. Is Jan adult? ($\rightarrow$ yes)
   2. Is Marc adult? ($\rightarrow$ no)
   3. Who drives a car? (*rightarrow* Jan, Piet)
   4. Does Marc drive a car? (*rightarrow* no)

In Prolog, this example could be formulated as follows :

```
adult (jan).
adult (piet).
a-child (marc).
car (X) :- adult (X).

? - adult (jan).
yes

? - adult (marc).
no


? - car (X).
X = jan
X = piet

? - car (marc).
no
```

The clause `car(X) := adult(X)` should be read as "in order that someone drives a car, that someone should be an adult".

The evaluation of a Prolog-questioning is done by plowing through the data-base (i.e. all predicates and rules) via a tree-searching-algorithm in order to *satisfy a goal*. This goal can be realized in several ways, depending on the questioning:

- by an affirmation ('yes') or a negation ('no');

- by looking for each instance of a variable (X) which leads towards an affirmation.

Normally, after a positive success of a goal, the data-base will be looked up further by returning back to a previous node (*Backtracking*) for eventual other and new successful instances. During the evaluation mechanism, the seeking for a goal mostly requires other *subgoals* first to be realized, resulting into deep nested searches in the data-base. The real power of Prolog comes out when diverse rules and facts are combined with each other, as in the next example.

Example 1 (following [CLOCK]).
   (comma inside a rule means 'AND')

16

```
/*1*/ thief (john).
/*2*/ likes (mary, food).
/*3*/ likes (mary, wine).
/*4*/ likes (john, X) :- likes (X, wine).
/*5*/ may-steal (X,Y) :- thief (X), likes (X, Y).
/*6*/ thief (marc).
/*7*/ likes (marc, X) :- likes (X, pizza).

? - may-steal (john, X).
X = mary
```

The reasoning behind this conclusion is as follows:

- In order that someone should steal something (rule /*5*/), that someone has to be first of all a thief; from /*1*/ follows that John is a thief and thereby satisfies this first condition.

- /*5*/ thus becomes:

```
may-steal(john, Y) :- thief(john), likes (john, Y)
```

- From /*4*/ we learn that John may like someone, if this one likes wine; i.e. the data base is scanned to satisfy a goal for:

```
likes (X, wine)
```

- This subgoal will be found in /*3*/; thereby succeeds the goal /*3*/ and also /*4*/ and a fortiori /*5*/; the answer to the original Prolog question will thus be **mary**.

- After this success, the data base will be searched further; thereby it will be found that marc is also a thief (rule /*6*/), and by rule /*5*/ marc may perhaps steal something; nevertheless, marc likes someone who likes pizza (rule /*7*/), and this condition is never satisfied; so, there is no more goal satisfaction.

In a similar way, the following question

```
? - may-steal (X, mary).
```

May give an answer

```
X = john
```

Example 2 (free following [CLOCK])

17

```
male (jan).
male (marc).
female (leen).
female (sonja).
parents (marc, mia, jos).
parents (leen, mia, jos).
sister (X,Y) :- female (X), parents (X,M,F), parents (Y,M,F).

? - sister (leen, marc).
yes

? - sister (leen, X).
X = marc
```

Prolog has several different structuring concepts, operators, variables, and mathematical operations:

- normal 'functors' are constructed as in the previous examples, but they can be nested; e.g.

```
reads (jan, book (de-witte, ernest-claes))
reads (jan, book (de-witte, author (ernest-claes)))
```

- arithmetic operators:

$$+ (X,*(Y,Z))$$

- lists and many functors for list processing (cfr. LISP): [   ]

Application Example: 'Towers of Hanoi'

This game can also be solved in Prolog. The following program from [CLOCK] is very short, but nicely does the job. It is started by entering

```
            hanoi(3)
```
for three discs.

```
        hanoi(N) :- move(N,left,centre,right).

        move(0,_,_,_) :- !
        move(N,A,B,C) :-
            M is N-1,
            move(M,A,C,B),
            inform(A,B),
            move(M,C,B,A).

        inform(X,Y) :- write([move,a,disc,from,X,to,Y]), nl.
```

18

When Prolog has reached a successful goal, it will undo this goal, and retry to satisfy the next goal. This mechanism is called 'Backtracking'. It can be done with a systematic tree-searching methodology of the data-base by means of place markers and alternative solutions. This property sometimes leads to inadmissible long processing times. This is sometimes unnecessary because Prolog searches parts of the data base from which the programmer knows in advance they will never succeed, or whenever the user is satisfied with one single solution. A cure for this problem is the so called *cut* (written as '!') mechanism: this puts a place marker in the tree, which never can be passed back again on backtracking. This mechanism enables the programmer to tell Prolog:

- "once you are as far as to reach a successful goal, you don't have to go any further, because there are no more";

- "once you are as far as to fail, you don't have to try any further: you will never succeed";

- "once you find this solution, I am satisfied: don't look for any other alternatives".

Programming in Prolog, just as in Lisp, is never pure logic nor functional. Both languages contain a number of structures which clearly go into the direction of imperative programming. A Prolog 'cut' may be seen in this context as an imperative indication from the user.

5.3 Smalltalk-80.

Smalltalk-80 (Adele Goldberg - Xerox Parc) is the most important language in the object programming style. In Object-Oriented Programming, the emphasis is not on procedures, nor on a functional specification, but on the data, called 'objects'. In the Smalltalk language, the abstract data typing is pushed as far as possible by introducing the concept of 'class'. A class contains the complete description of a type, both to its nature as to the possible operations which can be carried out. An object is just an instance of a class. A program proceeds by sending 'messages' towards objects. These messages tell the object which operation to carry out, and *not how:* this is determinated completely by the object itself and is *hidden inside the class* ('information hiding / data encapsulation'). In this way, identical operations on objects of different nature may be carried out in different ways. Each instance of a class delivers an object with an own 'private' memory. Modifications of this private memory can only be carried out by the object itself. By this mechanism it will be automatically guaranteed that the implementation of an object will not depend from internal details of other objects, but only from the messages to which it has to respond. This method also guarantees the modularity of the system.

As a generic example, consider the following Smalltalk-80 program. This program sets up a kind of file for managing household finances (following [ADGOL]).

```
class name                    FinancialHistory
```

```
instance variable names            cashOnHand
                                   incomes
                                   expenditures
instance methods

transaction recording

  receive: amount from: source
      incomes at: source
              put: (self totalReceivedFrom: source) + amount.
      cashOnHand <- cashOnHand + amount

  spend: smount for: reason
      expenditures at: reason
                  put: (self totalSpentFor: reason) + amount.
      cashOnHand <- cashOnHand - amount

inquiries

  cashOnHand
      ^cashOnHand
  totalReceivedFrom: source
      (incomes includesKey: source)
              ifTrue: [^incomes at: source]
              ifFalse: [^0]
  totalSpentFor: reason
      (expenditures includesKey: reason)
              ifTrue: [^expenditures at: reason]
              ifFalse: [^0]

initialization

  initialBalance: amount
      cashOnHand <- amount.
      incomes <- Dictionary new.
      expenditures <- Dictionary new
```

The above SMALLTALK-80 program (in fact a class definition) will be triggered by 2 actions:

- instancing this class to a real object;

- instancing of a number of private memory cells.

The required procedure is :

```
Huishoudkas <- FinancialHistory new initialBalance: 10000
```

This initializing step puts a start value to `CashOnHand` and creates 2 files (i.e. `incomes` for income / **expenditures** for outgoing). Several actions are possible on the object, which for convenience, can be grouped in 3 parts :

1. Transactions :

   - `receive: amount from: source`
   - `spend: amount for: reason`

2. Questioning:

   - `CashonHand`
   - `totalReceived From: source`
   - `totalSpent For: source`

3. Initializing:

   - `initialBalance: amount`

Examples of possible actions ('messages') are:

- `Huishoudkas spend: 135 for: food`
  this action increases the expenditures of the section 'food' with 135 and decreases the total 'CashonHand' with the same amount;

- `Huishoudkas receive: 2000 from: RUG`
  idem for the income of the section 'RUG'

- `totalSpent for: hobby`
  returns the total amount for the section 'hobby'.

Application Example: 'Towers of Hanoi'

The former mentioned problem can also be solved in Smalltalk. The following program is written in the *Little Smalltalk* implementation from [TIMBUD]. It is started by entering

```
                Hanoi new discs:3
```

for three discs.

21

```
Class Hanoi
[
    discs: numberOfDiscs        |i|
        i <- MoveDisc new.
        i moveDiscs:numberOfDiscs from:#A to:#B using:#C.
        i reportMoves
]

Class MoveDisc
| numberOfMoves |
[
    new
        numberOfMoves <- 0

|
    reportMoves
        ('There are ', numberOfMoves, ' moves done.') print

|
    moveOne:origin to:destination
        ('Move from ',origin,' to ',destination) print.
        numberOfMoves <- numberOfMoves +1

|
  moveDiscs:n from:origin to:destination using:buffer
      (n == 1)
        ifTrue: [self moveOne:origin to:destination]
        ifFalse:[self moveDiscs:(n-1) from:origin to:buffer
                    using:destination.
                self moveOne:origin to:destination.
                self moveDiscs:(n-1) from:buffer to:destination
                    using:origin]
]
```

Smalltalk-80 relies completely on the class / object mechanism. The notion class has been elaborated largely in a vertical hierarchy in the form of meta-classes (upwards) and of subclasses (downwards) among which exists certain inheritances. The development of programs is done on highly graphical interactive workstations, which themselves are completely driven by object-oriented software. The powerful operation of such workstations lies into the presence of very large libraries with primitives and application oriented definitions and objects, which, also due to an intrinsic modularity, are rather easy to handle.

The concept of Object-Orientation (OO) is very important not only in A.I. but more general in to-day's programming style. Therefore the user should study OO more in depth in order to learn the real issues about it.

6 Expert Systems.

6.1 Introduction.

Already in the beginning of this text, the radical change in the scientific research of A.I. during the 60's has been stressed out: research towards general basic principles of knowledge and intelligence did swap to the appreciation of specific domain knowledge as a base for practical intelligent behavior.

This change in attitude was caused by the development of the first expert system: DENDRAL, started in 1965 at Stanford. DENDRAL (Feigenbaum, Lederberg, Djerassi) was a program used as an assistant for the scientific reasoning in the search and tests of hypotheses of molecular structures, starting from chemical data. DENDRAL has been elaborated in such a way that it outperforms the human possibilities (even those of the designers).

An important number of expert systems at present time are in use inside the medical world. A well known example is MYCIN (Stanford): this system provides a diagnosis concerning hart and meningitis infections, and advises an antibiotic therapy treatment. The medical doctor provides MYCIN with all possible data and test results. When suggesting a possible diagnosis, the doctor may engage a conversation with MYCIN, and ask questions such as "why are you asking me this?", "how do you obtain this conclusion?". MYCIN may also inform the user why it rejected certain hypotheses. The inference engine of MYCIN has been carried later on to another knowledge base, this time for lung diseases, and has been applied with the same success.

6.2 Definitions of Expert Systems.

What is understood as an 'expert', has been carefully expressed by P. Johnson, in the following text:

> *"An expert is a person who, because of training and experience, is able to do things the rest of us cannot. Experts are not only proficient but also smooth and efficient in the actions they take. Experts know a great many things and have tricks and caveats for applying what they know to problems and tasks; they are also good at plowing through irrelevant information in order to get at basic issues, and they are good at recognizing problems they face as instances of types with which they are familiar. Underlying the behavior of experts is the body of operative knowledge we have termed expertise. It is reasonable to suppose, therefore, that experts are the ones to ask when we wish to represent the expertise that makes their behavior possible."*

An expert system is composed of 3 important elements:

23

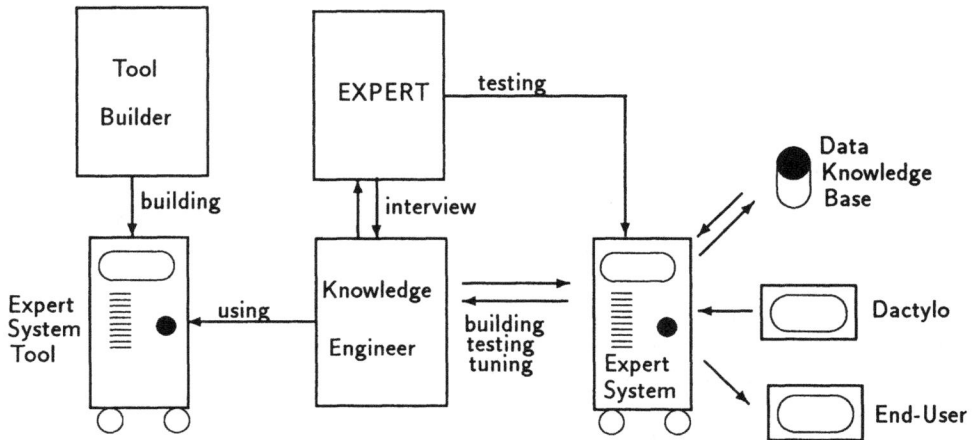

Figure 1: Expert System Building.

1. the knowledge of facts;

2. the knowledge of relations between the facts;

3. a heuristic, efficient method for storage and acquisition of this information.

In the 70's, especially the last item received a lot of attention in scientific research. The importance of this element is crucial for any practical use of expert systems, as is stressed out by Hayes-Roth:

> *"The central notion of intelligent problem solving is that a system must construct its solution selectively and efficiently from a space of alternatives. When resource limited, the expert needs to search this space selectively, with as little unfruitful activity as possible. An expert's knowledge helps spot useful data early, suggests promising ways to exploit them, and helps avoid low payoff efforts by pruning blind alleys as early as possible. An expert system achieves high performance by using knowledge to make the best use of its time."*

The construction of an expert system is called *knowledge engineering*. The knowledge engineer's job is to extract the knowledge out of human experts: i.e. procedures, strategy, rules of thumb and to transfer this into a data base (figure 1).

Besides a base for high value expertise, such data bases can be very useful for a number of other application.

24

- The knowledge base can be seen as a model for the validation of forecasts. The user has the possibility to confrontate the knowledge with eventual control actions, hypotheses, or modifications which he intends to carry out on a real system and may tune his actions more optimal by a deeper insight into the process behavior.

- The knowledge base can be seen as institutionalized archives: the enormous amount of knowledge in the base has been constructed and gathered gradually as a consensus between all possible experts. Although key persons disappear out of the plants, their knowledge will stay forever and will be enhanced by their followers.

- The knowledge base is also an ideal training facility for people who have to take important decisions in industry. In this case, the knowledge base has to be equipped with a suitable 'educational' interface.

When developing and using expert-systems, a number of key people (figure 1) are involved:

- the (human) expert of the specific application domain;

- the knowledge engineer;

- the person who builds the expert-system tools.

6.3 Human Expert versus Artificial Expert System.

There are a number of clear distinct reasons for introducing expert systems. The advantages can be summarized as follows.

Human expertise	Artificial expertise
perishable	permanent
difficult transferable	easy transferable
difficult documentable	easy documentable
unpredictable	consistent
expensive	affordable

Nevertheless, currently it is quite wrong to exclude completely the human expertise. In many cases it is useful to keep a 'moderate' human expert inside the loop, which may fill up the holes and the imperfectnesses of the expert system. The most important trumps of a human expert are summarized in the following table.

Human expert	Artificial expert
creative	uninspired
adaptive	has to be teached
sensorial observation	symbolic input
broad view	blinkers
common sense	technical sense

Some remarks in this context are:

- Symbolic input is inferior to sensorial observations, because the latter starts from a global view where a priori mutual relations can be captured; when converting an image to individual objects, much information can be lost.

- The notion *common sense* really is a huge trump for a human expert. This notion is very difficult to describe, but is a kind of broad view of general knowledge about the world in which we live, and about its mechanisms and relations. Common sense is both the awareness of a certain knowledge as the awareness that some knowledge is not present at all.

7   Knowledge Representation and Extraction.

The key issue in A.I., and especially in expert systems, is the way knowledge is stored and extracted from the knowledge base. This fact is not only related to the application field itself, but also to the feasibility of the application. How nice and intelligent an expert system may be, if the time to infer the required answer is too long, it is completely useless. This is especially true in process control environments where the real-time aspect is a severe requirement. Therefore, knowledge representation and extraction mechanisms not only have to fulfill their task, but they have to do it very quickly and efficiently, which is difficult to achieve because of the painful limitations of to day's computers! There is no general rule for knowledge representation and extraction. Some general schemes can be recognized and will be given here; two of them are mentioned mainly for reference only, but the third one will be explored more deeply because of its importance in to day's operational expert systems.

7.1   Frames and Scripts.

Frames gather inside one organizational entity, facts and rules which apply consistently to each element referring to it. The technique is mainly used in the U.S. (MIT / Stanford / Carnegie Mellon) and follows the previously introduced concept of objects. Knowledge is divided into so-called *clusters*, which are connected to each other in an hierarchical way. This can be compared to the principle of classes, metaclasses and subclasses of Smalltalk-80. The frames of a certain level inherit knowledge from the higher level, and in their turn put this knowledge at the disposal of the lower levels. This mechanism is supposed to be very close to the organization of the human brain. Frames can represent information about the normally expected characteristics of things, as they can indicate the exceptions. Besides the descriptive part, frames often contain so-called scripts, which represent the normal action sequences to take when an outside trigger to the object-instantiation arrives. In process control they are much like the black box system which is used in currently available process control systems: a specific control loop can be seen as an instantiation of some control's scheme 'frame' (e.g. PID), which in its 'script' represents the algorithm to calculate whenever a 'trigger' (sample) occurs. A.I. frames are much more closely linked to each other in a complex hierarchical scheme with strict inheritance rules.

7.2 Semantic Networks.

In semantic networks all knowledge is represented as a labeled graph. The nodes thereby describe objects, and the links (arrows) represent the relations in between. One should realize that, although the principle is fixed, there are many ways to use such graphs; one often gets the impression that they are just some free drawing of elements, but once inside a consistent scheme a great deal of subtle information can be represented. Semantic networks are very popular e.g. in natural language and grammar processing. The translation of such a graph to a computer suitable form can be done in many ways, Prolog e.g. can be a suitable formalism to host semantic networks.

7.3 Logic Inference and Production Rules.

Logic systems, especially Predicate Logic, are powerful tools to both represent and infer A.I.-problems. Whereas Prolog, as a direct implementation of Predicate Logic, is the most rigorous implementation, many practical systems use another related form of logic processing: *Production Rules* or *Rule-based Systems*. Production rules are very popular as an aid in representing and manipulating knowledge because many existing expert systems are based upon it.

Production rules originate from the popular IF ... THEN structure, which is known from traditional procedural languages.

In its most simple form, they just deals with single logic facts.

> *"If there is a furnace alarm, and there is no temperature alarm, then the thermocouple is defect"*

These IF...THEN structures are based on the *modus ponens* rule from *Propositional Logic*:

$$(X \wedge (X \Rightarrow Y)) \Rightarrow Y$$

I.e. when $X \Rightarrow Y$ is known to exist, and X is true, the Y will also be true. Note that if we think of X and $X \Rightarrow Y$ as two entries in a database, the modus ponens rule allows us to replace them with the single statement, Y, thus eliminating one occurrence of the connective '$\Rightarrow$'. Modus ponens is therefore called the '$\Rightarrow$'-elimination rule.

Although at first sight propositional logic does exactly what we need, it turns out to be rather limited. Therefore *Predicate Logic* - as is used in Prolog - is the base for Production rules. As already mentioned Predicate Logic adds a concept of a *variable* to logic systems. This enables more general pronounced statements: "if someone is ... then that someone will ...". Thereby that someone can be instantiated by any suitable individual from the facts data base. Predicate rules can be formulated as follows (the numeric value is the variable element).

- 'If the oven temperature becomes higher than 250 degrees Celcius, then the thermoswitch is defect'.

- 'If the oven temperature becomes lower than 50 degrees Celcius and if the light in the oven is on, then the thermoswitch is defect'.

- 'If the outside temperature is higher than 28 degrees Celcius, then beer consumption will rise'.

Rules always contain two parts:

1. a *condition-part*, sometimes called 'left-hand-side' (LHS),

2. an *action-part*, sometimes called "right-hand-side" (RHS)

One must however be careful with the indication 'left' and 'right', because in some syntaxes (e.g. Prolog) the 'left' (condition) is lexically written on the righthand side of the rule.

A set of rules can be evaluated in either two different ways: from left to right or vice versa. Those two approaches to the solution reflect two different reasonings and classes of problems. As an example, consider e.g. an expert system for medical diagnoses, with the following rules.

```
IF symptom(A) and symptom(B) and symtom(C)
        THEN measles.
IF symptom(A) and symptom(D)
        THEN measles
IF symptom(A) and symptom(E)
        THEN flu.
```

The two first rules together form an OR-clause (with inside each rule itself an AND-clause).

In a left-to-right reasoning, the doctor will ask the patient about all possible symptoms, write them on a scoreboard, evaluate all rules in sequence, and tries to find out the disease.

In a right-to-left approach, the doctor will *suppose* the patient has measles, scan the knowledge base to find the measles symptoms and then diagnoses the patient for a possible match.

Those two reasoning processes are called:

- **forward-reasoning** ('left-to-right') : it is often referred to as 'data-driven' or 'bottom-up';

- **backward-reasoning** ('right-to-left') : this is also called 'goal-driven' or 'top-down'.

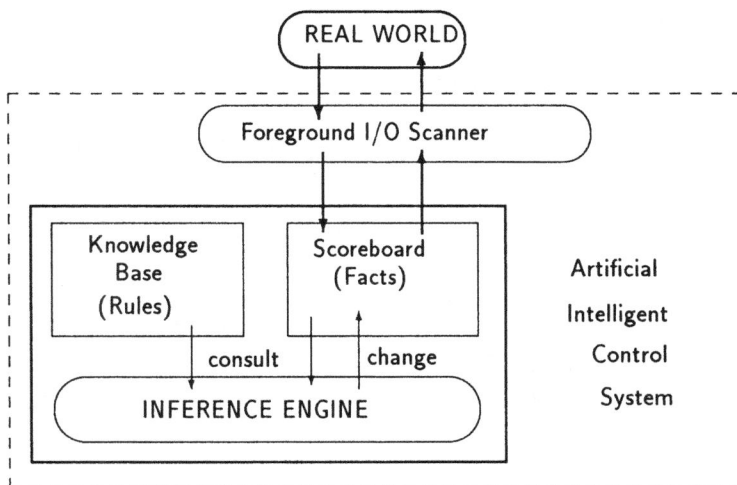

Figure 2: Real-Time Production Rule System.

The different elements of a production rule- based expert system are drawn in figure 2.

The knowledge base contains all the rules. The scoreboard contains all known facts: i.e. the world to which the system applies in a given state at a particular instant of time. As time goes by, both by external events, or by evaluating rules, the contents of this scoreboard may change. The interpreter acts on the given rule set and mainly has to match the lefthand sides in order to succeed the goals or sub-goals (action). One must however be aware that there is a fundamental difference with a classical procedural language: in a rule-based system, many rules (IF ... THEN ...) may be 'open' (i.e. may be executable) in parallel and at the same time. A special feature therefore is the *conflict resolution:* i.e. the strategy to follow when at a given instant of time more than one rule 'fires'. Several schemes are then conceivable: priority, selecting the first (as in Prolog), taking the most recent rule, the most important one, the most selective one etc... Finally many operational expert systems offer the user some extra facilities in the form of the explanation (tracing) of how a specific conclusion has been drawn. This is very important not only in the development phase to evaluate the system, but also in real operation in order to make the user confident about the suggested actions.

From the above it should be clear that Prolog is a specific implementation of a Production Rule System. The 'translation' of the above examples into Prolog is straightforward and much concise:

```
thermodefect :- temp(oven,X) , X>250.
thermodefect :- temp(oven,X) , X<50,
```

29

```
ovenlight.

raise beer :- temp(outside,X) , X>20.

measles :- symptom(A) , symptom(B) , symptom(C).
measles :- symptom(A) , symptom(D).

flu :- symptom(A) , symptom(E).
```

As Prolog contains in itself also an inference engine, Prolog can be seen as a production rule system. It employs the backward inference procedure. Prolog evaluation is a goal searching procedure, whereby a goal is put forward and then all necessary conditions ('subgoals') are searched in order to try to attain the main goal. E.g. the above mentioned example

```
thermodefect :- temp(oven,X) , X<50 , ovenlight.
```

is evaluated by asking Prolog

```
?-thermodefect.
```

Then Prolog will

1. measure the temperature;

2. compare this as ' < 50'

3. look at the oven light.

Only is if all sub-goals succeed ('and' clause) Prolog will answer 'yes'.

The action 'measure the temperature', which is typical for a real-time system, can be implemented in two ways :

- As a rule in the knowledge base, activating some special (non-Prolog) routine to actually measure the temperature;

- As a fact: i.e. the engine will look at the scoreboard for something like `temp(oven,40)`. This approach inherently supposes some other mechanism, probably a conventional foreground program in the computer, is present (see also figure 2) which periodically scans measurements end adapts the facts on the scoreboard correspondingly. The latter approach is substantially better, especially from real-time and performance constraints, although many existing tools often choose the first.

30

In order to have a full operational system, this is only part of the story. There should indeed by a kind of 'autopilot': i.e. a program which continuously and automatically questions the system for results. Just having a program which cyclically asks all possible questions would be highly inefficient. Some methodology therefore should be introduced. This could be some feed-forward mechanism which after changing facts on the scoreboard triggers all involved rules.

The reader should also realize that the above example is rather simple. In practice a questioning involves a complex evaluation procedure of many linked rules / facts. It can be seen as an (inverted) tree traversing mechanism, with many backtracking (i.e. on failure of a branch) before a successful answer eventually could be found.

8   Bibliography.

[CLOCK] W.F. Clocksin & C.S. Mellish, 'Progarmming in Prolog', Springer, Berlin (1981)

[LBOUL] L. Boullart (ed.), 'Monographic Series on Automatic Control, Vol. 1 Expert Systems', Belgian Institute for Automatic Control, Antwerp (1987)

[DOUGH] Edward Dougherty & Charles Giardina, 'Mathematical Methods for Artificial Intelligence and Autonomous Systems', Prentice-Hall, Englewood (1988)

[EDFEIG] Avron Barr & Edward Feigenbaum 'The Handbook of Artificial Intelligence, Vol. 1,2,3', Heuristech Press, Stanford (1981)

[ENISEN] Eli Nisenfeld (ed.), 'Artificial Intelligence Handbook, Vol. one & two', Instrument Society of America (1989)

[GLORIO] Robert Gloriosi & Fernando Colon Osorio, 'Engineering Intelligent Systems', Digital Equipment Corporation, Bedforf (Mass.), 1980

[SWAAN] H. De Swaan Arons & Peter van Lith, 'Expert Systemen', Academic Service, Den Haag (1984)

Appendix

A Rule based system
called "CLIPS".

Rule-based systems can be used to simulate the reasoning mechanisms of experts. They will most of the time explain their actions, decisions or advice, using certain conditions to be fulfilled before this result can be obtained. The rules therefore consist of a condition side (the *antecedent*) and an action side (the *consequent*). They are essentially of the following form:

```
IF <condition> THEN <action>
```

The antecedent is also referred to as the Left Hand Side of the rule (LHS) and the consequent part is also called the Right Hand Side (RHS). The antecedent consist of one or more conditions, checking the existence or non-existence of certain facts in a database. The consequent consists of several actions inducing certain changes in the database.

The database is the representation of the real world to the program and consists of facts, structured in a manner designed for the specific languages. This structuring varies from text strings to vectors with attribute-values, frames and objects in object-oriented languages. The facts can be changed during program execution. The database is often referred to as the *working memory* or the *scoreboard*.

The difference in rule-based programming as opposed to conventional programming, using languages like FORTRAN or C, lies in the fact that the statements in conventional programming languages are programmed in a predefined order. The instructions will be executed in the order designed by the programmer, and will never diverge from that strategy. The rule-based programming languages use an inference engine, which determines the rules to be fired only from the facts present in working memory. In this way the appearance of facts that were not expected by the programmer, can cause the program to proceed to other rules than those planned. This results, especially when the programs become very large, in less readable programs. Because of this, debugging is a rather difficult task.

Some expert system shells are created using rules as a programming paradigm. The difference between those shells and rule-based programming languages however, lies in the fact that in shells more effort has been put into a good user environment (explanation facilities and layout options for the expert system), rather than on the specification of variable types, computational aspects or the ability to use programming languages as well (the better shells excluded from this list). Rule-based programming languages often do offer those options, but do not have the mentioned extra's. In most cases, the languages they interface to, supply possibilities for that.

32

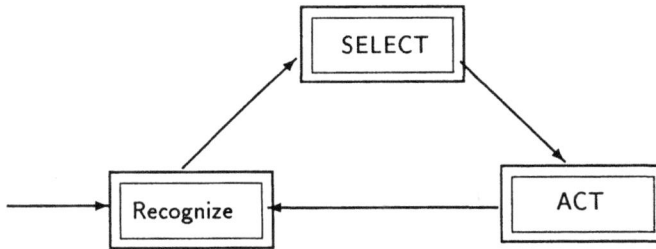

Figure 1: The Recognize-Select-Act cycle

Many explicit rule-based programming systems are commercially available on the market. Mostly they operate in a forward way, but backward evaluation can often be obtained by introducing extra facilities in the language to create goals, subgoals, e.o. A rather simple, however powerful rule based system, is 'CLIPS', developed at the Artificial Intelligence Section of the Johnson Space Center, and available on the public domain for many computer types (in fact, CLIPS is written in K&R-C and its source is available). It will be used in this text. CLIPS's structure is very similar to the OPS5 expert system language from DEC (although there are slight differences in the evaluation procedure in the conflict set).

CLIPS is in essence a forward chaining system, but it can be used for backward chaining as well. The real-world facts are stored in the working memory and the rules, which in CLIPS resemble IF-THEN constructions from some conventional programming languages, are stored in the *production memory*. The *recognize-select-act cycle* can be sketched as follows (figure 1).

Recognize: Matching of the facts on the LHS of the rules. Rules for which facts, matching the LHS, can be found, are said to be instantiated by these facts. These instantiations, together with the name of the rule, are stored in the *conflict set* or *agenda*.

Select: Selection of an instantiation to be fired. A simple stack based or a given priority scheme can be used.

Act: Executing the selected instantiation, modifying the knowledge base.

A CLIPS program can be divided into two main parts. The first part contains the declarations of the working memory elements, while in the second part the productions are specified. First, the representation of the facts will be discussed, followed by a description of the productions.

33

Facts.

The facts representing the real world are stored in the working memory. The working memory is the database with information representing the problem to be solved. The information is stored in elements, grouped into classes. These working memory elements are being matched with the conditions on the left hand sides (LHS) of the rules.

A working memory element in CLIPS consists of a class name and a number of attributes which can get values during the execution of the program. The attributes can be either a symbol (a 'word' of plain text) or a numeric value). A working memory element is specified by its class name and the list of attributes.

The class name of the working memory element indicates the class the element belongs to and often groups certain aspects of the problem to solve. The attribute names can indicate the characteristics of a certain behavior. An example of a working memory element is:

(Car Ford Mustang 1979 Jackson)

Thereby specifies:

Car	the class
Ford	the Mark
Mustang	the Type
1979	the year
Jackson	the owner

However this specification is only a matter of interpretation, and is the interpretation the designer of the rule attributes it. For CLIPS they are just a list of plain symbols, which in the recognize phase can be matched.

Within the working memory all the elements are assigned a time tag. This time tag, which is unique during the execution of the program, indicates the rank number in which the element has been created. An example of the representation of facts in the working memory:

<1> (Car Ford Mustang 1979 Jackson)
<2> (Car Rolls-Royce Silver-Shadow NIL NIL)
<3> (House Veldstraat 3 B-9000 Ghent)

The element with the highest time tag (<3>)is the most recent element. By removing elements from the working memory, the time tag is deleted, and it will not be used again within that run of the program.

Rules.

A rule, determining the actions to be taken when certain events occur, is called a production. The IF part of the IF-THEN rule, called the left-hand side of the rule (LHS), contains a number of conditions. The right hand side of the IF-THEN rule (RHS) contains the actions to be taken if the LHS of the rule is matched completely. A production in CLIPS consists of :

- 'defrule' followed by a production name

- a left hand side (LHS) containing the conditions

- a $\Rightarrow$ dividing the production in the LHS and the RHS

- a right hand side (RHS) containing one or more actions.

The production has the following standard form:

```
(defrule <name>
     ( cond-elem 1 )

     ( cond-elem n )
 =>
     ( action 1 )

     ( action m )
)
```

The production has to be enclosed by parentheses, as are the separate conditions and actions. There can be both positive and negative conditions. Positive conditions test the presence of a certain fact, while negative conditions only match if there is no element matching that condition at all.

An example of a production is :

```
(defrule Old-Car
     ( Car $? ?year ?name )
     ( test ( < ?year 1950 ) )
     ( test ( neq ?name Company ) )
 =>
     ( printout ?name "owns an old car" CRLF ) )
```

In this test, the data base will be scanned for cars built before 1950, with owners who are not companies (but persons, families, ...), and print the corresponding names.

In the conditions, some of the attribute values have been bound to so called *local variables* written as '?xxx'. In the example the year has been bound to ?year in order to test the value of year to the required value 1950 and the owners name has been bound to ?name, in order to be used in the second condition and in the action statement. The '$?' is a so called *wildcard* to skip the unused attributes 'mark' and 'type'.

The actions that can be done vary from simple messages to the user, by specifying a 'printout' statement, to making new working memory elements using 'assert', removing working elements from working memory using 'retract and reading text from the terminal, using 'read'. There are special facilities provided in CLIPS to call routines written in the C-language. This is especially useful when e.g. complex calculations are to be done (athough CLIPS can perform most mathematics by itself too). Also a CLIPS program can be called from another program as part of it execution process.

Matching of Facts on Rules in CLIPS.

In the match-phase, the working memory elements are compared to the conditions of the LHS of the rules of the production memory. For a rule each positive condition must be satisfied and elements specified in the negative conditions must not be in the working memory. Consider the following example rule:

```
(defrule Sell-Car
  ?car <-  ( Car ?mark ?type ?year ?name )
   (Sell-Car From ?name to ?new-name)
   (Pay money ?amount by ?new-name to ?name)
 =>
   (retract ?car)
   (assert (Car ?mark ?type ?year ?new-name)
   (printout ?name "sold his car to" ?new-name "FOR $"
                                      ?amount CRLF) )
```

and suppose the working memory contains the following facts:

1.   (Car Chevrolet Camaro 1975 Anderson)
2.   (Car Ford Mustang 1975 Jones)
3.   (Car Rolls-Royce Silver-Shadow ? McCenna)
4.   (Sell-Car From Anderson To Brown)
5.   (Sell-Car From Jones To Johson)
6.   (Pay money 18000 by Brown to Anderson)
7.   (Pay money 25000 by Johson to Jones)
8.   (Pay money 100000 by Rochin to McCenna)

In this example only positive conditions are being used. The first condition is matched by working memory elements 1, 2 and 3. Only the names of the original owners are being bound to the variable ?name. The second condition now has to be matched by elements from the class Sell-Car, with the sellers 'from' as the original owners (indicated by ?name from the first condition). We now see that from the match in condition 1 with element 1, element 4 matches the second condition. Also from the match of element 2 in condition 1, element 5 matches the second condition. For element 3 of the working memory, no matching element for condition 2 can be found. For the third condition elements have to be found from the class Pay with as payers the new owners, as specified in the second condition and as receivers the original owners as specified in condition 1. For the pair of elements (1, 4) matching the first two conditions, element 6 can be found that matches the third condition. Also for the pair (2, 5) matching the first two conditions, element 7 can be found that matches the third rule. Note that somebody (in this case Rochin) is paying McCenna a large amount of money, but because there is no match of the second condition with McCenna as seller and with Rochin as buyer, this transaction must have been on something else (maybe houses).

For this match the following instantiations are being placed in the *conflict set*.

```
Sell-Car 1 4 6
Sell-Car 2 5 7
```

The time tags of the elements from the obtained match of the LHS are used to create the conflict set. Within CLIPS, a Rete Net matching algorithm is used. On this very efficient matching algorithm discussions can be found in the literature.

Selection of Instantiations.

The rules that have obtained complete correct matches (together with the accompanying working elements that have caused the match) are stored in the conflict set or agenda by the Rete-Net algorithm. The next step in the cycle is to select an instantiation, which will be executed. In CLIPS two strategies are available for this selection of the instantiation: either by a LIFO-stack or by using a priority (*'salience'*). The agenda is essentially a stack. Rules are pushed onto the stack when they are selected in the previous phase. If the priority of the new rule is less than the rule currently on top of the stack, the new rule is pushed down the stack until all rules of higher priority are above it. Rules of equal or lower priority will remain below the new rule.

Other production-rule systems have more sophisticated mechanisms. OPS5 (from Carnegie Mellon, and commercialized by Digital Equipment Corporation) has two basic strategies.

LEX (Lexicographic-Sort): is a strategy consisting of three steps:

1. **Refraction.** The instantiations of the conflict set that have already been executed, are removed from the conflict set, so they will not fire again. It prevents the program from looping infinitely on the same data. Also the instantiations of which the elements have been changed by the firing of another rule, (so the match is no longer correct) are deleted from the conflict set. However, if an element is correct even after a MODIFY action, it will remain in the conflict set.
   If after this first step only one instantiation is left in the conflict set, that instantiation will be executed. If there are no instantiations left, the program halts. If more than one instantiation is left, the strategy proceeds to 2.

2. **Recency.** The elements of the instantiations will be tested on the time tags from the working memory elements that matched the rules. The instantiations with the highest time tags in the conflict set will be chosen, for the working memory elements with the highest time tags are the most recent data.

3. **Specificity.** When still more than one instantiation is candidate for being executed, the specificity of the match will have to be used for the decision. The specificity of the match is determined by the complexity of the LHS of the rule. The compiler will start looking for the total number of tests in the LHS of a production. The language specifies exactly which items are considered to be a test and which not. The production with the highest number of time tags will be executed.

4. If more than one instantiation remains candidate for being executed after these three steps, the choice will be made arbitrarily.

MEA (Means-Ends-Analysis) is an alternative strategy. MEA considers the first time tag in a series of conditions to be the most important when recency is being checked. This strategy allows the programmer to influence the order in which the rules are fired. The steps in MEA can be described as follows :

1. **Refraction.** (see LEX)

2. **First element.** Order the remaining instantiations in order of the first time tag only. If more than one instantiation has the highest level of recency, then proceed to 3.

3. **Recency.** (see LEX)

4. **Specificity.** (see LEX)

5. **Choice.** (see LEX)

In the literature [ ] and [ ] many examples are studied in detail to give a deeper insight into the mechanism.

Acting of the Rules.

Once an instantiation has been selected from the conflict set it has to be executed. In this phase the actions specified in the RHS of the rules are being performed. These actions can include writing output to the screen or to a file, changing elements from the working memory, removing rules from the production memory or adding rules, etc....

Application Example: 'Towers of Hanoi'

This game can also be solved in CLIPS by the following rule set. It can be started simply by stating '(reset)' and '(run)', after which the number of discs to move will be asked for. The example uses some features not yet explained in this brief overview, such as:

- CLIPS asserts by itself the fact (initial-fact) in working memory on a (reset) instruction. This is a handy way to start the whole process (here with the rule 'startup');

- the construct '$?' is a so called *wildcard*, which can be instantiated by any number of attributes in the fact; it is used when an attribute should not exactly match, i.e. is not needed in the rest of the rule;

- '(gensym)' generates a random symbol (each time a new one), which can be bind to a local variable by the 'bind' action; this is used here to uniquely identify each move-order;

- '=(...)' evaluates the expression between the parentheses (in prefix notation) and returns the result;

- the 'not' before a fact-clause matches the *absence* of the fact.

Notice that the program counts also the number of moves, which are reported afterwards. It should be equal to $2^n - 1$, n being the number of discs. Furthermore in the 'move-n-stones' rule, the sequence of the three asserts in the acting part is in fact the *inverse* of the required (procedural) execution order. This is necessary because of CLIPS's stack-oriented execution mechanism. In other rule based systems (e.g. OPS5), this could be different.

```
(defrule move_n_stones ""
    ?c1 <- (move_stone ?N ?origin ?dest ?buff $?)
    (test (!= ?N 1))
    =>
    (retract ?c1)
    (bind ?sym (gensym))
    (assert (move_stone =(- ?N 1) ?buff ?dest ?origin ?sym))
    (assert (move_stone 1 ?origin ?dest ?buff ?sym))
    (assert (move_stone =(- ?N 1) ?origin ?buff ?dest ?sym)))

(defrule move_1_stone ""
    ?c1 <- (move_stone 1 ?origin ?dest ?buff $?)
    =>
    (retract ?c1)
    (printout t "Move stone from " ?origin " to " ?dest crlf)
    (assert (count-now)))

(defrule startup ""
    ?c1 <- (initial-fact)
    =>
    (retract ?c1)
    (printout t "How many stones to move from pile A to B using C? ")
```

```
      (assert (move_stone =(read) A B C))
      (assert (count 0)))

(defrule done-message ""
   (not (move_stone $?))
   (not (initial-fact))
   ?c2 <- (count ?n)
   =>
   (retract ?c2)
   (printout t "All done; " ?n " moves" crlf))

(defrule count-move ""
   ?c1 <- (count-now)
   ?c2 <- (count ?x)
   =>
   (retract ?c1 ?c2)
   (assert (count =(+ ?x 1))))
```

Knowledge Representation by Logic

Prof.Dr.ir Luc Boullart
Automatic Control Laboratory, University of Ghent.

1 Introduction.

There exist a variety of ways to represent knowledge (facts,...) in A.I.-programs. When dealing with facts, there are actually two different aspects to consider:

- Facts themselves in some relevant world. These are the things we like to represent.

- Representation of facts in some formalism. These are the things we actually like to manipulate.

What we need therefore is a mapping mechanism (in both directions) between them. Furthermore, in any problem we need to input or output information for those facts. Consequently, regardless of the formalism chosen, we will have to deal also with natural (e.g. Dutch) language sentences (figure 1).

Usually mapping functions are not always one-to-one, but merely many-to-many relations: i.e. each object in the domain maps to several elements in the representation space, and vice versa. This may confuse the interpretation of the results of an automated reasoning system profoundly. The sentence "All mothers have children" could mean different things depending on the interpretation:

- each mother has one child;

- each mother has at least one child;

- each mother has several children.

This means the operation of each A.I.-program should be carried out in full accordance and consistency with both the designer's built in options, and with the ideas of the user having entered the knowledge data into the system.

One of the oldest representations methods is *logic*. It was developed by logicians as a means for formal reasoning, primarily in the area of mathematics. It is now widely used as one of the important schemes used in the A.I.-world. It even has found itself a protagonist in the language 'Prolog', which employs one of the logic formalisms, i.e. 'Predicate Calculus' as the underlying base.

Logic is a formal method for reasoning. Many concepts which can be verbalized can be translated into symbolic representations which closely approximate the meaning of these

41

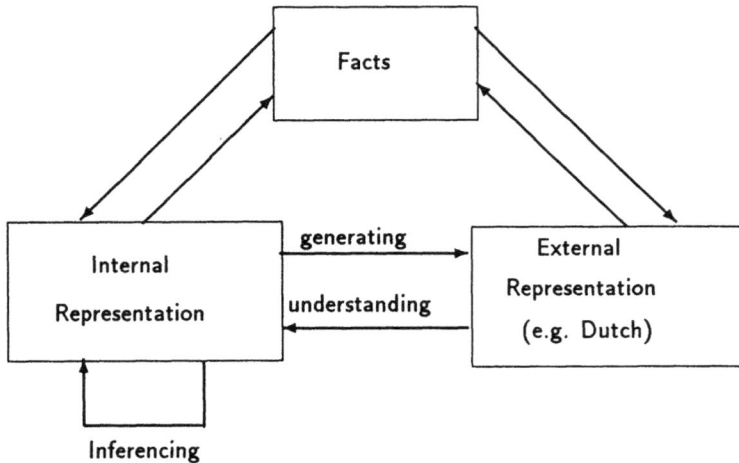

Figure 1: Mapping / Representation of Facts

concepts. These symbolic structures can then be manipulated in programs to deduce various facts, to carry out a form of automated reasoning.

In 'First Order Predicate Logic (FOPL)', statements from a natural language like English are translated into symbolic structures comprised of predicates, functions, variables, constants, quantifiers, and logic connectives. The symbols form the basic building blocks for the knowledge, and their combination into valid structures is accomplished using the syntax (rules of combination) for FOPL. Once structures have been created to represent basic facts or procedures or other types of knowledge, inference rules may then be applied to compare, combine and transform these 'assumed' structures into new 'deduced' structures. This is how automated reasoning or inferencing is acquired.

As a simple example of the use of logic, the statement "All Dutchmen are Europeans" might be written in FOPL as:

$$(\forall x) \ (DUTCHMAN(x) \Rightarrow EUROPEAN(x))$$

Here, $\forall x$ is read as 'for all x' and $\Rightarrow$ as 'implies' or 'then'. The predicates
$$DUTCHMAN(x), \text{ and } EUROPEAN(x)$$
are read as 'if x is a Dutchman,' and 'x is an European' respectively. The symbol x is a variable which can assume a person's name. Therefore, if it is known that Erasmus is born in the Netherlands:

$$DUTCHMAN(Erasmus)$$

42

one can draw the conclusion that Erasmus is an European:

$$EUROPEAN(Erasmus)$$

The example clearly shows that the transformation of simple English sentences into FOPL-statements is rather straightforward. If an appropriate software system would be at hand, those FOPL-statements could be typed in directly into the knowledge base, and the inference engine started.

In this text we will deal with the two following forms of logic:

- Propositional Logic;

- Predicate Logic.

It will be shown that, although the first form can be used to perform automated reasoning, it is rather harsh and unflexible. The second form will lead to more useful implementations in the field of Artificial Intelligence.

2 Propositional Logic.

Propositional Logic involves on the one hand the judgment of the *propositions (facts)* which are formulated by the language, and the other hand the *rules* to specify how to *deduct new propositions* from the former ones. Propositions are *elementary atomic sentences*, also called 'well-formed formulas (wffs)'. They may have either have the Boolean value 'true' or 'false', but no other value.

$$
\begin{aligned}
Erasmus \quad & is\ a\ Dutchman. \quad (\rightarrow true) \\
Erasmus \quad & is\ born\ in\ Brussels. \quad (\rightarrow false)
\end{aligned}
$$

More complex propositions are given by the logic connectives, formed with the following primitive operations:

And (conjunction)	$\wedge$ or &
Or (disjunction)	$\vee$
Not (negation)	$\neg$
Implication	$\Rightarrow$
Equivalence (double implication)	$\equiv$ or $\Leftrightarrow$

with the following truth tables.

X	Y	$X \wedge Y$	$X \vee Y$	$X \Rightarrow Y$	$\neg X$	$X \equiv Y$
T	T	T	T	T	F	T
T	F	F	T	F	F	F
F	T	F	T	T	T	F
F	F	F	F	T	T	T

43

A common precedence rule in propositional logic is from high to low:

$$\neg \quad \wedge \quad \vee \quad \Rightarrow \quad \equiv$$

Remarks:

- Implication means: '*if* we assume that X is true, *then* Y must be so'. I.e. the truth of X implies the truth of Y. This does not exclude of course that Y may be true when X is false, but it does exclude that Y would be false when X is true.

- The 'Or' is seen as 'inclusive': i.e. the result is also true when both X and Y are true. This is often in contradiction with usual English sentences where an 'exclusive Or' ($\oplus$) is meant. However the latter can easily be expressed as:

$$(X \vee Y) \wedge \neg(X \wedge Y)$$

- The 'Equivalence' ($\equiv$) can also be seen as a double implication, i.e. "if and only if ... then ...".

Propositional Logic draws logic conclusions based on the *modus ponens* rule,

$$(X \wedge (X \Rightarrow Y)) \Rightarrow Y$$

i.e. when $X \Rightarrow Y$ is known to exist, and X is true, then Y will also be true. Note that if we think of X and $X \Rightarrow Y$ as two entries in a database, the modus ponens rule allows us to replace them with the single statement, Y, thus eliminating one occurrence of the connective '$\Rightarrow$'. In what are called natural deduction systems of logic, there are typically two rules of inference for each connective, one that introduces it into expressions and one that eliminates it. Modus ponens is therefore called the '$\Rightarrow$'-elimination rule.

3   The resolution of Propositional Logic Connectives.

It would be useful from a computational point of view if we had a proof procedure that carried out in a single operation the variety of processes involved in reasoning with statements in predicate logic. Resolution is such a procedure, which gains its efficiency from the fact that it operates on statements that have been converted to a very convenient standard form.

Resolution produces proofs by refutation. In other words, to prove a statement (i.e., show that it is valid), resolution attempts to show that the negation of the statement produces a contradiction with the known statements (i.e., that it is unsatisfiable).

One of the steps which has to be carried out *before* the actual resolution process starts, is to convert all clauses to some standard form. This will be shown to be necessary too before the unification process in predicate calculus (see further) can take place. One form which is often used, is the so called '*conjunctive normal form*'. In this form the result is

44

Idempotency	$X \vee X = X$
	$X \wedge X = X$
Associativity	$(X \vee Y) \vee Z = X \vee (Y \vee Z)$
	$(X \wedge Y) \wedge Z = X \wedge (Y \wedge Z)$
Commutativity	$X \vee Y = Y \vee X$
	$X \wedge Y = Y \wedge X$
	$X \equiv Y = Y \equiv X$
Distributivity	$X \wedge (Y \vee Z) = (X \wedge Y) \vee (X \wedge Z)$
	$X \vee (Y \wedge Z) = (X \vee Y) \wedge (X \vee Z)$
De Morgan's Law	$\neg(X \vee Y) = \neg X \wedge \neg Y$
	$\neg(X \wedge Y) = \neg X \vee \neg Y$
Conditional Elimination	$X \Rightarrow Y = \neg X \vee Y$
Bi-Conditional Elimination	$X \equiv Y = (X \Rightarrow Y) \wedge (Y \Rightarrow X)$

Figure 2: Equivalence Laws

a set of clauses which are conjunctive (i.e. they all should be true in order to succeed), where each clause itself is a disjunction of literals. To convert whatever clauses to such normal form, a stepwise algorithm should be executed (see [ELRICH]), during which the equivalence laws of figure 2 have to be applied, although in many cases the conversion will be intuitive.

Example.

$$(rain \wedge umbrella) \vee (sun \wedge parasol)$$

becomes, after applying the distributive law twice

$$[rain \vee (sun \wedge parasol)] \wedge [umbrella \vee (sun \wedge parasol)]$$

or

$$(rain \vee sun) \wedge (rain \vee parasol) \wedge (umbrella \vee sun) \wedge (umbrella \vee parasol)$$

which gives the 4 following conjunctive clauses:

1. $(rain \vee sun)$

2. $(rain \vee parasol)$

3. $(umbrella \vee sun)$

45

4. (*umbrella* ∨ *parasol*)

The resolution procedure is a simple iterative process, at each step of which two clauses, called the parent clauses, are compared (resolved), yielding a new clause that has been inferred from them. The new clause represents ways that the two parent clauses interact with each other. Suppose that there are two clauses in the system:

winter ∨ *summer*

¬*winter* ∨ *cold*

Recall that this means that both clauses must be true (i.e., the clauses, although they look independent, are really conjoined).

Now we observe that precisely one of *winter* and ¬*winter* will be true at any point. If *winter* is true, then *cold* must be true to guarantee the truth of the second clause. If ¬*winter* is true, then *summer* must be true to guarantee the truth of the first clause. Thus we see that from these two clauses we can deduce

summer ∨ *cold*

This is the deduction that the resolution procedure will make. Resolution operates by taking two clauses that each contain the same literal, in this example, *winter*. The literal must occur in positive form in one clause and in negative form in the other. The *resolvent* is obtained by combining all of the literals of the two parent clauses except the ones that cancel.

If the clause that is produced is the empty clause, then a contradiction has been found. For example, the two clauses

winter

and

¬*winter*

will produce the empty clause. If a contradiction exists, then eventually it will be found. Of course, if no contradiction exists, it is possible that the procedure will never terminate, although as will be seen further there are often ways of detecting that no contradiction exists (e.g. look out for looping).

The Resolution Procedure [ELRICH].

In propositional logic, the procedure for producing a proof by resolution of proposition S with respect to a set of axioms F is the following:

1. Convert all the propositions of F to the standard clause form.

Original	Standard Form
p	p
$(p \wedge q) \Rightarrow r$	$\neg p \vee \neg q \vee r$
$(s \vee t) \Rightarrow q$	$\neg s \vee q$
	$\neg t \vee q$
t	t

Figure 3: Converting Propositions into Standard Clauses

2. *Negate* S and convert the result to standard clause form. Add it to the set of clauses obtained in step 1.

3. Repeat until either a contradiction is found or no progress can be made:

   (a) Select two clauses. Call these the parent clauses.

   (b) Resolve them together. The resulting clause, called the *resolvent*, will be the disjunction of all of the literals of both of the parent clauses with the following exception: if there are any pairs of literals L and $\neg L$, such that one of the parent clauses contains L and the other contains $\neg L$, then eliminate both L and $\neg L$ from the resolvent.

   (c) If the resolvent is the empty clause, then a contradiction has been found. If it is not, then add it to the set of clauses available to the procedure.

Let's look at an example with the axioms shown in the first column of Figure 3 and we want to prove r. First we convert the axioms to clause form, as shown in the second column of the figure. Then we negate r, producing $\neg r$, which is already in clause form. Then we begin selecting pairs of clauses to resolve together. Although any pair of clauses can be resolved, only those pairs that contain complementary literals will produce a resolvent that is likely to lead to the goal of producing the empty clause (shown as a box). We might, for example, generate the sequence of resolvents shown in Figure 4. We begin by resolving with the clause $\neg r$ since that is one of the clauses that must be involved in the contradiction we are trying to find.

One way of viewing the resolution process is that it takes a set of clauses all of which are assumed to be true. It generates new clauses that represent restrictions on the way each of those original clauses can be made true, based on information provided by the others. A contradiction occurs when a clause becomes so restricted that there is no way it can be true. This is indicated by the generation of the empty clause. The tree-representation in the Figure 4 indicates the whole procedure up to the final step where an empty clause is found, indicating thereby a contradiction. As this contradiction is between the *negated* proposition $\neg r$ and the rest of the connectives, this thereby proves r to be true.

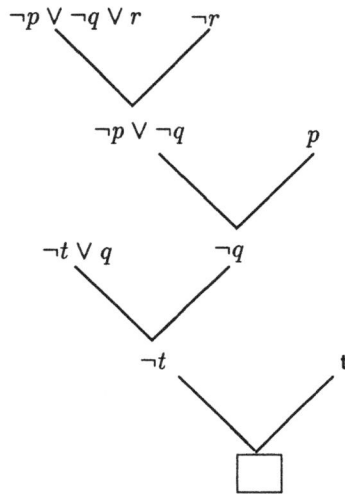

Figure 4: Resolution Tree

4 Predicate Logic.

For the purposes of AI, propositional logic is not very useful. In order to capture adequately into a formalism our knowledge of the world, we need not only to be able to express true or false propositions, but also to be able to speak of objects, to postulate relationships between these objects, and to generalize these relationships over classes of objects. We turn to the *predicate calculus* to accomplish these objectives.

The predicate calculus is an extension of the notions of the propositional calculus. The meanings of the connectives ($\wedge$, $\vee$, $\Rightarrow$, $\neg$, and $\equiv$) are retained, but the focus of the logic is changed. Instead of looking at sentences that are of interest merely for their truth value, predicate calculus is used to represent statements about specific objects, or *individuals*. The most important advantage compared with propositional logic is that *variables* can be used in clauses. Those variables are a kind of placeholders that can be instantiated (e.g. during the inference process) by any individuals matching the predicate.

Predicates: are statements about individuals both about themselves and in relation to other individuals. They are applied to a number of arguments and can have the value of either *true* or *false* after they have been instantiated by their arguments.

Example

$$Man(X)$$

48

$$Man(X) \Rightarrow Drivescar(X)$$

The first predicate may be

true for X=John
false for X=Lilian

which means that somewhere in the data base there should be a (direct or indirect) prove that $Man(John)$ is *true*, and that no prove can be found that $Man(Lilian)$ is *true*. The second predicate says that, if an individual exists for which $Man(X)$ is *true*, that for same individual $Drivescar(X)$ will be *true*. Probably this will express the intention of the designer that "if someone is a man, he drives a car". Those clauses thereby say that the individual Lilian will not drive a car, unless perhaps two other predicates would exist, stating:

$$Woman(X) \Rightarrow Drivescar(X)$$
$$Woman(Lilian)$$

Predicates can have more than one argument. So for example:

$$DrivesLimo(X,Y)$$
$$Greater-than(X,Y)$$

where X may stand for a person (e.g. John)
Y may stand for a car type (e.g. Rolls-Royce)
Greater-than(7,4) is *true*
Greater-than(3,4) is *false*

Each one-place predicate defines what is called a *set* or *sort*. That is, for any one-place predicate, all individuals X can be sorted into two disjoint groups, with those objects that *satisfy P* (for which $P(X)$ is *true*) forming one group, and those that don't satisfy P in the other. Sorts may include thereby other sorts.

Remark: For a matter of convenience, the last letters of the alphabet (X,Y,Z) are used for variables, while the first letters (A,B,C) are reserved for individuals.

Quantifiers: We shall often have occasion to refer to facts that we know to be true of all or some of the members of a sort. For this, we introduce besides a variable the notion of *quantifier*. A variable is a place holder, one that is to be filled in by some constant, as X has been used in this article. On the other hand there are two quantifiers, $\forall$, meaning *for all...*, and $\exists$, meaning *there exists...* The English-language sentence "All men are mortal" is thus expressed in predicate calculus, using the variable X as

49

$$\forall X.\ Man(X) \Rightarrow Mortal(X)$$

which is loosely rendered, "For all individuals X if X is a man (i.e., $Man(X)$ is *true*), then X is mortal." The English sentence "There is a course which is advanced and deals with AI" becomes the predicate calculus statement

$$\exists X.\ Course(X) \wedge Advanced(X) \wedge AI(X)$$

More complicated expressions, or *well-formed formulas (WFFs)*, are created with syntactically allowed combinations of the connectives, predicates, constants, variables, and quantifiers.

Inference rules for quantifiers: In a typical natural deduction system, use of the quantifiers implies the introduction of four more *inference rules*, one for the introduction and elimination of each of the two quantifiers. For example, the ∀-elimination, or *universal specialization* rule states that, for any well-formed expression Φ that mentions a variable X, if we have

$$\forall X.\ \Phi(X)$$

we can conclude

$$\Phi(A)$$

for any individual A. In other words, if we know, for example,

$$\forall X.\ Man(X) \Rightarrow Mortal(X)$$

we can apply this to the individual Erasmus, using the ∀-elimination rule, to get:

$$Man(Erasmus) \Rightarrow Mortal(Erasmus).$$

The rules of the propositional calculus, extended by predicates, quantification, and the inference rules for quantifiers, result in the *predicate calculus*. The inference algorithm itself is called *unification* (see further).

The rules of the propositional calculus, extended by predicates, quantification, and the inference rules for quantifiers, result in the predicate calculus.

First order logic: Predicate Calculus is very general, and often quite clumsy. Two other additions to the logic will make some things easier to say, without really extending the range of what can be expressed.

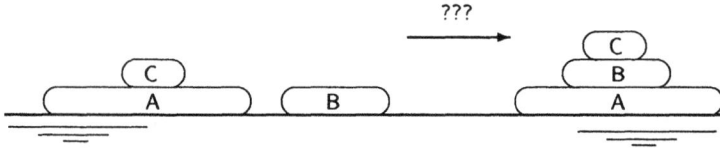

Figure 5: Stacking Disc Problem

1. *Operators or Functions:* Like predicates, they have a fixed number of arguments; but functions are different from predicates in that they do not just have the values *true* or *false*, but they "return" objects related to their arguments. For example, the function father-of when applied to the individual Mary would return the value John. Other examples of functions are absolute-value, plus, etc... Each of the arguments of a function can be a variable, a constant, or a function (with its arguments). Functions can, of course, be combined; we can speak of the father-of (father-of (John)), who would, of course, be John's paternal grandfather.

2. *Equals:* Two individuals X and Y are equal if and only if they are indistinguishable under all predicates and functions. More formally, $X = Y$ if and only if, for all predicates P, $P(X) \equiv P(Y)$, and also for all functions F, $F(X) = F(Y)$.

What we arrive at with these additions is no longer pure predicate calculus; it is a variety of *first-order logic*. (A logic is of first order if it permits quantification over individuals but not over predicates and functions. For example, a statement like "All predicates have only one argument" cannot be expressed in a first-order theory.) First- order logic is both *sound* (it is impossible to prove a false statement) and *complete* (any true statement has a proof).

Example.
   The problem in figure 5 requires to handle a number of discs from the starting point at the left to its final position as indicated on the right. The task is a sub-problem of the well known 'Towers of Hanoi' game.
   The initial situation can formally be described as follows:

```
ON(C,A)
FLOOR(A)
FLOOR(B)
FREE(C)
FREE(B)
```

51

The final goal (bottom up) on the other hand is:

```
FLOOR(A)
ON(B,A)
ON(C,B)
FREE(C)
```

Some predicates have to be declared more precisely:

$$FREE(X) \equiv \neg(\exists Y.\ ON(Y,X)) \tag{1}$$

$$ON(Y,X) \wedge REMOVE(Y,X) \Rightarrow FREE(X) \wedge \neg ON(Y,X) \tag{2}$$

$$FREE(X) \wedge FREE(Y) \wedge STACK(X,Y) \Rightarrow ON(X,Y) \tag{3}$$

The clause (1) defines when a disc is seen to be free, clause (2) defines the *action* 'REMOVE', while the clause (3) says how to 'STACK' a disc.

Inferencing Sequence.

1. The first goal to fulfill is:

   ```
   FLOOR(A)
   ```

   which is *true* in the initial state, and thereby fulfilled.

2. The next subgoal is:

   ```
   ON(B,A)
   ```

3. Inferencing between this subgoal and the 'STACK' definition in (3) gives another three subgoals:

   ```
   FREE(B)
   FREE(A)
   STACK(B,A)
   ```

4. The clause FREE(B) proves to be *true* (see initial state).

5. The clause FREE(A) raises two new clauses by inferencing with the 'REMOVE' definition from (2):

   ```
   ON(Y,A)
   REMOVE(Y,A)
   ```

6. The clause ON(Y,A) instantiates to ON(C,A) in the initial state.

52

7. The clause `REMOVE(Y,A)` instantiates then to `REMOVE(C,A)`, thereby ordering the action 'REMOVE!'.

8. The next subgoal is now `STACK(B,A)` which orders another action 'STACK!'.

9. As next in row, we have `ON(C,B)` which can be reached by the three next subgoals:

    ```
    FREE(C)
    FREE(B)
    STACK(C,B)
    ```

10. The clauses `FREE(C)` and `FREE(B)` prove to be true, while `STACK(C,B)` involves again an action 'STACK!'.

11. The last goal `FREE(C)` is also true, and thereby **the whole goal succeeds**.

The overall action list, thereby becomes:

Actions
REMOVE(C,A)
STACK(B,A)
STACK(C,B)

which indeed solves our original problem completely.

It should be clear by careful examination of this example, that the solution as proposed is rather 'ad hoc' and that e.g. changing the order of the final goals would lead to a different solution, if at all. However it will be shown in the next paragraph that there exists a generalized *Unification Algorithm* to deal with the generic solution of such goal-seeking problems.

5 Unification in Predicate Logic.

The basic idea of unification is very simple. Clauses which have the same predicate can be unified if the 'arguments' are either identical or can be substituted to make them identical. E.g.

$$Man(X)$$
$$\neg Man(John)$$

can be unified since the predicates ('Man') are identical, and since the substitution John/x can be carried out (and leads in this example to a contradiction). The only restriction thereby is that, when a substitution is carried out, it should be done consistently throughout the whole clause. E.g.

$$\neg Roman(X) \lor loyalto(X, Caesar) \lor hate(X, Caesar)$$
$$\neg hate(Marcus, Caesar)$$

53

requires a threefold substitution of Marcus/X, and thereby unifies to the following clause:

$$\neg Roman(Marcus) \lor loyalto(Marcus, Caesar)$$

From this example it should be clear that resolution in predicate logic is more complicated than in propositional logic. Suppose the following clause:

$$Man(John)$$
$$\neg Man(X) \lor Drivescar(X)$$

Based on the resolution algorithm in propositional logic, one would be tempted to reduce those two clauses to the following one:

$$Drivescar(X)$$

This is not correct, because the reduction is only valid under the assumption that John substitutes X. Therefore the correct reduction should be:

$$Drivescar(John)$$

The detailed unification algorithm can be found in [ELRICH], but the main idea of its operation is given below.

Given that that standardization has been done, it is easy to determine how the unifier must be used to perform substitutions to create the resolvent. If two instances of the same variable occur, then they must be given identical substitutions.

We can now state the resolution algorithm for predicate logic as follows, assuming a set of given statements F and a statement to be proved S:

1. Convert all the statements of F to clause form.

2. Negate S and convert the result to clause form. Add it to the set of clauses obtained in 1.

3. Repeat until either a contradiction is found, no progress can be made, or a predetermined amount of effort has been expended:

   (a) Select two clauses. Call these the parent clauses.

   (b) Resolve them together. The resolvent will be the disjunction of all of the literals of both of the parent clauses with appropriate substitutions performed and with the following exception: If there is a pair of literals $T1$ and $\neg T2$ such that one of the parent clauses contains $T1$ and the other contains $T2$ and if $T1$ and $T2$ are unifiable, then neither $T1$ nor $T2$ should appear in the resolvent. We will call $T1$ and $T2$ complimentary literals. Use the substitution produced by the unification to create the resolvent.

54

(c) If the resolvent is the empty clause, then a contradiction has been found. If it is not, then add it to the set of clauses available to the procedure.

If the choice of clauses to resolve together at each step is made in certain systematic ways, then the resolution procedure will find a contradiction if one exists. However, it may take a very long time. There exist strategies for making the choice that can speed up the process considerably.

Example. Consider the following predicates [ELRICH].

1. Marcus was a man.
   $Man(Marcus)$

2. Marcus was a Pompeian.
   $Pompeian(Marcus)$

3. All Pompeians were Romans.
   $\forall X \quad Pompeian(X) \Rightarrow Roman(X)$

4. Caesar was a ruler.
   $Ruler(Caesar)$

5. All Romans were either loyal to Caesar or hated him.
   $\forall X \quad Roman(X) \Rightarrow loyalto(X, Caesar) \vee hate(X, Caesar)$

6. Everyone is loyal to someone.
   $\forall X \quad \exists Y \quad loyalto(X, Y)$

7. People only try to assassinate rulers they are not loyal to.
   $\forall X \quad \forall Y \quad person(X) \wedge ruler(Y) \wedge tryassassinate(X, Y) \Rightarrow \neg loyalto(X, Y)$

8. Marcus tried to assassinate Caesar.
   $tryassassinate(Marcus, Caesar)$

After conversion to a normal clause form, we arrive at the clause-set as in figure 6.

Let's try to solve the proof of the statement "Did Marcus hate Caesar?'

$$hate(Marcus, Caesar)$$

We will therefore add to the 8 clauses from the figure 6 the 9th statement:

$$\neg hate(Marcus, Caesar)$$

and look if we run eventually into a contradiction as drawn in the tree of figure 7.

55

1	$man(Marcus)$
2	$Pompeian(Marcus)$
3	$\neg Pompeian(X1) \lor Roman(X1)$
4	$ruler(Caesar)$
5	$\neg Roman(X2) \lor loyalto(X2, Caesar) \lor hate(X2, Caesar)$
6	$loyalto(X3, Y1)$
7	$\neg man(X4) \lor \neg ruler(Y2) \lor \neg tryassassinate(X4, Y2) \lor \neg loyalto(X4, Y2)$
8	$tryassassinate(Marcus, Caesar)$

Figure 6: The Pompeian Story.

On the other hand, if we had to give a resolution proof of

$$\neg hate(Marcus, Caesar)$$

we had to look for a contradiction of the complementary clause

$$hate(Marcus, Caesar)$$

for which it is immediately clear that there exists no contradiction, for:

- there is no $\neg hate$ to eliminate;

- clause 5 is made true by substituting Marcus/X2, and there is no further inference.

In practice of course, this may not be seen at once, but only after some preliminary evaluation steps, as is the case when we try to give proof of the following:

$$loyalto(Marcus, Caesar)$$

by adding the complementary statement

$$\neg loyalto(Marcus, Caesar)$$

As can be seen in figure 8, after some steps we arrive at the clause

$$hate(Marcus, Caesar)$$

which cannot be resolved any further, therefore stands as it is and should be added to the set of connectives.

Sometimes, the conclusion can only be drawn by observing a loop in the resolution process, which happens when a previous clause suddenly reappears. Suppose for example we add to the connective set a statement that "People who hate each other, persecute each other and vice versa"

$$persecute(X, Y) \equiv hate(Y, X)$$

which can be converted to two extra normal form clauses:

56

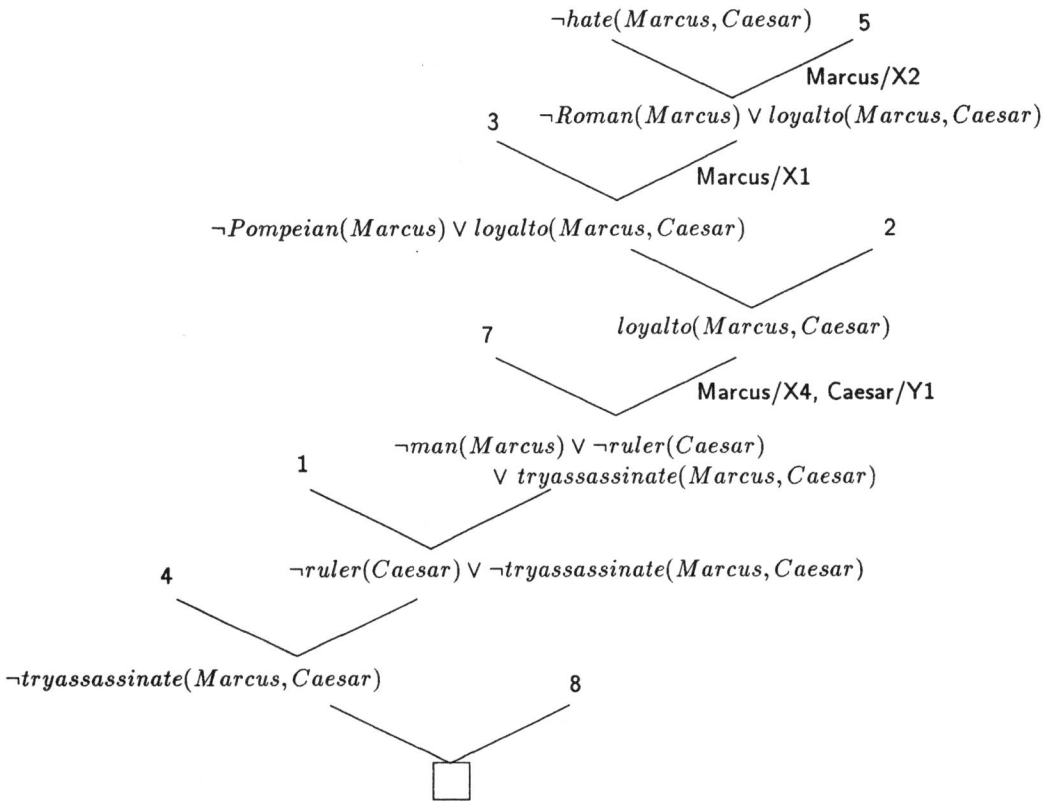

Figure 7: Solving the question whether Marcus hated Caesar

$$\neg loyalto(Marcus, Caesar) \qquad 5$$

Marcus/X2

$$\neg Roman(Marcus) \vee hate(Marcus, Caesar) \qquad 3$$

Marcus/X1

$$2 \qquad \neg Pompeian(Marcus) \vee hate(Marcus, Caesar)$$

$$hate(Marcus, Caesar)$$

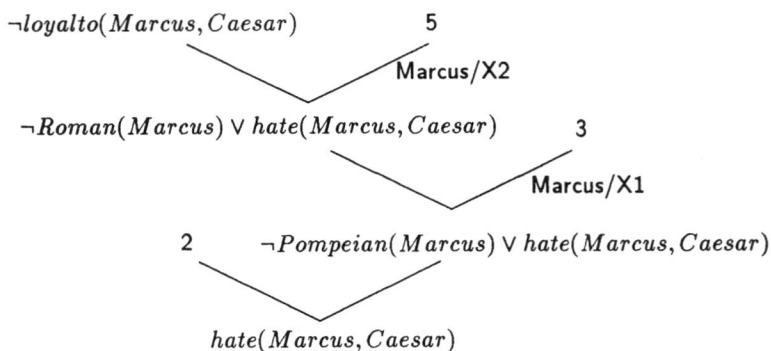

Figure 8: Testing Marcus' loyalty

9	$\neg persecute(X5, Y3) \vee hate(Y3, X5)$
10	$\neg hate(X6, Y4) \vee persecute(Y4, X6)$

If we now try to look for a contradiction for

$$hate(Marcus, Caesar)$$

we get the tree of figure 9. The same clause reappears after two resolution steps and cannot be resolved any more. Therefore again it stands as it is.

Another point to mention is that during the unification procedure, it may be necessary to try out several alternative paths. In the tree-structure this happens when at a certain point more than one clause matches. The correct way to deal with the situation, is to firstly make an arbitrary choice, and resolve it to the end of the branch. Afterwards one has to return to the culminating point to step into the other alternative(s). This mechanism is widely known as 'backtracking' and makes the whole resolution process often highly time consuming, even with modern high speed computers. In the Prolog language an ad-hoc (procedural) intervention can be programmed to deal with it ('cut'-mechanism).

The unification algorithm also enables to search into the clause space instantiations which fulfill a certain demand. This is kind of 'question answering' facility. As an example, suppose the following clauses are given:

$$Pompeian(Marcus)$$
$$\neg Pompeian(X) \vee died(X, 79)$$

and we like an answer to the question "When did Marcus died?"

$$\exists t \; died(Marcus, t)$$

58

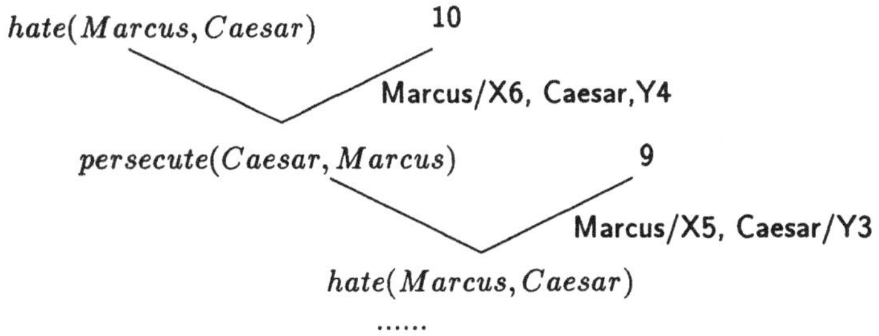

$$hate(Marcus, Caesar) \qquad 10$$

Marcus/X6, Caesar,Y4

$$persecute(Caesar, Marcus) \qquad 9$$

Marcus/X5, Caesar/Y3

$$hate(Marcus, Caesar)$$

......

Figure 9: Looping in the Resolution

Therefore, we will add the fact

$$\neg \exists t \quad died(Marcus, t)$$

or

$$\forall t \quad \neg died(Marcus, t)$$

and prove this to be in contradiction with the given clauses. The figure 10 gives the resolution tree, which shows that a contradiction is found for a substitution of 79/t and Marcus/X1. Once this contradiction has been found, in order to give the required answer, one has to grasp back to the original substitution (79/t), which should have been memorized somehow by the system.

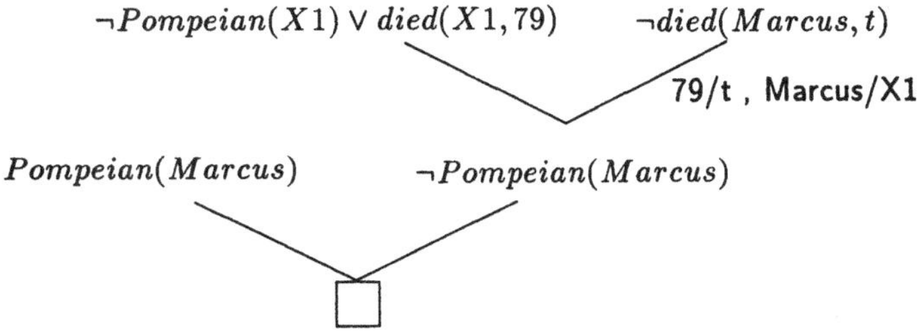

$$\neg Pompeian(X1) \lor died(X1, 79) \qquad \neg died(Marcus, t)$$

79/t , Marcus/X1

$$Pompeian(Marcus) \qquad \neg Pompeian(Marcus)$$

□

Figure 10: When did Marcus died?

59

6 Conclusion.

This text has discussed two important forms of logic: propositional and predicate. Such logic is used in many forms of A.I. inference machines, such as Prolog based systems and production rules. Nevertheless, there are still other forms which are also important in this perspective.

Nonmonotonic Logic: thereby facts and statements can be deleted or added to the knowledge base, e.g. to enforce proving capabilities or to eliminate inconsistencies between such statements. Also this type of logic enables the introduction of 'belief', when no real prove but only some general indication but no contradiction can be found for a certain statement.

Probabilistic Logic: thereby the likelihood for the belief or non-belief of statements can be accompanied with probability figures. Combination of statements should consequently then also take into account the probabilistic mathematics.

Fuzzy Logic: to represent properties of objects which are fuzzy or have a continuous nature. Fuzzy logic has it own mathematical base, and is of great importance, especially in process control matters.

For the detailed study of those other types of logic, the reader is referred to the literature (e.g. [ELRICH]), or to other course texts.

7 Bibliography.

[BONNET] A. Bonnet, 'Kunstmatige Intelligentie', Addison Wesley Nederland B.V. (1987)

[ELRICH] Elaine Rich, 'Artificial Intelligence', Mc.Graw Hill, London (1983)

[PATTER] Dan Patterson, 'Introduction to Artificial Intelligence and Expert Systems', Prentice Hall, Englewood (1990)

Object Orientation and Object Oriented Programming.

Prof.Dr.ir Luc Boullart

Automatic Control Laboratory, University of Ghent.

1 Introduction.

Object Orientation is a formalism which resembles strongly to normal human behavior. In the surrounding world exists a continuous interaction between human beings ('objects'), which often results in a scene where one person ('the boss') requests a service from another one ('the servant'). For example: "Bring all books from the second floor downstairs". Thereby the requester of this job is not at all interested in the detailed procedure of the executor to carry out his task: as long as everything runs smoothly and efficiently he won't worry. This shows that the abstraction of each object is fixed by itself at the correct necessary level. The book carrier has to know exactly how a book caddy and an elevator operate, but he is probably not at all interested in the detailed operation of the caddy itself, nor of the elevator, and certainly not in the contents of the books. The description of a task in an object oriented perspective clearly has an anthropomorphic aspect, and may be seen as a world ruled by a 'client/server' model (see figure 1).

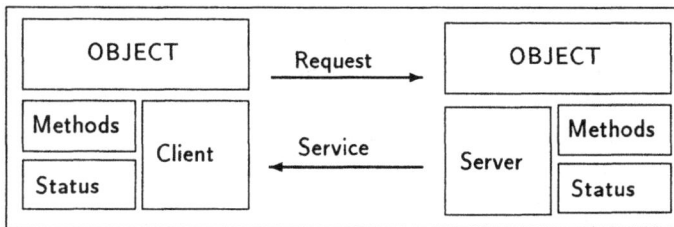

Figure 1: Object Orientation as a Client/Server model

In this client/server model, a *client*-object may issue a *request* to a *server* for the execution of a task: i.e. to deliver a *service*. Each object disposes thereby of a number of *methods*. Some of these methods are visible to the client ('public'), and may be activated as such by the requester. Other methods, which are only important for the object itself (the details) are consciously hidden from the outside world ('private'). Finally each object is also responsible for the consistency of its internal state, and has to take care of its own

61

domestic duties. The server-object by this philosophy shows to the outside world a certain intelligent behavior: by its internal tricks and treats it can execute tasks independently, without any need for outside help.

This approach shows a number of advantages.

- the identity of an object to the outside world can remain unchanged for its full life cycle;

- the internal functionality ('internal kitchen') can undergo important changes when necessary.

Suppose some object to represent a table, which an end-user may modify by inserting, deleting,... elements. In such case, the internal representation of the table may well change fundamentally during its life (e.g. an indexed table instead of a linked list because of efficiency), without noticing the outside user at all. The object oriented model is thus very *robust*. Re-usability is stimulated and the maintainability of the software increases considerably.

Object Orientation also decreases the *conceptual gap* for the end-user, for the following reasons.

- Conventional software performs a translation between our living world full of concrete objects, and an obscure world full of bits, variables, boolean values, loops, go-to's,...

- Object Orientation on the other hand does a translation between the same world-objects and software-objects which they resemble strongly, thereby hiding all internal details (bits,...) as much as possible.

The essence of object orientation is the creation of an *illusion for the end-user*.

An excellent example is how computers are used in process control. In the fifties and sixties industrial processes were controlled by a bunch of measuring and controlling equipment: operators had a direct impact on these concrete objects. This equipment can be replaced without any problem by a process computer, which uses control software to perform the task. When this is done by a usual (e.g. Pascal) program, the end-users (process analyst and operator) will undoubtedly loose their track in the labyrinth of program instructions and constructions. If, on the other hand, a formalism is carefully built to encapsulate the early day's equipment in software objects, which communicate and interact by request/service procedures, then the process operator will find himself comfortable in this illusion, even when behind the smoke curtain the same algorithms are being executed.

2   Basic Principles of Object Orientation and Object Oriented Programming.

An object consists of a number of data encapsulated inside procedures. Those procedures are called *methods*, and are the only way to get access to these data (figure 2). Methods are formally defined in a *class* (see further), and can have two basic forms:

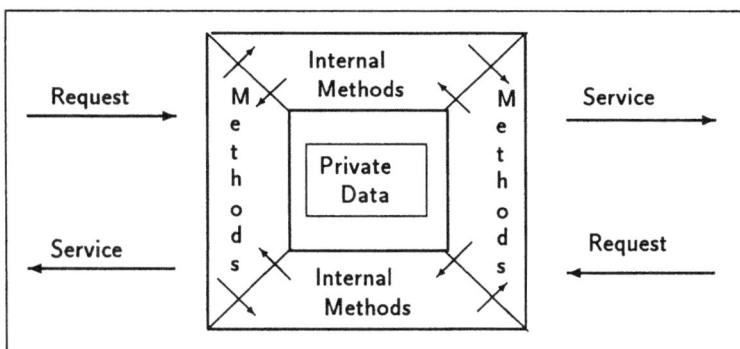

Figure 2: An Object and its interface

- they may be *public accessible*: i.e. they may be called by any other object for a service;

- they may be *private accessible*: i.e. they only may be called by the own object itself (to provide internal services).

Some object oriented programming languages still have other alternative and more subtle mechanisms for access by methods. C++ has the possibility to declare other classes as 'friend', whereby methods of that foreign class can be defined for accessing private data. C++ can also declare a method as *protected*, whereby it can be considered as public by the classes from the same hierarchy (see further), but as private by all other classes. Even for data, sometimes the difference is made between public and private, although in principle data should only be accessed via methods. All these object formalisms guarantee both hiding of useless details, as fine tuning and real protection of - possible dangerous - privileged operations on the data.

Object Oriented Programming (OOP) differs in many ways from conventional programming. Where in a classic program a procedure (by calling) is told on which data to operate, in an OO-environment an object will be told which methods to carry out. After completion the method will send back a value or an object. In OO-terminology this is called *exchanging messages*. In contrast with 'old fashioned' instruction flow programs, an OO-program operates rather by data flow. This enables possibly a stronger parallel execution, when the necessary equipment and underlying software for task division and scheduling would be available.

From a semantic viewpoint objects are introduced as follows.

- All data and methods are defined in an *object type* (in the same way as 'integer', 'float', 'record',... in a normal language). This object type is called a *class*.

63

- Afterwards variables of this class can be created. The class is thereby instantiated into a real object to which the methods of that class can be fired.

A class contains (figure 3):

- The definition and the code for all methods of that particular class;

- a framework or layout of all data of that class.

The instantiated object on the other hand contains:

- the necessary space for the data;

- pointers to the different methods.

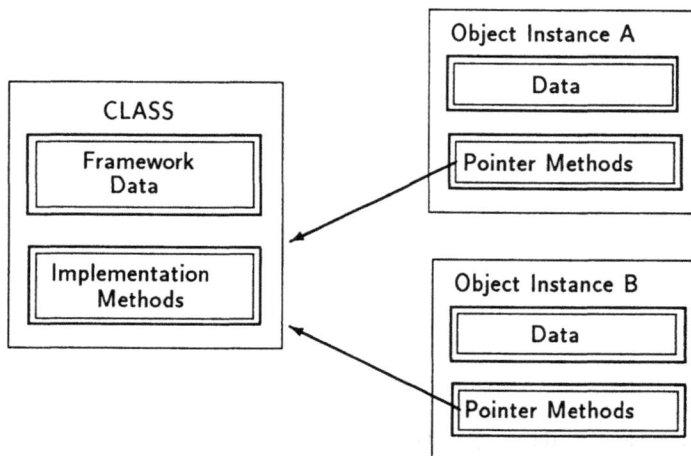

Figure 3: Software implementation aspects of OO

This strong separation of the context (program code versus data) offers a normal and useful optimization. In some OO-languages or dialects (e.g. C++) the formal definition and the implementation itself are still more separated (see example in next paragraph). This stimulates the use of finer protection mechanisms in order to enhance the outside coupling /interfacing (formal definition) of the classes.

3 Programming Example in Object Orientation.

OO-languages can grosso mode be divided into two groups:

Pure OO-languages: These languages are erected exclusively on an OO-formalism. Objects are the only working elements, and they all are deduced hierarchically from one fundamental object (*The* object). Furthermore all objects live completely dynamically in a free allotted space: there is no real compilation of static defined objects and the implementation is interpreter oriented. Smalltalk is a typical example of such an OO-language.

Hybrid OO-languages: Thereby, starting from well known common languages, extra OO-facilities are added. This can of course most easily be done in languages which are no too far away from an OO-concept: Pascal (record) and C (struct). In these languages the normal procedural formalisms can still be used, and if necessary one can switch to Object Orientation. Typical examples are: Objective-C, C++, Object Pascal. Because those languages are implemented with compilers, there is a fee to pay: less dynamic behavior of the objects, and more explicit programming by the programmer.

Follows a simple introductory programming example. A class to deal with complex mathematical numbers is constructed. To fix the idea, only the addition of two complex numbers, complemented with the necessary domestic methods (conversions,...) is given. The example is written in three different implementations: C++, Object Pascal and (Little) Smalltalk.

C++

```
#include <stream.hpp>

class complex
{
 float rpart;
 float ipart;
public:
 complex (float r=0.0, float i=0.0)  {rpart=r; ipart=i;}
 ~complex ()                          {};

 getr ()                       {return rpart;}
 geti ()                       {return ipart;}
 void add (complex& a);
};

// 'add' class method

 void complex::add(complex& a)
 {
 rpart += a.rpart;
```

65

```
      ipart += a.ipart;
      }

  // main test program

  main()
  {
  complex c1, c2(1.1), c3(2.2, 3.3);
  c1.add(c2);
  c3.add(c1);
  cout << "Real: " << c3.getr()
            << " * imaginary: " << c3.geti() << '\n';
  }
```

Object Pascal

```
  { Object definition }

  type
  complex = object
  rpart, ipart: real;
  constructor init0;
  constructor init1(r: real);
  constructor init2(r,i: real);
  { destructor done();  no destructor necessary here }
  function getr: real;
  function geti: real;
  procedure add(a: complex);
  end;

  { Methods implementation }

  constructor complex.init0;
   begin
   rpart := 0.0;
   ipart := 0.0
   end;

  constructor complex.init1(r: real);
   begin
   rpart := r;
   ipart := 0.0
   end;
```

66

```
constructor complex.init2(r,i: real);
 begin
 rpart := r;
 ipart := i
 end;

function complex.getr: real;
 begin
 getr := rpart
 end;

function complex.geti: real;
 begin
 geti := ipart
 end;

procedure complex.add(a: complex);
 begin
 rpart := rpart + a.rpart;
 ipart := ipart + a.ipart
 end;

{ Main test program }

var
 c1, c2, c3 : complex;
begin
 c1.init0;  c2.init1(1.1);   c3.init2(2.2, 3.3);
 c1.add(c2);
 c3.add(c1);
 writeln('Real: ',c3.getr, ' * imaginary: ',c3.geti )
end.
```

(Little) Smalltalk

```
Class Complex
| rpart ipart |
[
new
 rpart <- 0.0 .
 ipart <- 0.0
|
new: r
 rpart <- r .
```

```
      ipart <- 0.0
      |
      setimag: i
       ipart <- i
      |
      getr
        rpart
      |
      geti
        ipart
      |
      add: a
       rpart <- rpart + a getr .
       ipart <- ipart + a geti
      ]

      Class Main
      | c1 c2 c3 |
      [
      main
       c1 <- Complex new.
       c2 <- Complex new: 1.1 .
       c3 <- (Complex new: 2.2) setimag: 3.3 .
       c1 add: c2 .
       c3 add: c1 .
       ('Real: ', c3 getr, ' * imaginary: ', c3 geti) print
      ]
```

Although in the above example similarities can be found between C++, Object Pascal on the one hand, and Smalltalk on the other hand, clearly there are important differences. Because Smalltalk deals only with objects, also the 'main program' should be defined as a class (i.e. as a type):

Class Main

with only one method, namely:

main

(class names are written with a capital letter, while method names start with a small letter). The start of this little Smalltalk program has to be done by giving an *interactive* command:

Main new main

which in fact means:

Main new: creates a new object of the class 'Main'; the result is an object in dynamic memory;

...main: to this created object - stored in a variable - the message 'main' is fired.

The formal definition and the implementation code should both be written in Smalltalk inside the class definition, while in C++ this can optionally be done at once, such as:

```
complex (float r)        (rpart=r; ipart=0.0;}
```

If it should have be done afterwards, a special notation (::) is required to indicate to which class the method belongs to:

```
void complex:: add(complex& a)
{ ....}
```

Each OO-definition contains further a so called *constructor* and a *destructor*. A *constructor* is a method which by the creation of an object (instantiation) automatically (C++) or explicitly (Object Pascal and Smalltalk) will be executed. Its task is the necessary initialization of the (private) data of the object. In C++ these methods carry the same name as the object itself. In our example of complex number processing, there is only one constructor necessary. Because of the 'default' properties of C++, this enables three different variants for the programmer. They illustrate the three different ways how the instantiation of a complex type can be done, with or without standard values (zero):

```
complex c1              -> default real = imaginary = 0.0
complex c2(1.1)         -> real = 1.1 ;   imaginary = 0.0
complex c3(2.2, 3.3)    -> real = 2.2 ;   imaginary = 3.3
```

In Object Pascal these names may be freely chosen. Only the keyword 'constructor' or 'destructor' has to be written explicitly:

```
constructor init0();
constructor init1(r: real);
constructor init2(r,i: real);
```

In Smalltalk this new constructor has the standard method name 'new'. Two variants are possible (being with the specific Smalltalk syntax), and for the third variant (explicitly entering the imaginary part), an extra method ('setimag') is executed on the object being created.

```
c1 <- Complex new.
c2 <- Complex new: 1.1 .
c3 <- ( Complex new: 2.2 ) setimag: 3.3
```

69

A *destructor* is a method which automatically is executed just before an object disappears at the end of its life time out of a program. This method is necessary especially in the hybrid languages, because the programmer has himself to take care of dynamic memory management of the objects. In pure OO-languages this management is done by the underlying system. A typical example is returning the 'heap'-space to the garbage collector service just before the object gets out of scope. In the C++ example this destructor is given as

```
~complex()
```

Furthermore the given examples clearly illustrate messages are effectively being sent to the object, eventually with extra arguments. The formal definition of a method implicates the mean object ('carrier') is always present, and extra arguments always should be mentioned by the programmer. The reader will notice also the difference between the annotation of C++, Object Pascal and Smalltalk to access (private) data of other objects of the same class (type):

```
rpart <- rpart + a getr        (Smalltalk)
rpart += a.rpart               (C++)
rpart := rpart + a.rpart       (Object Pascal)
```

In C++ and Object Pascal this is a (static) annotation comparable with calling a 'struct'- or 'record'-element, while in Smalltalk (dynamically) a real method-message to the object is being sent. For the rest the Object Pascal version in the example resembles the C++ one (with its specific syntax), and needs no further comment. However, as will be illustrated further, there are other very important differences between them (e.g. operator overloading).

The foregoing simple examples are given for the seek of illustrating Object Orientation concepts, and are in no way an image of the power and elegance of the languages themselves. C++, Object Pascal and Smalltalk are all full fledged programming languages, although with differences in purpose and implementation. This is reflected also in their appearances to the end user. C++ and Object Pascal have classical implementations, i.e. are batch oriented by means of the usual path of editing, compiling, linking, debugging and executing. Smalltalk on the contrary is a strong interactive language, which is used at powerful multi-window stations with integrated tools for editing ('browsing') and debugging (powerful 'trace-back' facilities). Furthermore Smalltalk installs itself in the user environment: all classes, methods,... which are being developed, are permanently added to the environment as an extension.

4 'Abstract Datatype'.

The Object Orientation model is a quasi perfect image of what computer scientists call an 'abstract data type'. The notion 'type' is common knowledge, even from the era that Fortran ruled the (computer) world. In stronger typed languages, such as Pascal, this concept is further elaborated: the types 'integer, real, boolean, char, record, ...' are introduced. Such a type, which can be instantiated in a program variable, is in fact characterized by the following:

1. the way the information is represented internally in the computer (e.g. binary values, floating point, ...);

2. the operations which instantiations of this type may carry out; they may be different from one type to the other:

   - integer: $+ \ - \ / \ *$ succ pred mod ...
   - real: $+ \ - \ / \ *$

3. the underlying algorithms which are used by these operations (Fourier series, logarithms, ...).

The end user in principle doesn't care a bit how this information is internally structured, and which algorithms are exactly used. These are annoying *implementation details*. The only things which he cares about, are the possible operations he can use.

The concept 'abstract data type' is a generalization of this 'type', and can be defined as follows.

- Building modules at a higher level until types with the same style as the basic types.

- Defining these types (object classes) only on the basis of *their functionality*: i.e. the service they can provide ('abstract') and the formal properties. The implementation details are released completely to the responsibility of the type designer, and are even not absolute. A type may - depending on the application - receive different internal structures without loosing its service or functionality.

The concept is called *information hiding* and *data encapsulation*. In the literature [BMEY] abstract data types are extensively discussed in OO-perspective. An important aspect thereby is the way to exactly describe their service and properties via formal methods (e.g. mathematical).

5 Properties of Object Oriented Systems.

Following Bertrand Meyer [BMEY] there are *seven important rules (conditions)* which an object oriented system should fulfill. Its is clear that, whether or not a certain rule is fulfilled, a 'level' of OO-approximation arises. These rules thereby offer the reader the possibility to recognize real and pseudo OO-software.

This is the aspect of 'data flow' as has already been described.

This was explained thoroughly in the previous paragraph. Programming languages such as Ada and Modula-2 provide this mechanism. Also Fortran can, by clever using its subroutine facility, give some relief. Standard Pascal with its straight forward structure cannot deliver this feature properly. Some Pascal variants (with modular compilation) go somewhat in the right direction.

Many 'classical' languages fall out the boat by this rule. Object Pascal e.g. doesn't have a domestic garbage collecting service. Following this rule Pascal will never completely be object oriented. C++ attacks this problem (preceded by ANSI C) somewhat by the introduction of the so called *scope*. A scope can be limited to a file, function or class, but also to an instruction sequence delimited by a {} ('compound statement'). Objects, which are defined within a certain scope, are not visible outside. Whether their garbage is really cleaned up can be expected, but is not obvious. An example:

```
main()
{  // local scope of level #1
      int a,b;
      ...
      while ( ....)
      {  // local scope of level #2
         for (int i=0; ...)
            ...
      }  // end of level #2
   ...
}  // end of level #1
```

72

In the above example the 'integer' object 'i' is no longer visible outside the scope of level 2. Normally one would expect the garbage collector to clean up the object after leaving the scope. One should notice that in C++ this construction is pure *static*, and is dealt with by the compiler. In natural OO-languages (Smalltalk) this is done completely dynamic during the execution, which of course is much more powerful.

Rule 4

Classes. Each non-simple type is a module, and each high-level module is a type.

Here the difference is made between real and pseudo object oriented languages, by postulating the following equivalence:

$$module \equiv type$$

This means data abstraction and encapsulation alone are only syntactical make-up, while in real OO-languages the module should be defined as an abstract data type with an own semantic denotation. In principle both Smalltalk, C++ and Object Pascal satisfy this rule, although in Pascal this is rather poor (no automatic execution of a constructor for example). The terminology 'non-simple' allows still simple existing types (such as integer, real, ...). The term 'high-level' enables normal procedures can be used in a type definition.

Rule 5

Inheritance. A class can be defined as an extension or as a restriction of another existing class.

Here the re-usability of software is envisaged by the following provisions.

- An object can rely in its functionality onto other already defined objects.

- The principle of *subtype* or *subclass*: a new object class is defined in function of an existing one, however with extra modifications such as:

    extensions: offering extra services which the antecedents don't;

    restrictions: abolish certain services which are offered by the antecedents;

    modifications: change services as offered by the antecedents ('overriding', 'overloading').

73

Semantically this comes to redefine a new class as a *subclass* of an existing one. Thereby in the new subclass all methods which have been defined in the higher hierarchy are at disposal. The above example of complex numbers may e.g. be extended to a class of complex arrays. Thereby minimal a new constructor and a new destructor shall be defined, and probably also some new methods. On the other hand all methods for mathematical operations on complex numbers remain in effect via inheritance.

C++

```
Class ComplexArray: public complex
{
    int size;
    complex* arrptr;
public:
    ComplexArray(int = 100);
    ~ComplexArray();
    . . .
};
```

Object Pascal

```
type
    ComplexArray = object(complex)
        size: integer;
        arrptr:  complex
        constructor initcarr(i:integer);
        . . .
    end;
```

The other aspect, i.e. overloading or overwriting in the new subclass of methods from a hierarchical higher class, can semantically dealt with as follows. In the OO-languages the principle holds that the name of a method needs *not be unique*, but may be identical to the name from the higher hierarchy. This enables thereby to overwrite that method. Another way to express this rule, is that not the name of the method, but its *signature* should be unique. The signature is seen as the name, *plus* the arguments, including the class (type). When an object receives a message, an up going hierarchical way will be followed: first its own class, then its superclass (just above), and so on until eventually the highest class has been reached. If at that point no method is found, an error message will be given

---

Rule 6

Polymorphism and dynamic binding: Program instructions should have the possibility to refer to objects of more than one class, and the operations thereby may have different realizations in different classes.

74

At the first sight this is a logical consequence of rule 5, but it goes much further. Polymorphism is the mechanism for one single operation to have effect on objects of different classes. Although not widely understood, this occurs in all languages, especially for mathematical expressions. The following C-example shows how.

```
main()
{
    int i, j, k ;
    float a, b, c ;
    . . .
    i = j * k;
    a = b * c;
    . . .
}
```

In both multiplications the method '*' refers in fact to two completely different object types (classes): 'int' on the one hand, and 'float' on the other. Obviously the underlying implementation of the multiplying operation of integer numbers differs considerably from floating point numbers. Polymorphism is an extension of this simple principle to all possible classes.

The effective implementation of polymorphism is not so evident. As long as *static* defined objects are being used in a compiling language (C++, Object Pascal), the compiler can exactly sort out the exact class to which an argument belongs to, and generate the correct calling sequence. When dealing with *dynamic* objects, things become more difficult (e.g. 'pointers' in C or Pascal). In such case only at *run-time* can be found to which class from the hierarchy the object belongs to, and the *binding between the method and the corresponding instruction code has to be deferred*. This is called dynamic or 'late' binding. In Smalltalk, which operates in full dynamics, this is a normal procedure, but in other languages, such as C++ and Object Pascal, a 'hint' has to be given to the compiler to defer the binding. This is done by adding the keyword 'virtual'. As a practical example (figure 4) a class hierarchy is proposed to draw on a screen objects of the following form:

$$location \rightarrow point \rightarrow circle$$

The framework of this task looks in C++ as follows.

```
class location
{   . . .
    int xloc, yloc;
public:
    location(int x=0, int y=0);
    . . .
}

class point: public location
```

75

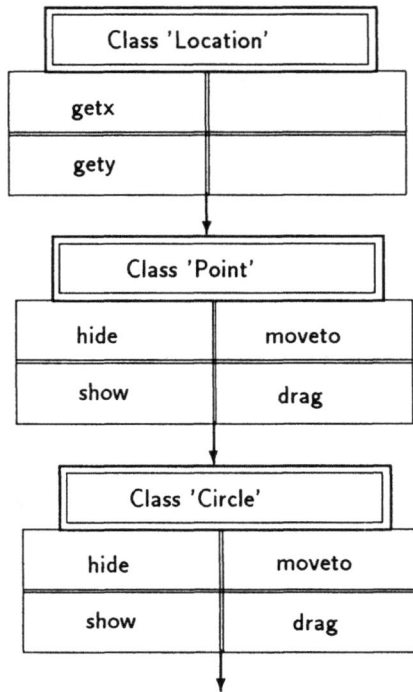

Figure 4: Polymorphism and Late Binding.

```
{
    int visible;
public:
    point(int x=0, int y=0);
    ...
    virtual void show();
    virtual void hide();
    ...
}

class circle: public point
{
    int radius;
public:
    circle(int x=0, int y=0, int r=1);
```

76

```
                    virtual void show();
                    virtual void hide();
                        . . .
            }
```

This series of classes contain two virtual methods: 'show' and 'hide', because in practice the underlying implementation of the object 'point' is quite different from 'circle'. When a dynamic object then executes the method 'show' or 'hide', at run-time the question has to be solved which underlying class in meant. The true fundamental reason is the principle of extended type compatibility, by which a circle type in fact is also a point type. This is comparable to the 'subrange' type of Pascal, which also keeps its scalar mother type (e.g. integer).

Rule 7

Multiple and repeated inheritance: It should be possible a class inherits from more than one class, and also multiple times from the same class.

By this principle an object can refer to methods of more than one parent class. This way a new class can be composed by combining methods of other classes. A typical example is a class which enables to draw graphical objects apart and within each other on a computer screen. This class can recall thereby (see figure 5, but without the superclass 'point' at the top) *three* other classes, which furthermore have nothing else in common.

An extra problem (ambiguity) raises when the direct parent class descend again from a common grand parent (superclass), and the mother types have some methods with the same name, but a different implementation (figure 5). In such case again 'hints' should be given to the compiler with indication via which path (the end user wants) to look for the ambiguous methods.

It has to be clearly stated that not everyone is very happy with the principle of multiple inheritance. The most important objection which arises is, it would be against any natural behavior and would break with the anthropomorphic aspect of an object: 'no child can have more than one father'. These computer scientists state the need for multiple inheritance only exists with badly analyzed systems.

6 Overloading of (internal) Primitive Methods.

Once the principle of polymorphism is accepted, immediately raises the question if it wouldn't be possible / desirable to overload the standard original methods of the language. One could easily think of redefining input, output but also addition ('+'), subtraction ('−'), multiplication ('*'), division ('/'), indexing ('[ ]'), assignment ('='), . . . In the earlier example with complex numbers, this could mean complex numbers would be absorbed as e

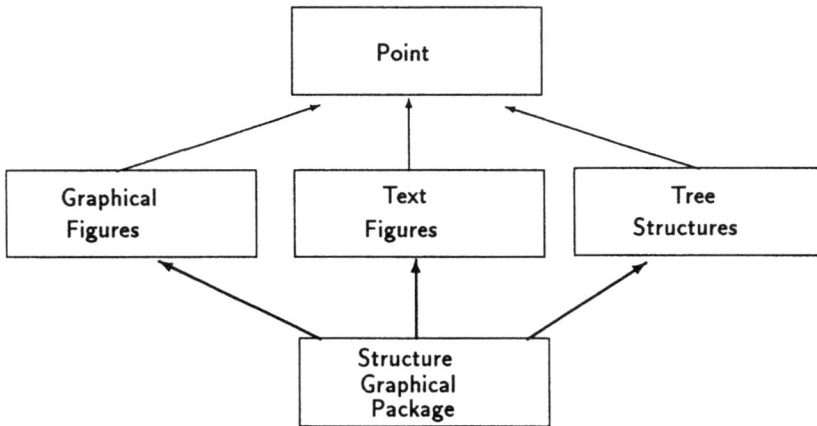

Figure 5: Multiple Inheritance.

new intrinsic type in the language. The end user wouldn't notice any difference any more between the primitive and new complex types. In pure OO-languages, such as Smalltalk, there is fundamentally no difference between primitive and new added types: each new definition installs itself in the environment and enriches it. Applying polymorphism on primitive types is there the most natural thing. In more ''usual' languages, such as C, Pascal, ... this is not obvious at all. Nevertheless C++ offers this total polymorphism to a rather high extent. Object Pascal on the other hand fails dramatically. It is out of scope of this text to give all details of the semantic and syntactic implementations, which at the start looks rather cryptic. To fix the idea, the example of complex numbers is given here again in C++, but now with the addition and the subtraction really implemented as primitive types, i.e. with the '+' and '−' operators. Also output is done with the stream operator '≪'. It should be noticed overloading of primitive methods is strongly connected with the underlying structure of these methods inside the system. Thereby the formal definition will not be identical in all C++ implementations.

```
#include <stream.hpp>

class complex
{
    float rpart;
    float ipart;
public:
    complex (float r=0.0, float i=0.0)      {rpart=r; ipart=i;}
    ~complex ()                      {};
```

78

```
                complex& operator=(const complex& x);
                complex  operator+(const complex& x);
                friend ostream& operator<<
                        (ostream& s, complex& x);
        };

        // class methods

                complex& complex::operator=(const complex& x)
                {
                    rpart = x.rpart;
                    ipart = x.ipart;
                    return *this;
                }

                complex complex::operator+(const complex& x)
                {
                    complex sum;
                    sum.rpart = rpart + x.rpart;
                    sum.ipart = ipart + x.ipart;
                    return sum;
                }

                ostream& operator<<(ostream& s, complex& x)
                {
                    s << x.rpart << " * " << x.ipart << '\n';
                    return s;
                }

        // main test program

                main()
                {
                    complex c1, c2(1.1), c3(2.2, 3.3);
                    c1 = c1 + c2;
                    c3 = c3 + c1;
                    cout << c3;
                }
```

In the above example, we find besides the well known constructor and destructor operators

```
        complex (float r=0.0, float i=0.0)
                {rpart=r; ipart=i;}
```

79

```
~complex ()                          {};
```

three new classes to overload the standard operators. The formal definition

```
complex  operator+(const complex& x);
```

overloads the addition ('+') operator and specifies a formal parameter 'x' of the type 'constant reference (&) to a complex type' (the notation 'constant' inhibits the programmer to modify inadvertently the value of x). The class returns a complex number. One should be aware of the fact that inside the method the parameter **x** also intrinsically is present. Thereby the corresponding method can make the sum as follows:

```
rpart + x.rpart;
ipart + x.ipart;
```

This method returns the value of the local variable (also a complex type) 'sum':

```
complex sum;
...
return sum;
```

Declaring overloading of the assignment operator ('=') is very similar. Only a *reference* to the resulting object is returned:

```
complex& operator=(const complex& x);
```

The implementation of this assignment method differs only by the fact that on returning (the internally in C++ defined) pointer 'this' is used, which always points to the own instantiated object:

```
return *this;
```

One should remark that returning this pointer is not strictly necessary for a simple assignment as in the example

```
c1 = ... ;
c3 = ... ;
```

but on the other hand would enable linked assignments as in:

```
c4 = c1 = ... ;
```

The overloading of the stream operator (ANSI C) '≪' is done via two parameters: a reference to the output device **s**, and a reference to the value **x** to be output:

```
friend ostream& operator<<
        (ostream& s, complex& x);
```

The word 'friend' is rather strange here, but is imposed to enable the basic method 'ostream' to use the private newly defined method.

7   Conclusion.

New tools which offer themselves to the world of information systems, should indeed bring forward effective progress. Fancy constructions, bells and whistles are totally irrelevant when they do not contribute in ameliorating the *software quality*, be it by changing the underlying methodologies, be it by increasing the functionality.

Software quality is characterized by a number of external and internal factors [BMEY]. The external factors are important for the user (end user and software managers). internal factors - such as modularity of software in all its forms - are important for the software developers.

The most important external quality factors are the following.

Correctness: the software product should do exactly what has been described in the requirements and the specifications.

Robustness: software systems are supposed still to function in abnormal situations.

Extensibility: software should be easily adaptable to changing specifications.

Re-usability: software should easily be re-used - totally or partially - in the development of new applications.

Compatibility: software should easily be combined with other software.

Object Orientation is a software formalism which clearly encounters many of these essential factors. In the introduction the anthropomorphic aspect of object orientation and the creation of an illusion for the end user was already pointed out. Thereby OO is an incitement to make the structure of the software more clearly resembling the real world model. Grouping all forces of an object and the hiding of the internal details blur the conceptual gap between a problem and its solution. Thereby a new structure arises which is easier extendable and reusable. The software developer must however be vigilant. Using object orientation needs an important swing in mentality: as long as it remains a linguistic make-up, it will miss its goal. Object Orientation starts with the *design* of software, and not with the programming. It is necessary to *think object oriented*. This suggests the need for object oriented design methodologies too. An example of such software is the HOOD-methodology as developed by the ESA ('**H**ierarchical **O**bject **O**riented **D**esign'). It is a hierarchical decomposition design methodology based on identification in the problem phase of objects and their mutual relations. Nevertheless the developer may employ an object orientation approach without using formal OO-design tools, in order to have a smooth transition to the programming phase. This maturity he shall probably only reach after having gone through many projects in a *real OO-spirit*.

8 Bibliography.

A reference book on object orientation in general is without any doubt *Object-Oriented software construction* from Betrand Meyer. All aspects are explained by means of the *Eiffel*-language and methodology. It contains also an excellent and extensive bibliography.

For Smalltalk the reference book is of course *Smalltalk 80* from 'Smalltalk-lady' Adele Goldberg. It is an extensive book, maybe too voluminous for those who won't use Smalltalk. Nevertheless it is of the highest importance for a thorough OO concept understanding to look at the basic constructs of Smalltalk. In such case the book *A Little Smalltalk* from Timothy Budd may be an adviser. It contains the most important classes (with the main methods) in a good educational style. Furthermore the underlying software 'to try it out' is available in the public domain for all kinds of PC's.

For Object Pascal we can easily refer to the manual of Turbo Pascal V5.5 and later. It contains – in good tradition – an excellent tutorial on the language. The reader should be warned however that the developments around Object Pascal are not yet fully stabilized.

For Object-Orientation in the C-language, there are two different streams: on the one hand there is 'Objective C', and on the other 'C++'. 'Objective C' is described in the book *Object-oriented programming: an evolutionary approach'* from Brad Cox, but is de facto overwhelmed by the organization backing C++ (AT& T Bell Laboratories), whereby C+ clearly is on its way to a standard. The basic book on C++ *The C++ Programming Language* is not an easy book to read. Another one, which also can be used for self-study, is *C++ Primer* from Stanley Lippman.

For the reader interested in object-oriented data base systems, the work *Object-Oriented database management systems: concepts and issues.* is a good basic paper.

[**AGOL**] Adele Goldberg: "Smalltalk 80":
> part 1: the language and its implementation (with Robson D.);
> part 2: the interactive programming environment.
> Reading, Addison Wesly (1984)

[**BCOX**] Brad J. Cox: "Object-Oriented Programming: an Evolutionary Approach" Reading, Addison Wesley (1987)

[**BMEY**] Bertrand Meyer: "Object-Oriented Software Construction"
New York, Prentice-Hall (1988)

[**BSTR**] Bjorne Stroustrup: "The C++ Programming Language"
Reading, Addison Wesley (1986)

[**EBER**] Elisa Bertino, Lorenzo Martino: "Object Oriented Database Management Systems: Concepts and Issues"
IEEE-computer, p. 33-47 (april 1991)

[**SDEW**] Stephen Dewhurst, Kathy Stark: "Programming in C++"
New York, Prentice Hall (1989)

[**SLIP**] Stanley B. Lippman: "C++ Primer"
Reading, Addison Wesley (1989)

[**TBUD**] Timothy Budd: "A Little Smalltalk"
Reading, Addison Wesley (1987)

Expert System Case Study:
The Chocolate Biscuit Factory. [1]

J. Efstathiou

Queen Mary College, London, England.

Summary

The purpose of this paper is to demonstrate the technology of expert systems through the example of diagnosing faults on the production line of an imaginary chocolate biscuit factory. The basic principles of expert systems are explained together with some of the problems of setting up an expert system, such as choice of knowledge representation method and knowledge acquisition. Using the example, forward chaining, backward chaining, mixed initiative interaction and control strategies are explained. The importance of knowledge as a resource within an organization is emphasized and it is argued that constructing an expert system can help use the existing knowledge more effectively as well as helping to obtain new knowledge.

1 Introduction.

The Chocolate Biscuit Factory is an imaginary example designed to illustrate some of the uses of expert systems. The reader is asked to imagine a chocolate biscuit factory which suffers some elderly and unsatisfactory plant. By contrast, the latest thing in expert system technology is about to be installed.

2 Knowledge representation.

The expert system shell which the factory plans to use has a rule-based knowledge representation format. It has been tailored in some ways to suit domains which require the diagnosis and repair of faults. This is because the operators commonly observe something going wrong, which suggests several possible causes. These can be checked out until one or more breakdowns have been identified. This format of knowledge representation has other benefits because it makes explanations easier and more intelligible. This is hoped to lead to a better trained and more capable workforce.

An operator on one of the machines in the chocolate biscuit factory occasionally notices something going wrong with the chocolate or the biscuits. These observations are called triggers, because they trigger a suspicion in the operator's mind that something is going wrong. Some of the operators are quite good at realizing what it is that could have caused these triggers, so a class of rule is constructed which have the following format:

---

[1] Reprinted (with permission) from Journal A, Vol. 27, no 2, 1986, pp. 62 - 68, Belgian Institute for Automatic Control (BIRA), Antwerp, Belgium

```
IF <trigger> THEN SUGGEST <hypothesis>
```

Hypotheses here are things like 'oven too hot', which link a fault in a piece of equipment to an observed effect, such as runny chocolate. Some triggers suggest a single hypothesis, but others could suggest more than one. So, the operator observes a trigger, which suggests one or more possible hypotheses. The hypotheses have to be caused by something in the plant, so another class of rules is needed:

```
<hypothesis> CAUSED BY <fault>
```

For example,

oven too hot CAUSED BY gas temperature too high

The operator needs a way of checking whether the fault has occurred or not, so a test has to be applied. The proper test is given by a rule of the format:

```
<fault> CONFIRMED BY <test>
```

If the test fails, then that fault is ruled out. But if the test succeeds, then some action must be taken. The action might be a simple repair or, in the case of something more serious, calling a skilled repairman or reporting the fault to the shift supervisor, before discarding a batch of product. Thus, another type of rule is needed to provide the appropriate action:

```
IF <fault> THEN DO <action>
```

The knowledge representation format outlined above uses five different kinds of object in the knowledge base. The trigger is the observation that first causes the operator to notice something is going wrong. This could suggest several hypotheses to the skilled fitter or repairman. Each hypothesis could be caused by one or more faults. The hypothesis is the mechanism which links a fault on the plant to the trigger. If someone asked "Why does high gas temperature make the chocolate runny ?", the answer lies in the hypothesis "Because the oven is too hot". Each fault can be checked by a test and if the fault is confirmed, then some action needs to be taken. These five objects are linked together by four kinds of rule:

```
IF <trigger> THEN SUGGEST <hypothesis>
<hypothesis> CAUSED BY <fault>
<fault> CONFIRMED BY <test>
IF<fault>THEN DO<action>
```

The four different kinds of rules are needed to represent the different kinds of knowledge that exist. The first rule uses data to suggest a hypothesis, but the second uses a hypothesis to suggest data. The third kind of rule provides basic information on how to test for a particular fault, but the fourth says that in the circumstances where a fault is confirmed, then a particular action should be done (figure 1).

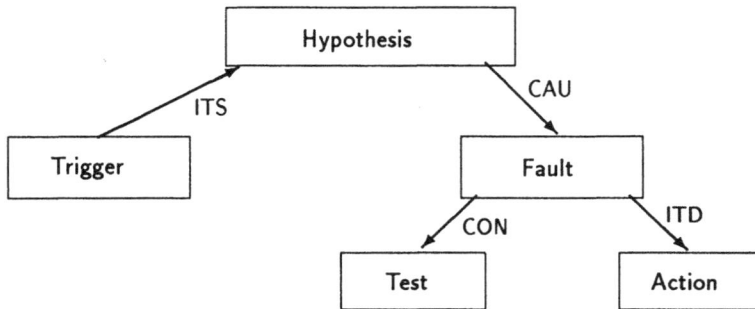

Figure 1: Rule-based knowledge representation.

3 Knowledge elicitation.

The knowledge engineer feels happy that the knowledge representation format and inference mechanisms are suitable for the chocolate biscuit factory and so embarks on a process of knowledge elicitation. This is not a straightforward task, because so many different people have pieces of relevant knowledge. The operators can describe most of the triggers, but sometimes leave it to the fitters to find out what has really gone wrong. The first list of triggers is usually incomplete, but after talking to some of the other people involved, other items can be added.

The next job is to compile a list of faults. If a log of faults is kept, then this simplifies the task, but otherwise it is a matter of interviewing the repairmen and fitters to find out what jobs they have to do.

Given these two lists, the rules can be elicited by asking for the links between the two. This can turn out to be an exercise in knowledge refining, because the experts find that having to sit down and think and explain what they do helps them perhaps to find new links or think more carefully about the old ones.

The next stages are fairly straightforward. Each fault can be tested in some way or another. Sometimes, it is quite simple, but other times it might involve laboratory tests or queries to suppliers. Whatever the case, some action must be taken. The shift managers can be helpful here. They might have been trying to impose some discipline and consistency on the operators' response to faults and this gives them a chance to review their recommendations and put them down as rules.

The above account of a knowledge elicitation task makes it sound very simple and easy. In reality, all sorts of things can go wrong. Operators might be resentful of a computer program taking over the skills which they are proud to protect. Knowledge might be difficult to elicit from the old hands who feel they have nothing to learn from a computer and, besides, they do it all by feel anyway. Such obstacles require patience and tact to

86

overcome. Observing the expert at work, if possible, might suggest what the problem solving strategies are. Even if knowledge elicitation is made into an arcade game, expert systems will be no use if people resist and don't use them.

Assuming the knowledge elicitation proceeds successfully, the knowledge engineer could end up with four sets of rules which look like the following tables.

IF <trigger> THEN SUGGEST <hypothesis>		
Rule number	trigger	hypothesis
ITS 1	biscuits pale	oven too cool
ITS 2	biscuits not risen	oven too cool
ITS 3	biscuits burnt	oven too hot
ITS 4	biscuits burnt	too much cocoa
ITS 5	biscuits burnt	too much sugar
ITS 6	biscuits not risen	too much cocoa
ITS 7	biscuits not risen	too little raising agent
ITS 8	biscuits not risen	faulty flour
ITS 9	biscuits pale	wrong sugar
ITS 10	biscuits pale	faulty flour
ITS 11	chocolate sticky	not enough air in mixture
ITS 12	chocolate sticky	faulty cocoa
ITS 13	chocolate bubbling	oven too hot
ITS 14	chocolate runny	rancid butter
ITS 15	chocolate runny	factory too warm
ITS 16	chocolate too dark	too much cocoa
ITS 17	chocolate too light	wrong sugar
ITS 18	bad smell	rancid butter
ITS 19	bad smell	CHECK biscuits burnt

The trigger-hypothesis rules, ITS

	<hypothesis> CAUSED BY <fault>	
Rule number	hypothesis	fault
CAU 1	faulty flour	change of source
CAU 2	faulty flour	old stale consignment
CAU 3	too little raising agent	someone forgot to add it
CAU 4	too much sugar	nozzle flowing freely
CAU 5	too much cocoa	hopper faulty
CAU 6	too little cocoa	hopper faulty
CAU 7	faulty cocoa	change of source
CAU 8	wrong sugar	faulty batch
CAU 9	not enough air in mixture	dust cover blocked
CAU 10	not enough air in mixture	fan slow or stopped
CAU 11	oven too cool	gases too cool
CAU 12	factory too warm	warm temperature outside
CAU 13	oven too hot	gas temperature too high
CAU 14	oven too hot	cooling failed
CAU 15	too little sugar	nozzle opening restricted
CAU 16	faulty flour	faulty batch
CAU 17	rancid butter	warm temperature outside
CAU 18	faulty cocoa	faulty batch
CAU 19	oven too cool	low temperature gases

The hypothesis-fault rules, CAU

Rule number	fault	test
	<fault> CONFIRMED BY <test>	
CON 1	dust cover blocked	remove back from dust cover and look
CON 2	fan slow or stopped	shine strobe light and the steady black line is not visible
CON 3	faulty batch	sending sample to laboratory
CON 4	warm temperature outside	taking reading of wall thermometer. Should be less than 80°
CON 5	hopper faulty	measure time for level in hopper to drop by one gradation. Should be 4 minutes
CON 6	change of source	check bags
CON 7	old consignment	check date on bags
CON 8	someone forgot	check log OR ask supervisor
CON 9	nozzle flowing freely	see if sugar flowing steadily from nozzle, without breaking into droplets
CON 10	nozzle opening restricted	see if sugar dripping at less than 10 drops per minute
CON 11	cooling failed	tap turned on AND water jacket cool to touch AND pump not vibrating
CON 12	cooling failed	tap turned on AND water jacket warm to touch
CON 13	gas temperature too high	white color at viewing window
CON 14	low temperature gases	black color at viewing window
CON 15	low temperature gases	red color at viewing window
CON 16	low temperature gases	flickering light at viewing window

The fault-test rules, CON

	IF <fault> THEN DO <action>		
Rule number	fault	action	lost production
ITD 1	dust cover blocked	clean and replace	some
ITD 2	fan slow or stopped	call fitter	major
ITD 3	faulty batch suspected	report fault to supervisor	major
ITD 4	warm temperature outside	report fault to supervisor	some
		If butter rancid THEN discard batch	major
ITD 5	hopper faulty	call fitter	major
ITD 6	old consignment	report fault AND discard batch AND	major
	OR change of source	check faulty batch	
ITD 7	someone forgot	add ingredient AND report fault	major
		AND discard batch	
ITD 8	nozzle flowing freely	turn down sugar supply	some
ITD 9	nozzle opening restricted	switch off sugar supply AND	some
		remove nozzle AND replace with spare	
		AND send faulty nozzle for cleaning	
ITD 10	cooling failed AND	call fitter	major
	water jacket warm to touch		
ITD 11	cooling failed AND	call electrician	major
	pump not vibrating		
ITD 12	cooling failed AND	turn on tap	some
	water tap turned off		
ITD 13	gas temperature too high	turn down oxygen input	some
ITD 14	low temperature gases	restart override	some
	AND black color		
ITD 15	low temperature gases AND	raise alarm AND evacuate factory	factory
	flickering light		shutdown
ITD 16	low temperature gases AND	turn up oxygen supply	some
	red color		

The fault-action rules, ITD

4 Forward Chaining.

Forward chaining (figure 2) is the simplest way to use rules such as these. Each rule consists of an antecedent and a consequent. The antecedent is the IF part and the consequent is the THEN part. Forward chaining goes from antecedent to consequent, from the IF to the THEN. As an example of forward chaining, let us suppose that an operator sees that the

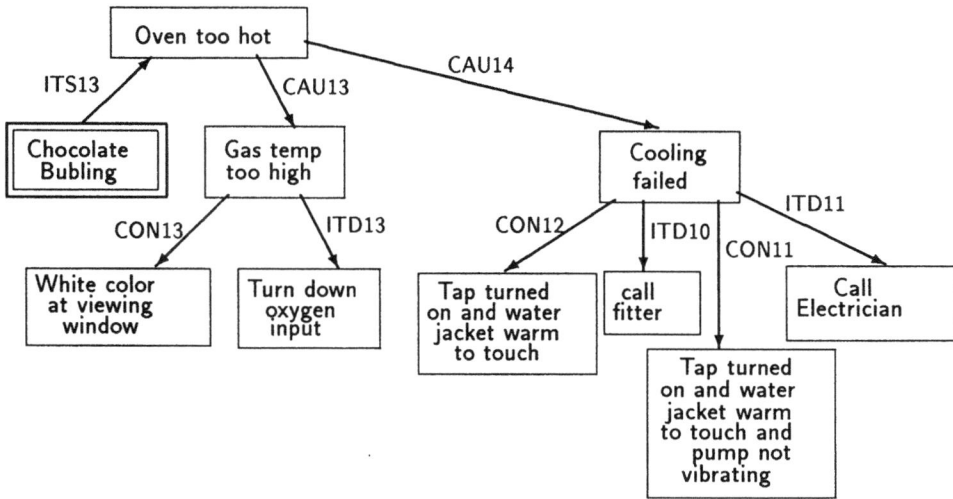

Figure 2: Example with forward chaining.

chocolate is bubbling. Reasoning forwards, we use Rule ITS-13 to suggest the hypothesis 'oven too hot'. Scanning on down the list of ITS rules, we see that there are no others with 'chocolate bubbling' as the antecedent. So, this is the only hypothesis. Turning to Rule CAU-13, this suggests the likely fault 'gas temperature too high'. Rule CAU-14 also has 'oven too hot' as the antecedent, so we add 'cooling failed' to the list of possible faults. Next, the faults have to be tested. Rule CON-13 suggests a test for 'gas temperature too high', i.e that there is a white color at the viewing window. Suppose the operator goes to check this and sees that the gas color is perfectly normal, at an orange glow. That rules out 'gas temperature too high' as a fault, but 'cooling failed' still needs to be checked. Two tests exist to check whether the cooling has failed, supplied by Rules CON-11 and CON-12, because there are different ways in which the cooling could fail. This is reflected in the actions that are recommended by Rules ITD-10, ITD-11 and ITD-12. The operator is requested by the expert system to check several items and depending on the answers given, the system can infer the particular fault and recommend the appropriate action. In the example above, the reasoning proceeds forwards from antecedent to consequent each time. The control strategy of the inference engine had to cope with the possibility of two faults (which could have occurred simultaneously) and be able to recommend tests for the second fault once the first one had been discounted.

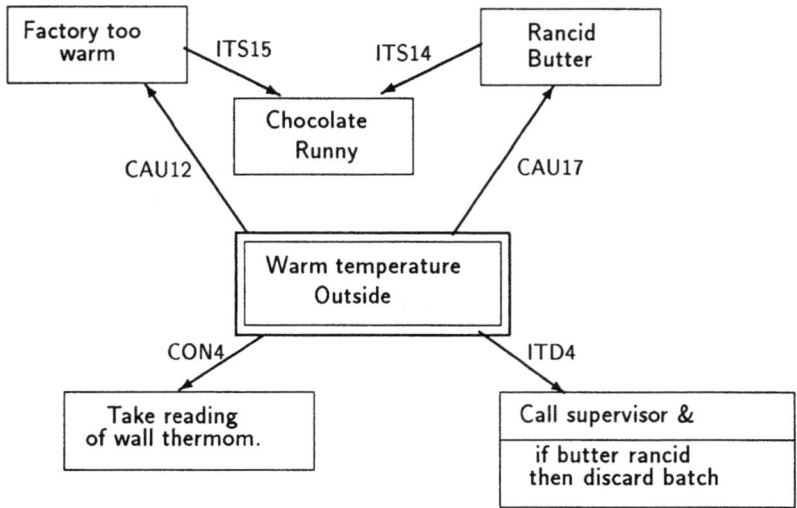

Figure 3: Example with backward chaining.

5 Backward Chaining.

Forward chaining with such a well ordered set of rules can produce something like expert behavior quite well, but it is rather rigid. The reasoning process must always start with the trigger and go forwards. If there were some way of starting with the hypothesis, forward reasoning would not let us infer which triggers would be manifest under that hypothesis, because that would require reasoning from consequent to antecedent under the ITS rules. That would be Backward Chaining. To illustrate the usefulness of backward chaining (figure 3), let us do another example. Suppose on the way in to work one day, an operator notices that the weather is unusually warm. This suggests to the alert person that some faults could probably occur because of the heat and that it might be worth consulting the expert system to find out what to watch out for.

'Warm temperature outside' is one of the faults that the system recognizes. Rule ITD-4 reasons forward to recommend that the shift supervisor should be advised and the batch discarded if the butter is rancid. This could lead to a major production loss, so it would be a fault worth avoiding if possible.

Rules CAU-12 and CAU-17 have 'warm outside temperature' as their consequents, so reasoning backwards suggests the hypotheses 'factory too warm' and 'rancid butter'. Reasoning backwards with the ITS rules reveals that the triggers are 'chocolate runny' (ITS-14 and ITS-15) and 'bad smell' (lTS-18). So, the operator is forewarned to watch out for the chocolate becoming runny or a bad smell developing. Incidentally, reasoning forwards with rule CON-4 suggests the test of taking the reading on the wall thermometer. By volunteering the information that the temperature outside is warm and using a mixture of forward and backward reasoning, the user of the expert system is able to discover the seriousness of such a fault developing and finding out the triggers and tests which should be monitored during the day.

6 Control Strategy.

We have already seen a simple example where more than one fault could occur. In the simplest case, the expert system chooses to test the faults in the order in which they are suggested, but this could lead to problems. If the biscuits are burnt, rules ITS-3, ITS-4 and ITS-5 all suggest possible hypotheses. Each of these hypotheses generates possible faults that would all need to be tested before the true cause (or causes) could be discovered. It would seem to be a good idea to assign some sort of priority to the testing procedures.

There are many candidate criteria which could be used for determining priority. In this simple case, the expert system has used the potential production loss associated with each fault. It would seem a good idea to rank the possible faults according to the amount of product that would be lost, taking care to identify the most expensive faults as quickly as possible, so as to minimize the production loss.

Other possible criteria that a more sophisticated control strategy could take into account are:

1. probability of the fault occurring;

2. cost of performing the test;

3. if more than one trigger is observed, whether the fault is suggested by more than one hypothesis;

4. possible damage the fault could inflict.

If several faults were suggested, the human expert would presumably be prudent in performing tests to check the most common faults before proceeding to check the least likely. Similarly, if a test is difficult or expensive to perform, then that would be left until later as well The interplay of these criteria has not received much study yet, but expert systems are being developed which do take such factors into account.

7 Other Enhancements.

Some of the rules will be observed to contain other items apart from those defined in the earlier section on Knowledge Representation. We have already added a cost to the ITD rules so as to improve the control strategy. Several of the rules contain AND or OR operators, which serve to keep down the number of rules that need to be maintained .

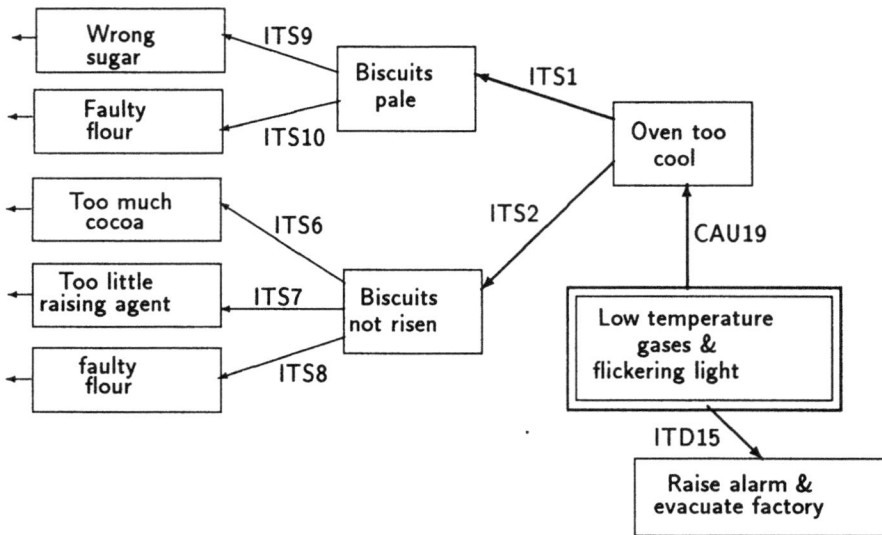

Figure 4: Example with backward chaining.

Rule ITD-6 contains an instruction to 'Check faulty batch'. Although the fault has been identified, the rule has this extra addendum which adds 'faulty batch' to the list of possible faults. This fault triggers CON-3 which requires sending a sample to the laboratory for tests. However, sending material to the laboratory can mean a long wait for the result of the test. All the other tests can be performed by the operator and the results obtained fairly quickly. There often is not time to await results from the laboratory, so ITD-3 has the rider 'Suspected' attached to the antecedent. Before obtaining the test results, the operator should proceed with the action recommended.

This carefully elicited and refined knowledge can be put to other uses too. The most expensive fault is shown at ITD-15, when low temperature gases are suspected and a flickering light is observed at the viewing window. In these cases, the alarm must be raised and the factory evacuated. Reasoning backwards, we find that low temperature gases are suspected by the hypothesis 'oven too cool'. This hypothesis is suggested by the triggers 'biscuits not risen' or 'biscuits too pale', reasoning backwards with ITS-1 and ITS-2. However, these triggers suggest many other hypotheses also, as can be seen by reasoning forwards from ITS-6, ITS-7, ITS-8, ITS-9 and ITS-10. If these triggers were seen, without a good control strategy, time would be wasted searching relatively trivial hypotheses (figure 4). Apart from using a better control strategy, the factory planners could decide that since they now know that this is the only test for an expensive fault, it should be changed from the status of test to that of trigger, i.e. something which captures

94

the operator's attention immediately. This could be done by implementing some sort of automatic alarm system which would be sprung immediately the temperature began to drop, so that conditions on the other side of the viewing window could be monitored more carefully.

Furthermore, it could be shown that sending material to the laboratory for tests occurs so frequently that the time for receipt of results should be cut. A faulty batch is deemed responsible under rules CAU-8, CAU-16 and CAU-18, and ITD-6 suggests it too. This is particularly important because so many triggers could be due to faulty materials that it is worth doing lab checks as a matter of course, perhaps regularly sampling the ingredients of the biscuits.

8   Conclusions.

We have attempted to introduce some of the basic techniques of knowledge representation and inference in expert systems, through an example of the Chocolate Biscuit factory. Expert systems are still in their early days and we can expect to see improvements in the quality of software over the next few years. The systems which are currently available are primitive by comparison with what is being developed within universities and software houses, but should make some impact soon.

Part of the reason for the shortage of expert systems in practical installations is the problems and difficulties of knowledge engineering. There are many ways of avoiding this bottleneck and we must hope that they are developed soon. Otherwise, expert systems are not going to live up to their promise and will be dismissed as a flash in the pan.

Another problem is the perennial gap between what interests the researchers at the universities and what the people in industry actually want. Current techniques, while they might appear passe in Stanford or MIT, could solve plenty of genuine problems if they were honestly applied .

Expert systems will not win many converts until there are some strong, commercial or industrial applications. Expert systems research would benefit from such ventures, both in terms of its publicity and its credibility.

Using A.I.-Formalisms in Programmable Logic Controllers

Prof.Dr.ir L. Boullart

Automatic Control Laboratory, University of Ghent.

1 Summary

Programmable Logic Controllers are already being used for some decades in automation applications. They provide a way of cost-effective technology for automating all types of processes ranging from the very simple ones to parts of rather complicated systems. The field has made important progress during the last decade, but only in a technological way: i.e. just the hardware itself has kept up with new components, such as microcomputers. The principles of operation however and its very peculiar way of programming never changed considerably. The advent of Artificial Intelligence stimulated new languages and tools. Thereby a whole set of new directions for programming PLC's open up. This section presents some ideas and possible ways to go, in order to stimulate new developments for the PLC application user, who has been deprived too long from up-to-date programming methodology and tools.

2 Programmable Logic Controllers: state-of-the-art

Programmable logic controllers are basically controlling devices intended for applications where both input and output are pure boolean values: i.e. each input/output value can only be "true" or "false". In the newer systems this has been extended to binary words and also some arithmetic has been included. The latter obviously goes more and more into the direction of common process computers, and at a certain point the question undoubtedly arises whether the device is still a PLC or can be considered as a general purpose computer. However in programming methodology, the difference is more clear: PLC devices mostly are programmed in the old-fashioned "assembly-like"-way, even when – as can be seen with some manufacturers – complete PID-type controllers are included. Clearly one may observe it would far be better to use an ordinary computer with its advanced and powerful software system instead of pinching high-level algorithms into stone-aged programming systems. The discussion here will not deal with those "hybrid" kind of systems, but will focus on pure boolean-like functions.

As in all boolean systems, two types of PLC applications exist. The first group comprises the combinational applications, in which the output values at each instant of time are determined by the input values alone. Those are the easiest to deal with. In the second group one finds the sequence control systems, in which the logic travels through a number of states. Here, not only the inputs, but also the current state (which has to be memorized) determinates the output. Where in PLC's of the first group boolean functions are fully parallel, the second

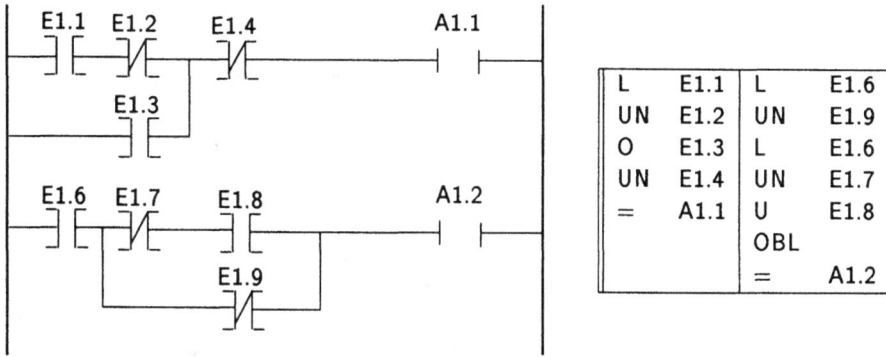

Figure 1: Ladder diagram of a combinatorial function.

group contains a mixture of parallelism and sequencing. In contrast with one would expect, the implementation of the second group into computers (which happen to be sequential machines) raises more problems, and leads to awkward structures.

Historically, programmable logic controllers emerged from the need in U.S. automobile assembly lines for a fast switching of the logic behavior of the machines and manipulators in the end of the 60's. At that time, relays were the state of the art in industrial technology, and instead of a fixed wired scheme, the concept of making it flexible and "programmable" was introduced.

In the first age, PLC manufacturers provided systems which were a simple functional replacement of the classical relay-like control. By offering devices with a kind of "central unit" (a boolean machine) implemented with semiconductor devices, "programming" of the input/outputs (mostly relays) became possible. However the programming methodology offered no new functionalities compared to a hard-wired system. Only a new way of describing the system had to be introduced, and following the concepts of that time the so called "ladder diagrams" and assembly-like notation was accepted (figure 1). Those syntactical constructs have to reflect both the parallel and the sequential elements in the application.

Representing sequence control by ladders, which are combinatorial primitives, was rather awkward because of the need of introducing feedback ("hold relays"). Therefore a new programming methodology was introduced, sometimes called "Grafcet". Such a scheme in fact draws a classical state transition diagram where state transitions are conditioned by "receptivities" (figure 7). The really new idea from the user's point of view was that inside one sequence loop only one state could be active at each instant of time. But one "program" could contain more such loops which had to be evaluated in parallel. Notations of all PLC formalisms have now been standardized in Germany by the DIN (19237, 19239, 40719) and the VDI (2880). The figures 1 and 7 represent some typical examples of PLC artwork.

97

3 Appreciation of the state-of-the-art PLC's

Giving a general appreciation on to-day's PLC's, a number of conclusion may be drawn.

1. Technologically PLC's are up-to-date and keep track of new progress.

2. Programming methodology has been frozen at the first (stone) age of its roots.

The latter requires some extra explanation. When looking at the old fashioned mechanical relay networks, they had at least one superb quality: it was true parallel processing. Introducing new semiconductor technology pinched this parallel system into a sequential schema, thereby hoping that when sequencing fast enough at the lowest level, the macroscopic behavior would look (quasi-) parallel. When looking even deeper into it, even when programming a sequence control automation (with state-transitions), this had to be "simulated" in order not to loose parallelism within other sequence loops. One would believe operating system principles never had been discovered. In this perspective, applying new technology seems more like a step backward. Worst of all, also in the programming methodology, a sequential procedural scheme is imposed, which hides the true parallel behavior of the controller.

The argument often heard, that this is the most "logical" way to program PLC's for non-technical people, is hard to keep up. Such systems really need a declarative, functional description, leading to a higher level of abstraction, enabling thereby better formal design, programming and debugging.

4 Using the PROLOG formalism in PLC's

In view of the class of problems dealt with in PLC's, the Prolog formalism by its predicate structure is very well suited to be the driving force for a PLC problem description. It can be shown that not only the formalism itself is an ideal tool, but that A.I.-methodology can be applied to enhance and ease the employment of PLC's in specific industrial situations.

Remark

Certain parts of this paper require a thorough understanding of the PROLOG-language. If necessary, a suitable book – such as the basic book "Programming in Prolog" by Clocksin and Mellish – could be consulted.

4.1 The Prolog declarative formalism in PLC programming

The ladder network of fig.1 could easily be written as follows:

```
a1_1 :- e1_1, not (e1_2), not (e1_4).
a1_1 :- e1_3, not (e1_4).
```

or, by decomposing:

98

```
a1_1 :- u, not (e1_4).
   u :- e1_1, not (e1_2).
   u :- e1_3.
```

This mechanism, not only formally describes the correspondent part of the ladder, it is also functionally correct: e.g. in order that the output a1_1 should be true (= the goal), the conditions e1_1, etc . . . , or the conditions e1_3. . . should be fulfilled.

The next obvious step to take, is making the predicate names more "user friendly":

```
furnace_ok :- temp_ok, not (press_low).
   temp_ok :- temp_fuel_ok, not (temp_water_low).
   temp_ok :- manual_switch.
```

In many situations, a more "numerical" formalism should be more convenient. However, a predicate such as

```
not (press <= 100)
```

is no correct Prolog, and should be transformed into something like

```
press(X), X <= 100
```

thereby leading to a full description of

```
furnace_ok :- temp_ok, press(X), X <= 100.
   temp_ok :- temp_fuel(X), X < 50, temp_water(Y), Y > 70.
   temp_ok :- manual_switch.
```

All the previous statements deal with rather specific situations. However the Prolog formalism is strong enough to climb an abstract level higher and to formulate more generic rules.

```
tree(_,_,X,Z) :- manual(X), not(Z)
tree(X,Y,_,Z) :- X, not(Y), not(Z).
```

Such rules can be the root for any specific declaration of the kind

```
furnace_ok :- tree((temp(fuel), temp(water), switch,press(water)).
```

Application example

Suppose there is a need for switching on and off a furnace following a number of conditions with numerical values involved. This could be described by the following declarations.

```
set(furnace, on_cond).                              (1)
reset(furnace, off_cond).                           (2)

on_cond :- temp(fuel,X), X < 50,                    (3)
           press(water,Y), Y > 100.

off_cond :- temp(fuel,X), X > 70.                   (4)
off_cond :- press(water,X), X < 80.                 (5)
```

Furthermore there is a need for declarations, which when asserted, involve a change of the state of the numerical predicate. E.g.

```
raise(temp(fuel), 40).                              (6)
raise(press(water), 110).                           (7)
```

How those assertions are dealt with in a real situation will be described in the next paragraph.

The generic rules to drive a furnace (or any other similar device) in such a way are given below.

```
set(X,Y)  :- Y, asserta(X).                         (8)
reset(X,Y) :- Y, retract(X).                         (9)

raise(X,Y) :- arg(1,X,A), functor(X,F,N),
              abolish(F,2), Z=...[F,A,Y],
              asserta(Z),!.                          (10)

raise(X,Y) :- arg(1,X,A), functor(X,F,N),
              Z=...[F,A,Y], asserta(Z).              (11)
```

A short note on how those generics operate is perhaps noteworthy.

(8): if a condition Y is satisfied, the fact X is asserted; e.g. if condition (3) is true, then (1) will assert the fact "furnace"

(9): if a condition Y is satisfied, the fact X is retracted; e.g. if one of the conditions (4) or (5) is true, then (2) will retract the fact "furnace" from the data-base;

(10) "arg(1,X,A)" instantiates the first argument of X into A; e.g. with raise (temp(fuel),40), A will be instantiated with "fuel"; "functor(X,F,N)" instantiates F with the functor of X; e.g. F will be set to "temp()" "abolish(F,2)" will erase e.g. the (old) fact temp(fuel, 70) from the data-base; "Z=...[F,A,Y]" will construct a new fact temp(fuel,40) into Z; "asserta (Z)" will assert the latter fact, thereby making further evaluation by (3), (4) and/or (5) possible; remark that the rule (10) completely succeeds only when a fact of the same kind already exists in the data-base; otherwise "abolish (F,2) will fail and the whole clause will fail;

(11): therefore this fact is similar to (10), except that it will be fired the very first time, i.e. when no fact such as temp(fuel,...) yet exists in the data-base, and rule (10) has failed.

The generic rules such as (8) → (10) are rather complicated for a typical PLC end-user. However they are of no concern to him and should be made available by the system programmer (or manufacturer). A library of many such generic rules could then be gathered, and the end-user only has to write into very simple and easy to read formalisms such as (1) → (5).

4.2 Prolog as an operational system for PLC's.

Writing formalisms for PLC's in Prolog is one thing; keeping Prolog continuously running as the driving force for the PLC is another. There are indeed a number of problems to solve before a classical Prolog environment actually can drive an operational system.

1. In a classical view, Prolog is an interactive environment intended to communicate with a terminal. In process control however Prolog should continuously run in the computer without any need for a user asking all kind of questions at his keyboard. The easiest way to solve this problem, although not the most performant, is to write a higher level interpreter which continuously asks a series of questions by itself. This automatic question/answer mechanism has the useful effect to maintain the working memory (facts-base, blackboard, scoreboard) at the exact state reflecting input/output elements. It further indirectly invokes the action onto the output steering elements. This could be called the "Prolog run-time subsystem".

2. Yet another important piece of software should be introduced at this stage. This is what could be called the "Prolog real-time subsystem". The task of this subsystem is to interact on one side with the real outside world an on the other side with the Prolog working memory. This is done – for input – by a real-time scanner which scans all input channels and writes any changes in the working memory. Furthermore it has to notify the Prolog run time system whenever anything changes. On the other hand – for output – a real time output module looks for any notification coming from the run-time system to output new values onto output channels.

The overall system is schematic depicted in the figure 2.

4.2.1 The run-time Prolog subsystem.

As already stated, the purpose of this subsystem is to keep the Prolog inference engine continuously running. A first approach consists of an outer program loop which repetitively executes all existing Prolog clauses. This loop evaluates all existing "measurements" in the facts-base, and put new output facts in the same facts-base. The measurements should have been written previously inside the facts-base by the measurement system, while the new output facts will be sent to the actual outputs by the steering system. Both steering and measurement system are part of the above mentioned real time subsystem (see figure 2) and will be discussed in the

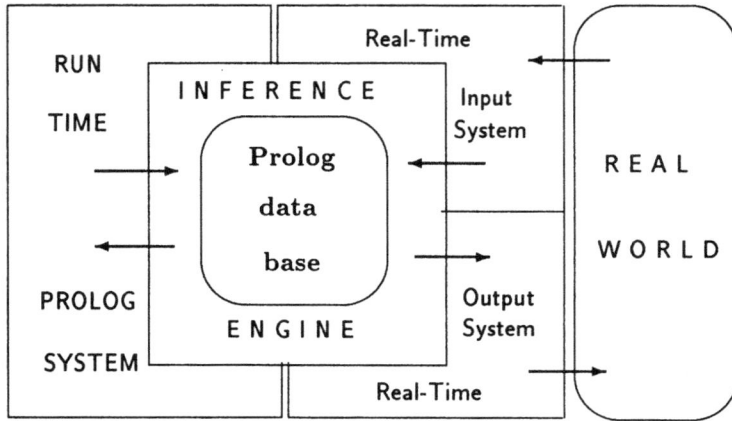

Figure 2: The PLC run-time and real-time subsystem.

next paragraph. A simple Prolog sequence to implement the run-time loop could be written as follows.

```
loop :- set(furnace, furnace_on_condition), fail.
loop :- reset(furnace, furnace_off_condition), fail.
loop :- set (pump, pump_on_condition), fail.
...
loop :- !, loop.
```

Such Prolog loop may be rather inefficient because of the time consuming task to check all (possible complex) conditions for setting, resetting, ... To overcome this, many enhancements are possible. One such optimization is to combine each Prolog statement of the loop with a trigger, which indicate that the following clause may be successful, and therefore has to be evaluated. Although each Prolog statement of the loop is still executed, most of them will immediately fail unless the trigger has been set previously by the measurement system.

```
loop :- trigger_1, retract (trigger_1),
        set (furnace, furnace_on_condition), fail.
loop :- trigger_2, retract (trigger-2),
        set (furnace, furnace_off_condition), fail.
loop :- trigger_3, retract (trigger_3),
        set (pump, pump_on_condition), fail.
...
loop :- !, loop.
```

102

Trigger	Rules
Trigger-1	rule-1,rule-2,...
Trigger-2	rule-3,rule-1,...
Trigger-3	rule-r,rule-p,...
...	...
Trigger-n	rule-e,rule-1,...

Figure 3: Relation-table "trigger"/"rules".

A more general way to construct such system would be to use a table, called a "connection-base" to indicate the relation between each trigger and the corresponding rules to execute (figure 3).

This table could e.g. be implemented by a list, with each element consisting of the trigger and another list containing the rules involved.

```
connect_base(
[[t1,[rule_A, rule_B, ...]]
 [t2,[rule_A, rule_C, ...]]
 ...
 [tn,[rule_X, rule_Z, ...]]]).
```

A suitable driver for this relation table is straightforward in Prolog.

```
evaluate (Y) :- X=..[Y,A],X, repeat, run(A).
run([]) :- fail.
run([X|Y]) :- run_head(X), run(Y)
run_head([X|Y] :- trigger(X),!, retract(trigger(X)),
                  run_rules(Y).
run_head(_) :- true.
run_rules([]) :- true.
run_rules([X|Y]) :- X,run_rules(Y).
```

The driver must be started by the command

```
evaluate (connect_base).
```

For this mechanism to work, another suitable relation table should exist for the measurement system to consult which trigger has to be set when a new measurement-value is noticed at the inputs (see figure 4).

Obviously many other optimization schemes can be thought of, depending on the skill and experience of the system designer.

103

Measurement	Trigger
temp-1	trigger-1
press.2	trigger-2
contact-4	trigger-1
sensor-4	trigger-p
...	...
course-1	trigger-n

Figure 4: Relation-table "trigger"/"measurement".

4.2.2 The real-time subsystem.

As already stated, this subsystem interfaces the facts-base, containing measurements and steering values, with the real world input/output. It is itself composed of two independent operating parts:

- the measurement system: this system scans the real world inputs and if any change in state is noticed, the relation-table with the triggers is consulted and the appropriate trigger asserted;

- the steering system: likewise this system inspects the facts-base for any changed output value and steers the real world output accordingly.

Although parts of the real-system could be written in Prolog, the communication with the real world input/output will require special attention and appropriate routines. Therefore probably the best way to implement the real-time system will be a program in a procedural language with suitable low-level facilities (e.g. C) and to interact with the facts-base by appropriate calls to Prolog primitives. Most professional Prolog-implementations (e.g. Quintus Prolog on Sun, Vax, etc...) provide such facilities to the user.

4.3 Application Example: a simple Robot manipulation.

To illustrate the above principles in a more practical environment, a simple Robot manipulation will be designed. Such a system is typical of a sequential nature (in contrast with previous combinatorial ones) : i.e. from the end-user's point of view there are a number of distinct states to recognize which the system sequences through, although at the programming level the complete program continuously keeps looping. The states therefore are kept inside internal locations for remembering.

The first item to understand, is the typical notational and terminology in sequential PLC-programs. Following the DIN-standard (widely used in PLC's), each succession in such a system is drawn following figure 5.

One recognizes the following elements:

104

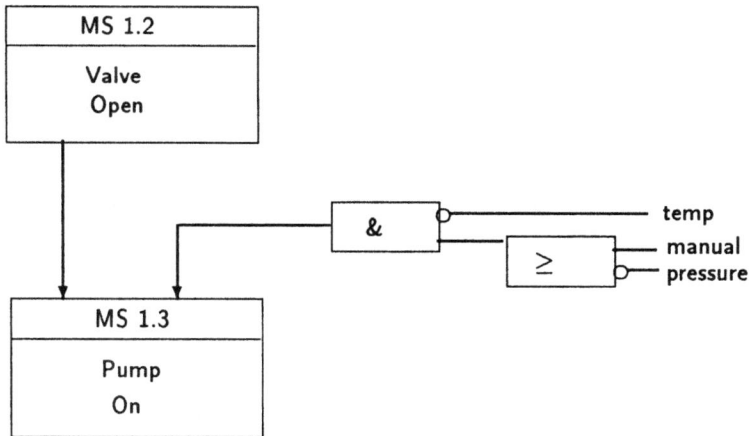

Figure 5: DIN-Notationals & terminology in sequential systems.

- the step itself (MS = "Merk-Schritte")

- the action at the step (e.g. "=", "s", ...)

- the condition to transit from one step to the next; this is called the "receptivity", and may be some (complex) boolean function.

Such formalism can easily be converted to Prolog as follows:

```
transit :- ms(valve_open), not(temp), (manual;not(pressure)),
           retract (ms(valve_open)), assert(ms(pump_on)), fail.
transit :- ..., fail.
...
action :- ms(valve_open), activate(valve), fail.
action :- ms(pump_on), set(pump), fail.
action :- ... , fail.
...
```

Obviously to make it more convenient for the end-user, a more generic formalism should be introduced. This is done by designing a table with at each entry the following items

    from_state/to state/receptivity/action

As always this is most practical by a list

105

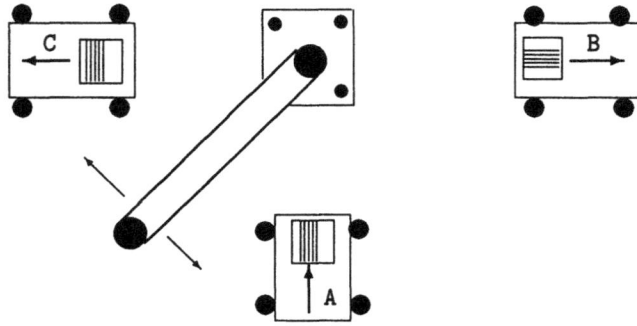

Figure 6: Simple Robot Manipulation.

```
state_diagram (
[[from_1, to_1, receptivity_1, action_1],
 [from_2, to_2, receptivity_2, action_2],
 ...
 [from_n, to_n, receptivity_n, action_n]]).
```

The interpreter for this list is the following program, which should completely be hidden from the user.

```
evaluate(Y) :- X=..[Y,A], X, repeat, run(A).

run([]) :- fail.
run([X|Y] :- action(X), run(Y).
action([From,To,Recept,Act]) :-
        ms(From),Recept,!,retract (ms(From)),assert(ms(To)),Act.
action(_) :- true.
```

The interpreter should be started with

```
evaluate(state_diagram).
```

As an example, take a robot situated as in figure 6.

The robot has to move objects fed by conveyer belt A towards the removing belts B and C, and the selection thereby is based upon the recognition of a product type_1 or type_2. The gripper is supposed to be of a very simple type (e.g. magnetic). The necessary contact sensors are given below.

106

prod_type(X)	product_type (X is 1 or 2)
pos (a)	positioning above A (true/false)
pos (b)	positioning above B (true/false)
pos (c)	positioning above C (true/false)
prod_at (a)	product_present on belt A (true/false)
prod_at (b)	product_present on belt B (true/false)
prod_at (c)	product_present on belt C (true/false)

The following output-signals are necessary:

gripper	gripper command (true/false)
motor (left)	motor-left command (on/off)
motor (right)	motor-right command (on/off)

The belts themselves are supposed to be completely self-controlled. The functional scheme of the overall solution for this illustrative example is depicted in the figure 7.

Following the functional methodology mentioned above, this solution can be offered to the interpreter with the following list. This list can be seen as the data describing functionally the complete robot movement.

```
robot_state_diagram(
[[initialize, go_to_start, not(pos(a)), set(motor(left))],
 go_to_start, wait_A, pos(a), reset(motor(left))],
[initialize, wait_A, pos(a), nothing],
[wait_A, grip_prod, prod_at(a), set(gripper)],
[grip_prod, go_to_B, prod_type(1), set(motor)(left))],
[go_to_B, wait_B, pos(b), reset(motor)(left))],
[wait_B, release_B, not(prod_at(b)), reset(gripper)],
[release_B, return_B, always, set(motor(right))],
[return_B, wait_A, pos(a), reset(motor(right))],
[grip_prod, go_to_C, prod_type(2), set(motor(right))],
[go_to_C, wait_C, pos(c), reset(motor(right))],
[wait_C, release_C, not(prod_at(c)), reset(gripper)],
[release_C, return_C, always, set(motor(right))],
[return_C, wait_A, pos(a), reset(motor(right))]]].
```

with the following clauses added to the earlier described interpreter.

```
set(X)    :- assert(X)
reset(X) :- retract(X).
always.
nothing.
```

The robot will start its operation after entering the following command:

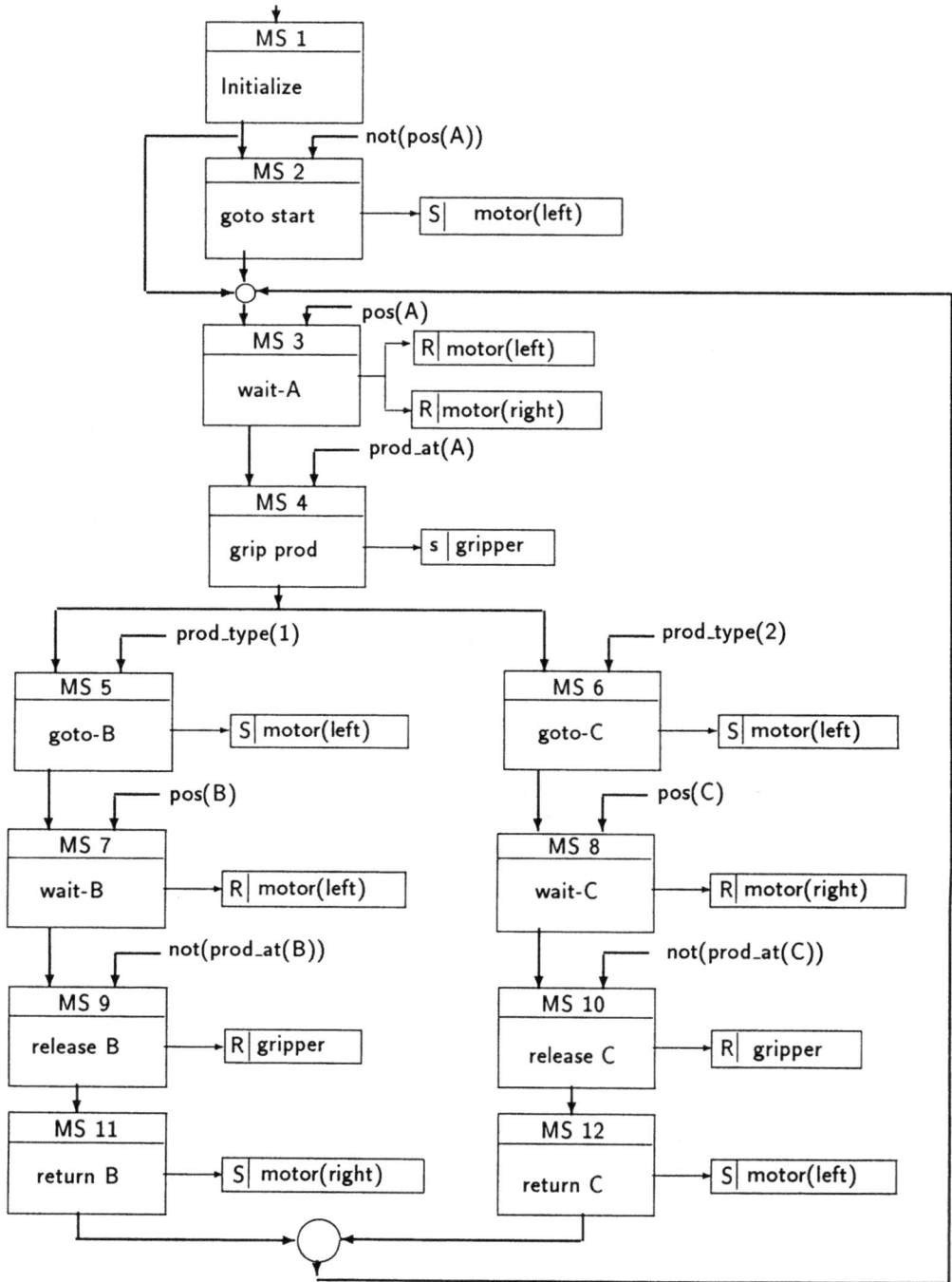

Figure 7: Functional Solution for the Robot-problem.

```
?- evaluate (robot_state_diagram).
```

The asserting and retracting of sensor-facts is – as mentioned before – the responsibility of the real-time subsystem.

5 Using Production Rules in PLC's.

Another formalism which has interesting properties to be used for PLC's are *Production Rules*. Production rules are as set of 'If-Then' clauses, where each rule is composed of two parts.

Condition part (or 'left-hand side'), which specifies all conditions which should be fulfilled in order for the rule to be triggered.

Action part (or 'right-hand side') which sums up all actions which have to be carried out once the rule fires.

Remark: one should be aware that the indication 'left' and 'right' is rather confusing, because in production rules - compared to Prolog - left and right are interchanged.

The driving force in rule based systems is done via the 'match-select-act' cycle (figure 8), which is executed as long as matches can be found.

Match: all rules are checked whether their condition parts are fulfilled or not. The rules which satisfy, and thereby are possible candidates for execution are placed on the 'agenda'.

Select: a special algorithm is executed to select out of these rules *one single* rule to execute. The strategy which is applied by the algorithm to make the choice can vary substantially from one system to an other.

Act: the selected rule is executed. During this execution many facts in working memory (the 'knowledge base') will probably be changed.

Afterwards the match phase starts over again, whereby, due to the changed contents of the working memory (the facts), rules which were fulfilled before could be no longer, and new rules could be added.

The interesting point about this driving mechanism, is that the execution is in no way procedural: the execution sequence is not fixed beforehand, but depends completely on the availability of the data in working memory. The behavior is pure data driven. Thereby it is very well suited as a possible underlying methodology for PLC's, which also (especially in the old relay's days) have a similar operation. In the next paragraphs this will be shown with 'CLIPS' (free domain rule based software from Johnson Space Center).

The furnace-problem, given at the beginning of this text, can easily be expressed in rules as follows. The first rule 'furn_go' says when the furnace should start, and put eventually the fact 'furnace_ok' in working memory. The second rule 'furn_stop'shuts it down by retracting the same fact. Of course the system supposes the real-time I/O-system will turn the fact 'furnace_ok' in a real output signal.

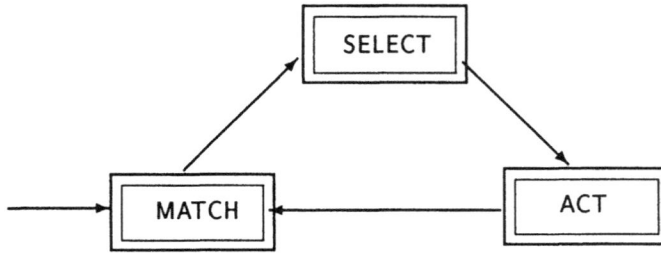

Figure 8: The Production Rule Driving Force.

```
(defrule furn_go                         (defrule furn_stop
   (not (press_low))                        (not (manual_switch))
   (or                                      (or
      (and                                     (not (temp_fuel_ok))
         (temp_fuel_ok)                        (temp_water_low)
         (not (temp_water_low)))               (press_low))
      (manual_switch))                      ?furnfact <- (furnace_ok)
=>                                       =>
   (assert (furnace_ok)))                   (retract ?furnfact))
```

The decision can also be drawn by facts based on numerical values. The example below uses three rules. One to switch on, and two to switch off. One of these two switches off when the fuel-temperature fails, while the other switches off when the water-pressure fails.

```
            (defrule furn_on
               (temp_fuel ?tf)
               (test (< ?tf 50))
               (press_water ?pw)
               (test (> ?pw 100))
            =>
               (assert (furn_on)))
```

```
(defrule furn_off_1                      (defrule furn_off_2
   (temp_fuel ?tf)                          (press_water ?pw)
   (test (> ?tf 70))                        (test (< ?pw 80))
   ?furnfact <- (furnace_ok)               ?furnfact <- (furnace_ok)
=>                                       =>
   (retract ?furnfact))                     (retract ?furnfact))
```

110

Also the programming of PLC sequential problems can be formalized by production rules. The example of the figure 5 can be written as:

```
(defrule transit_valve_to_pump
    ?msfact <- (ms valve_open)
    (not (temp))
    (or
        (manual)
        (not (pressure)))
=>
    (retract ?msfact)
    (assert (ms pump_on)))
```

This solution is rather 'ad hoc', but there is no problem at all to replace it by a more generic set of rules. As an example the moving robot problem of the figures 6 and 7 can be presented by:

• a set of facts stating all the receptivities;

• four generic rules to execute the transitions:

  − a positive 'set' transition: i.e. an action will be 'set' after a measuring signal is present;

  − a negative 'set' transition: i.e. an action will be 'set' after a measuring signal disappears;

  − a positive 'reset' transition: idem, but to 'reset' an action on a positive signal;

  − a negative 'reset' transition: idem, but to 'reset' an action on a negative (absent) signal.

The facts-base with the state transitions is given in the following default facts-list ('deffacts'). . Each **transition**-fact specifies

• from-state,

• to-state,

• 'negative'/'positive' (i.e. presence or absence),

• measurement,

• 'set' / 'reset',

• action

111

```
(deffacts robot-move
   (state initial)
   (transition initial go-to-start negative pos-a set motor-left )
   (transition go-to-start wait-A positive pos-a reset motor-left )
   (transition initial wait-A positive pos-a reset nothing )
   (transition wait-A grip-prod positive prod-at-a set gripper )
   (transition grip-prod go-to-B positive prod-type-1 set motor-left )
   (transition go-to-B wait-B positive pos-b reset motor-left )
   (transition wait-B release-B negative prod-at-b reset gripper )
   (transition release-B return-B positive always set motor-right )
   (transition return-B wait-A positive pos-a reset motor-right )
   (transition grip-prod go-to-C positive prod-type-2 set motor-right )
   (transition go-to-C wait-C positive pos-c reset motor-right )
   (transition wait-C release-C negative prod-at-c reset gripper)
   (transition release-C return-C positive always set motor-right )
   (transition return-C wait-A positive pos-a reset motor-right )
   (measuring always)
   (do-action nothing) )
```

The four generic rules are, considering their power, amazingly simple. The reader should try to 'play the inference engine' in order to see how the diverse rules are fired in the right sequence, only by the availability of the data in working memory. Again the real-time I/O subsystem is supposed to assert the measuring ... facts in working memory according to real-world facts.

```
(defrule pos-set-transition              (defrule neg-set-transition
   (transition ?from ?to                    (transition ?from ?to
  positive ?recept set ?act)                   negative ?recept set ?act)
   ?c1 <- (state ?from)                     ?c1 <- (state ?from)
   (measuring ?recept)                      (not (measuring ?recept))
=>                                       =>
   (retract ?c1)                            (retract ?c1)
   (assert (state ?to))                     (assert (state ?to))
   (assert (do-action ?act) ) )             (assert (do-action ?act) ) )


(defrule pos-reset-transition            (defrule neg-reset-transition
   (transition ?from ?to                    (transition ?from ?to
      positive ?recept reset ?act)             negative ?recept reset ?act)
   ?c1 <- (state ?from)                     ?c1 <- (state ?from)
   (measuring ?recept)                      (not (measuring ?recept))
   ?c2 <- (do-action ?act)                  ?c2 <- (do-action ?act)
=>                                       =>
   (retract ?c1)                            (retract ?c1)
   (assert (state ?to))                     (assert (state ?to))
   (retract ?c2) )                          (retract ?c2) )
```

6   Using Object-Oriented Programming in PLC's.

Object-Orientation uses a message sending mechanism between objects, which are instantiations of classes (types) to drive the execution of a program. This is also a sort of data-flow, because each message sends data to another object in order to perform a task on. After execution a value (or an object) is returned to the originator of the message. Nevertheless, the real program flow is still procedural. There is only a strong information hiding / data encapsulation. Object-oriented programming for PLC's can easily be done, but gives less benefits compared with the other mentioned A.I.-formalisms.

7   Conclusion.

When Artificial Intelligence tools should be used in PLC's instead of the usual paradigms (ladders, grafcets,...), it would mean an important step forward:

- *programming* can be clear, straightforward and supported by many usual computer tools, thereby freeing PLC development from any need for special development stations;

- *real program design* can be employed, enabling CASE tools with possible automatic program generation;

- *maintenance and debugging* of PLC programs can be much better formalized and unambiguously documented;

- *expert systems* can be developed in order to assist the designer as a process engineer, rather than as a technician;

- *a more flexible and soft migration between PLC's and process computers* can become feasible; this would certainly be beneficial for the designer, because his choice becomes neither limited, neither once-and-for-all; also he would be liberated from control algorithms cryptically pinched into PLC's (and which are just not made for it).

AN INTRODUCTION TO EXPERT SYSTEM DEVELOPMENT

Mateu Tomás Saborido

Instituto de Cibernética, Univ. Politécnica de Cataluña
Diagonal, 647, 08028 Barcelona, Spain.

Part I

Software life cycle.

The whole process of developing and maintaining a software product is called software life cycle. Different models of software life cycle have been devised, which structure the process in several phases, each one having a definite starting and ending point.

A software life cycle model defines all the activities required to define, develop, test, deliver, operate and maintain a software product. Different models emphasize different aspects of the life cycle and are appropriate for a range of situations. In any case, activities must be well defined, and some metrics should be available in order to evaluate if they have been completed successfully.

Only three of the multiple life cycle models and variations of models will be shortly reviewed in this document :

- **Waterfall.** This model and its descendants are the most classical ones. They are widely known and used (and criticized). Waterfall is the paradigm of conventional software engineering model. It is highly structured and pays special attention to analysis and documentation.

- **Prototyping.** Prototyping is very extended in AI research. It is an iterative model, where successive approaches to the final product are developed and evaluated.

- **Incremental development.** This model tries to integrate the aforementioned approaches. Incremental development may be specially appropriate for expert system projects, which come from AI research and try to become another software engineering technology. Nowadays, people from both communities cooperate in the same projects and a framework comfortable for everybody should be very useful.

114

1 The waterfall model.

This is one of the most commonly accepted phased models. At each phase, developers must verify that the application is being built correctly, meeting the specifications developed in the previous phases and satisfying all the requirements. If such verification succeeds, the next phase may be initiated, or else the work of the current or previous stages must be revised. This is a sequential model, where going backward to previous stages is a symptom of a bad design.

The waterfall model is structured in the following phases:

- **Study of feasibility.** The objective of this phase is to define an overall acceptable concept for the application, assuring its feasibility over the life cycle.

- **Requirements analysis.** At this stage the required functions are completely specified. Interfaces and performance requirements are also defined. The result must be validated, confirming that the system to be designed meets the customer's needs and wishes.

- **Product design.** Comprises the complete specification of control and data structures and overall hardware and software architectures, plus the necesssary documentation. The general specifications are verified in order to assure that they satisfy all the requirements defined at the previous phase.

- **Detailed design.** During this phase, all the control and data structures, all the input/output parameters and the algorithms for each program component are specified and verified against the previous specifications.

- **Codification.** At this stage, the different modules of the system are coded and tested.

- **Integration.** The tested components of the system are integrated and verified as a whole.

- **Implementation.** This phase refers to the full software and hardware installation, user training and customer acceptance.

- **Operation/Maintenance.**

This model is very extended, and a large set of design techniques, notations and, more recently, CASE tools support different formulations of the underlying ideas (i.e. Structured System Design, Jackson Structured Programming, etc).

A major difference between conventional approaches and expert systems is that in the former, requirements and specifications are essentially static, while in the latter many requirements may be discovered during design and implementation phases. On the other hand, deferring testing and acceptance to the later phases does not seem a good idea in this kind of projects, where users are closely involved in all stages of the development. All these considerations imply that developers will be often required to revisit prior phases and that is not in the waterfall spirit.

2 Rapid prototyping.

Rapid prototyping is a technique in which a simplistic model of the system is devised to demonstrate some functionality, to experiment with different approaches or to evoke user feedback. This methodology inherently forces multiple iterations through the knowledge acquisition, coding and testing phases until a final prototype is obtained.

This method has some troubles. For instance, people can extrapolate from early prototypes and think that the effort to complete the system will scale linearly from the effort to encode the partial prototypes. Another problem is scalability. Methods that perform well with simple prototypes may be inefficient in real cases. Moreover, as complexity grows, questions that were easily solved in the first prototypes may become unmanageable. Even if prototypes are scalable, as their complexity increases time and costs per prototype will grow.

Rapid prototyping may be used as a life cycle model itself when the problem is sufficiently small so that one person or a very small team can understand and code it directly, a tool is available to develop the prototypes easily and no maintenance or modification will be required. But it is more often used as a part of a more complex model; for example Bohem proposes a spiral life cycle where subsequent prototypes are developed until the full system is defined by an operational prototype, then waterfall detailed design, coding, testing, etc of the final system follow.

3 The incremental development model.

Incremental development represents a refinement of the waterfall model where prototyping is integrated in a top-down structured development.

In this model, there are a study of feasibility, a requirements analysis and a product design phases as in the waterfall model, but detailed design, coding, integration, and implementation are splitted in successive increments of functionality. Maintenance and enhancement may be splitted or not depending on the coupling between the different increments.

An increment differs from a prototype in that it results in a pre-release version of a part of the sytem and not a "throwaway" system. An increment must adhere to the specifications generated at the product design phase and it should have a fully specified detailed design. It should be as packed as possible (both in its code an documentation) and fully tested. After completing an increment, the development team should refine product design, incorporating all new features.

Determining the scope of an increment is not always easy, it must be small enough to be completed in at most some months but also large enough to represent a significant progress on the final system.

When all increments have been performed, system specifications are frozen. In some cases the final system release will require only a little integration work and the global testing, but in others, detailed design and coding of some or even all increments will have to be revisited in order to obtain a fully integrated and functional final system.

As will be later commented, one of the key points of incremental development is related with depth and breadth. System specification must cover the full breadth of the system and be deep enough to select good, easy-to-integrate increments. Increments must control both their breadth, in order to avoid overlaping with others, and their depth, in order to avoid cost desviations caused by enhancements which are not significative.

This model, if well used, allows to make use of the advantages of rapid prototyping and monolithic development without many of their disadvantages. It allows developers to incorporate new user requirements as the project develops, enables parallel development and eases incremental validation and transference.

Further reading : [MHB89], [Ros67], [Boe81],[Fox90],[Mag88]

Part II

Knowledge acquisition

4 Kinds of knowledge.

It is difficult to define, identify and isolate what the knowledge engineer is looking for during knowledge acquisition. The objective of this task is to conceptualize what the expert knows and how he or she solves problems. Thus, before being able to acquire knowledge from the domain experts, knowledge engineers should have a basic understanding of what knowledge is, how it is stored and organized and how experts access it. It is beyond the scope of this document to extend about this topics, but at least a brief categorization of some kinds of knowledge may be given :

- **Procedural knowledge.** Its main characteristic is that it is of such a low level that the expert has many troubles to explain it. It involves an automatic response to a stimulus. This kind of knowledge appears when the same knowledge is used over and over in a procedure, so the expert loses access to it, and as a consequence the ability to report it. A typical example is the knowledge of how to ride a bicycle. Procedural knowledge refers to *"know how"*.

- **Declarative knowledge.** This kind of knowledge represents surface information, that experts verbalize easily. It is referred to as *"know that"*. Declarative knowledge must be managed carefully because it generates a lot of information but usually does not express the deep knowledge that the expert uses, possibly unconsciously. Heuristics are a good example of this kind of knowledge and first generation expert systems contain only this kind of shallow knowledge.

- **Episodic knowledge.** This category refers to experimental information that the expert has grouped by episodes. These episodes are composed of events and temporal and spatial relationships among them.

- **Semantic knowledge.** Semantic knowledge is the deep knowledge of the expert. Concepts, facts, definitions, relationships among them and algorithms for manipulating symbols, concepts and relations compose the models that the expert

117

uses when reasoning. This is the richest knowledge and flexible expert systems are those based on it.

- **Metaknowledge.** It is concerned with the conscious awareness of the expert about what he/she knows and how he/she uses knowledge. Metaknowledge lets the expert to decide how to solve a problem, to evaluate the reliability of the generated solution and to detect unsolvable problems. This kind of knowledge is very important in order to avoid the *"cliff effect"* that appears when the expert system deals with problems that it is not capable of solving.

5 Knowledge acquisition techniques.

There are many knowledge acquisition techniques, many of which are derived from psychology practice. Some of the most extended are:

- **Unstructured interview.** Given the general session goals, the expert functions as a lecturer. The knowlege engineer asks questions to clarify his or her understanding. This method is often used in initial stages and is very dependant on the expert's ability to teach. This dependence and the inherent lack of focus may lead to an inefficient use of time and efforts.

- **Structured interview.** The knowledge engineer outlines specific goals and questions for the session and the expert is usually informed in advance about them. Specific lists of questions may be used. This is the most extended technique in knowledge acquisition.

- **Domain and task analysis.** Knowledge is obtained from the observation of the expert performing his or her task. All the analysis and conceptualization effort is performed by the knowledge engineer.

- **Process tracing and protocol analysis.** These techniques require that the experts report or demonstrate how they solve a specific problem. Their effectiveness depends on the ability of the experts to report their cognitive processes and the capacity of the knowledge engineer to analyze, structure and represent this information.

- **Simulations.** Computer-based simulations may be useful because they provide a fully controlled environment where the expert can demonstrate his/her expertise and the knowledge engineer may be released from part of the recording task. On the other hand, analyzing simulation results may be a hard task.

- **Automated tools.** Some tasks in the knowledge acquisition phase may be automated. For example, during the knowledge formalization phase, special-purpose editors or documentation managers may be used. Knowledge engineers may use CASE tools and some expert-system oriented utilities (i.e. KADS Power Tools) in order to make their work easier. Another kind of tools are those oriented to perform automated knowledge acquisition. One of the earliest attempts was TEIRESIAS, which was designed to enable automated acquisition of new knowledge for the MYCIN expert system. Other tools like AQUINAS, KNACK or TDE have continued this approach.

- **Multiple expert techniques.** When multiple experts are involved, some techniques for knowledge elicitation (i.e. brainstorming) or for obtaining consensual solutions may be useful.

5.1 Technique selection.

When deciding which technique or combination of techniques will be used, there are many variables to keep in mind, some of them are:

- **Project size and goals.** Large-scale projects need more structured techniques than smaller, more informal ones. The final system required functionality also determines which techniques must be used. For example, if the expert system must aid in aircraft steering, simulations may be basic.

- **Knowledge acquisition phase.** In the first stages, unstructured interviews are the usual. While acquired knowledge consolidates, structured interview is the most used. In the final steps it may be recommendable to make the expert interact with a tool to solve problems.

- **Availability of experts.** Structured techniques make a more efficient use of expert time than unstructured ones.

- **Type of knowledge to be elicited.**

  - Procedural: Structured interview, process tracing, simulations.
  - Episodic: Simulations, process tracing.
  - Declarative: interviews.
  - Semantic: Structured interview, process tracing, task analysis, repertory grids.
  - Metaknowledge: Structured interview, process tracing, task analysis, simulations.

- **Knowledge engineers availability and skills.**

6 Interacting with human experts.

Knowledge acquisition is not merely the application of some techniques that may be studied in textbooks; there is a human interaction and knowledge engineers are responsible of enabling and maintaining a good relationship between them and the experts. It is difficult to define how to do this, ut some critical concepts may be identified:

- **Expert motivatio** Domain experts should feel part of the development team. It is useful to let them now what happens with the information they provide, and how does it evolve until oming part of the final system.

- **Expert availabilit** Because the expert is an expert, day to day operation at the customer organi tion will demand his/her attention. Whenever possible it is better to get expert way from their workplace. Scheduling a time for knowledge

119

acquisition and another for their job, and trying to keep these activities separated, both expert and knowledge engineers will perform their task more efficiently.

- **Knowledge acquisition team.** Even in small projects, it is very difficult for a single person to carry on with the full knowledge acquisition effort. When using interviews it is hard to manage the overall flow of the conversation and record the information obtained. Therefore, it is better to split this work between two knowledge engineers.

7 Record keeping.

After knowledge acquisition techniques have been selected, record-keeping procedures must be established in order to:

- Reduce development costs by using expert and knowledge engineer's time effectively, for example making inconsistencies and knowledge gaps easy to detect, avoiding redundant work, etc.

- Organize the acquired information, allowing fast recovery and use, both in development and maintenance phases.

- Provide an audit trail for the acquired knowledge, so that a documented history of each piece of information from acquisition to implementation may be obtained easily.

Some methodologies define knowledge acquisition databases, used like project repositories in conventional developments. From the point of view of knowledge acquisition, information should be kept about:

- Session identification (date, number, etc).

- Knowledge engineer(s).

- Topic, subsystem or function of the expert system covered.

- Session source(s) (e.g. forms, reports, manuals, experts).

- Session documentation (notes, audiotape, videotape, etc.)

- Knowledge acquisition technique(s) used.

- Session review (information elicited, future actions, etc) with clear references to the documentation such as tape counter, page, etc.

- Expert verification of session notes.

Two components of the project database that will be extremely useful in all phases of the project are knowledge and data dictionaries. The knowledge dictionary contains the compilation of (nearly) all the terms used by the experts plus those "generated" during the project, providing a description of each term as common dictionaries do. This tool must provide a common language for everybody involved in the project, and will be invaluable for later enhancement and maintenance. Data dictionary is a more technical tool, it should be built and developed during system specification and design. For every entity in the system design, the data dictionary must provide a description of its meaning, use, attributes, format, restrictions, etc.

How this information will be processed until its implementation determines other structures in the project database. For example, if knowledge is codified as rules, standard rule headers may be defined. These headers may include information about when and by whom the knowledge was elicited, who formalized the rule and when, cross references, etc. Appending this header to the coded rule allows a trace of the knowledge that may be very useful later.

Any example case described during knowledge acquisition or testing should be carefully documented (problem description, solution, justification, required information, performance rates, etc) because it can be very useful both in system validation and as a regression test library. When significant changes have been done, these cases may be rerun in order to ensure that the system still handles them correctly and to evaluate how the system performace evolves as the system grows.

In large projects this record-keeping task may justify the figure of the project librarian and/or the knowledge acquisition coordinator, whose function is to ensure that all this information is managed properly.

Whenever possible, all record-keeping procedures and formalisms used during the project should be chosen and/or adapted in order to allow the use of commercial CASE tools. Alternatively, *ad-hoc* tools may be developed. This will ease the management of the great flows of information that this kind of projects generate.

Further reading : [MHB89], [CGJ89], [Efs89], [Sac91], [RU91], [Anj87]

Part III

Conventional Expert System development.

8   Introduction.

Expert systems are very similar to other information systems. In fact, much of the work in their development is conventional software engineering. The main difference is that conventional systems usually perform well-defined operations on data, while expert systems try to emulate human behavior in front of a given problem. Human expertise is difficult to formalize and, in general, expert system developers start coding without a fully detailed specification. This is not a caprice, allowing the expert to interact with prototypes and partial implementations enhances significantly knowledge elicitation and feedback. This advantage would be lost if complete knowledge acquisition was performed prior to design and codification. Moreover, paper-based-only validation of knowledge is impossible in nontrivial systems; it is necessary to develop prototype versions based on partial specifications in order to validate and refine them with expert's feedback.

Incremental development adapts well to the special characteristics of expert system projects and it will be used as a framework to present some general ideas about them.

The development of an expert system may be splitted into these phases:

- **Assessment and scoping.** Once a possible expert-system-based application has been identified, the first step is to define concisely the system concept, the key requirements and the business objectives and evaluate if such system has a good potential to be developed and used successfuly.

- **System specification.** Knowledge acquisition starts here. The outputs of this phase are:

  - The analysis of the expert's knowledge and environment concerned with the future system (conceptual model).

  - The functional, behavioural and architectural design of the future system (design model).

  - A prototype that reflects both models is an easy-to-understand manner.

  Conceptual and design models must be validated, i.e. customers must accept them as the formalization of their wishes and a solution to their needs (expressed in the previous phase) and developers must verify that all requirements are fulfilled by the proposed design.

- **Incremental development.** During this phase the global specifications are divided into increments of functionality of the initial prototype. It should be noted that some increments may be knowledge-based, while others may imply conventional techniques, or both. Each increment follows a particular life cycle:

  - Detailed design. If the increment has a knowledge component, specific knowledge acquisition is performed. In any case, the increment is fully detailed (algorithms, data structures and formats, etc).

  - Codification. The detailed design is translated to executable code.

  - Verification and validation. The increment is tested in order to ensure that it performs the right task in the right way. If this step fails, previous phases are reviewed, using the feedback information elicited during validation. In some cases, this review can affect the global system specifications.

  - Integration. Once the increment has been accepted, system specifications are adapted in order to acommodate the lessons learned during the development of the increment. Consistency with previous and next increments must be checked.

  It is possible to develop some increments in parallel, especially those that have no knowledge component (system interfaces, simulators, etc). On the other hand, knowledge engineers and other specialists, such as conventional analysts or programmers, will make a better use of their time if they can work in another increment when they are not necessary. For example, when an increment is in its codification phase, knowledge engineers can focus on another increment which is in the detailed design or validation phase. Later, they will participate again in the increment validation and integration. Good project management (scheduling, resource allocation, etc.) is essential to take advantage of this parallel incremental development.

- **Final system integration and testing.** When the full system has been developed and all specifications, and requirements are known, some refinement is performed, ranging from little integration work to modifications in the detailed design and recoding of the full system. In this extreme case, the final system may be developed using other tools or even by another team. Some authors state that this fully specified system should be reimplemented using conventional software engineering tools.

  Once the final system is developed, global verification, validation and tuning are performed. Except if the full system has been recoded after incremental development, this global testing should not be very difficult, because complete unit testing had been performed previously.

- **System tranference.** When the final system is implemented, tested and tuned, it is transferred to its operating environment and future users are properly trained.

- **Maintenance, enhancement and support.**

9 Assessment and scoping.

This is the first phase in the development of an expert system. The main steps are:

- First requirements definition, based on the needs and wishes expressed by experts, potential users and customer management.

- Study of feasibility. During this step requirements may be redefined attending to technical, financial, social or strategic reasons. Finally a set of requirements must be formulated. All parts involved in the development must agree that the system defined [Fab90]:

  - Is justified by real needs.
  - Is properly expressed from the right point of view.
  - Is possible to develop in the given technical and financial conditions.
  - Has a high potential for success.

  Some projects will be abandonned or entirely reformulated at this stage.

- If the previous steps are succesfully accomplished a project leader is appointed. A project coordinator, preferably at the senior management level, should also be appointed at the customer organization.

- Team selection.

9.1 Study of feasibility.

This phase is not more relevant than the others, but it will be seen in more detail because examining some significant concepts in the study of feasibility will provide a good profile of the environment of expert system projects and the troubles that may arise.

The most extended technique for estimating the potential for success of an application is checklists. Although they are rough measures and they work in a "quick and dirty" manner, experience shows that checklists are a good method because reliable estimates are obtained without spending resources and they can be revised and as the experience increases.

Different people have devised checklists for expert systems. Here is one due to T. Beckman [Bec91].

This checklist is organized in categories of criteria. Each criterion is a yes/no question and has a weight assigned according to its importance. Every category sums the weights of the affimative answers. The participation of each category in the final evaluation (the sum of all weights) expresses its influence on the project potential for success.

The checklist proposed by T. Beckman is composed of six categories with their importance expressed by the maximum points that each one can provide to the final evaluation. They are:

1. Characteristics of the task to be performed, 25 points.

2. Future system payoff, 20 points.

3. Customer management commitment, 20 points.

4. System designer skills, 15 points.

5. Domain expert characteristics, 10 points.

6. User characteristics, 10 points.

The first two categories are essential for success, and it is required that the application scores at least 50% in each one (13 points for category 1 and 10 for the second). The other four categories contribute in lesser degrees. Scores below 50% in any of them do not imply that the application should be discarded but indicate potential difficulties. It is also required to score at least 50% (50 points) overall. All the weights and many criteria are subjective and, therefore, may be discussed.

9.1.1 Task characteristics.

Clearly, the task to be performed by the expert system is the most critical topic. Not only must it be technically feasible, but use of expert sytems must also be necessary and appropriate.

124

	weight	concept
1	2	Task is primarily cognitive, requiring analysis, synthesis or decision making rather than perception or action.
2	2	Involves primarily symbolic knowledge and reasoning, rather than numerical computation.
3	2	Is complex, involving many parameters.
4	1	Involves chains of reasoning on multiple levels of knowledge.
5	2	Uses heuristics and requires judgement or reasoning about subjective factors.
6	1	Can't be solved using conventional computing methods
7	2	Often must be solved with incomplete or inaccurate data.
8	2	Often requires explanation, justification of results or reasoning.
9	1	Is at an intermediate stage of knowledge formalization that uses heuristics and classification rather than search or algorithms.
10	1	Task knowledge is confined to a narrow domain.
11	1	Task knowledge is stable.
12	1	Incremental progress is possible; task can be subdivided.
13	1	Does not require reasoning about time or space.
14	1	Is not natural-language intensive.
15	1	Requires little or no common sense or general-world knowledge.
16	1	Does not require the system to learn from experience.
17	1	Is similar to one in an existing expert system.
18	1	Data and case studies are available.
19	1	System performance can be accurately and easily measured.

- **Criteria 1-9 and 13-16:** Good tasks to be performed by expert systems are those wich involve essentially reasoning and manipulation with symbols. If there are many, say hundreds, parameters and chains of reasoning are deep, with a high level of heuristics and subjective factors, expert system technology can show all its power. Applications that involve a mostly numerical computation or that can be easily implemented with conventional data-processing techniques should be solved in this way. On the other side, if the desired task is too ambitious or risky it should be redefined. The underlying idea is to use expert systems in the problems that they can solve and only if there is no better way (the same can be said of data bases, parallel programming and any other technology).

- **Criteria 10-12 and 17-19:** Some task characteristics increase the probability of a successful project. If knowledge is well-bounded, its elicitation will be easier. If it is stable, system maintenance will be minimized. Incremental progress reduces risk and complexity and allows customers to follow the progress of the project, involving them in the development. Data and case studies available on magnetic media are invaluable both in knowledge acquisition and in testing.

9.1.2 Future system payoff.

The need of the expert system must be expressable in terms of benefits relevant to the customer management. Compared to conventional applications, expert systems are usually be expensive so that a good benefit/cost ratio is crucial.

	weight	concept
1	2	System would significantly increase revenues.
2	2	Reduce costs.
3	2	Improve quality.
4	2	Capture undocumented expertise or expertise that is perishable or in short supply.
5	1	Distribute accessible expertise to novice users.
6	1	Provide training effect on users through usage.
7	1	Raise barriers to future market entrants.
8	1	Require no or minimal more data entry than current system.
9	2	Be developed using commercial shells; little customized coding needed.
10	1	System maintenance would be low.
11	2	System would be delivered on an affordable desktop computer or workstation or integrated in preexistent hardware.
12	1	Could be phased in; partial completion would still be useful.
13	2	Would result in benefit/cost ratio of al least 10:1.

If the benefits are too low (criteria 1-7), the task probably does not provide sufficient value to the user organization. If the system costs too much to deliver (criteria 8-11), either the system needs to be recoded in a simpler environment (shell and/or hardware), custom coded in a compact language such as C or reduced in functionality.

9.1.3 Customer management

Even in the case that technical feasibility of a certain expert system is demosntrated, and payoff is ensured, the project can fail due to management problems or simply lack of interest. Hardware purchases, expert availability, access to documentation, etc. may be very difficult if customer executives are not supportive and interested in the project. Managers interest must have a realistic base, otherwise troubles will be merely postponed.

	weight	concept
1	4	Senior management is willing to commit significant funding resources to develop and deploy the system.
2	2	Willing to commit significant staffing resources to develop and deploy the system.
3	1	Supportive, enthusiastic, and has appointed a project coordinator.
4	1	Receptive to innovation and new technologies.
5	3	Site management has committed staffing resources for acquiring knowledge, preparing test cases and validating the system.
6	1	Is supportive, enthusiastic, and has appointed a project contact from management.
7	2	Management accepts primary responsibility for maintaining the installed system.
8	1	Use of the system will not be politically sensitive.
9	1	Either system requires minimal changes to existing procedures, or if it requires substantial changes, agrees to changes and recognizes the need for user training on the system.
10	2	Management understands that estimates for resources and deadlines are difficult and probably will not be met.
11	2	Management realizes the system will make mistakes and may perform no better than a moderately proficient user.

9.1.4 System designer(s).

Technical success of the system depends chiefly on the design team and secondarily on the domain expert. A solid background in AI and conventional techniques enables designers to recognize potential problems and use appropriate tools to solve them, rather than relying only on the functionality supplied by a given expert system shell. Knowledge engineers must be able to learn about the subject area in order to interact well with the domain experts and design a competent system.

	weight	concept
1	2	Designers have experience in designing and developing expert systems.
2	1	Know how to use a development tool appropriate for the system and have used the chosen tool or shell.
3	2	Are experienced in acquiring and eliciting knowledge from written sources and domain experts.
4	2	Have A.I. background to recognize which techniques will be useful in developing the system.
5	1	Understand cognitive psychology.
6	2	Have managed and developed more traditional computing applications.
7	2	Are knowledgeable or an expert in the domain.
8	1	Have hardware and software available for development.
9	2	Can commit enough full-time effort for developing, testing and implementing the system.

9.1.5 Domain expert(s).

In some circumstances it is possible to develop an expert system without a true domain expert, specially if there are knowledgeable practitioners or a good written source. If expertise exists and recognized experts are available, they should be cooperative, communicative and accessible for extended periods. Experts must feel the project as their own. If they are afraid of losing control or deskilling their work, they won't cooperate.

Some authors argue that to work with more than one domain expert leads to problems, because they will probably disagree. The fact is that if these discrepancies can be solved, the breadth and quality of the obtained knowledge will be better. There are few situations where consensus is impossible, and in this case a third party or the knowledge engineer himself must decide, and eventually reduce the experts group.

	weight	concept
1	2	Recognized experts exist.
2	2	Expert performance is provably better than that of amateurs.
3	1	Task is routinely taught to beginners.
4	1	Experts are accessible for extended periods of time.
5	1	Are cooperative.
6	1	Communicate well.
7	2	Are available to develop test cases and help evaluate the system.

9.1.6 User(s)

The final success of an aplication depends absolutely on the final users. If they feel threatened by the expert system or simply do not need it, or the interface is inadequate, they won't use it and the project will fail, despite its technical performance.

128

	weight	concept
1	2	Users feel a strong need for the system.
2	2	Won't be deskilled or unfavorably displaced as a result of implementing the system.
3	1	Want to be involved in the system's development.
4	2	Do not have unrealistically high expectations.
5	3	Have roughly the same level of expertise.

Criterion 5 is heavily weighted because when most users have approximately the same level of expertise, it is much easier to design an effective interface, to use standarized terminology as well as explanation and help facilities. When different kinds of users exist, different interfaces should be developed or some of them will feel unsatisfied.

9.2 Team selection.

Once it is decided to start the project and its scope is clear, the development team can be selected taking into account the project characteristics. In many cases, this team already exists so only expert selection and specific training have to be done.

9.2.1 Project leader.

The project leader must have a solid experience in management of information system projects (either conventional or expert systems). He or she is not required to be an AI expert if other members of the team can complement this weakness.

In many cases, two cultures will be present in the team: conventional software engineering an people who come from AI research. It is the project leader's responsibility to make them coexist and cooperate in a productive manner.

Not all project managers can lead an expert system project, they must tolerate uncertainty, and be willing to see their plans revisited frequently. Incremental development means that information required for planning a project is discovered when it is in an advanced state.

9.2.2 Knowledge engineer(s).

The essential attributes of good knowledge engineers are the same as those of good system analysts: they must work fluidly with different kinds of people, be inquisitive and able to assimilate information for a while without imposing a rigid structure on it too early. The main difference is that knowledge engineers usually begin codification when analysis is far from being complete.

9.2.3 Expert(s).

Usually there is not much choice about expert or experts. If it is possible, the most motivated and comunicative ones should be selected. If there are multiple experts, one of them may be appointed as main contact, in charge of obtaining consensus solutions if the experts disagree.

9.2.4 Software engineer(s) and programmers.

Much of the work in developing an expert system is conventional software engineering (i.e. interfacing the expert system with other information systems). In small projects, the knowledge and the software engineer may be the same person, but in medium size and big ones there will be enough work for two or more specialists.

A critical task for software engineers will be the development of the user interface.

9.2.5 Apprentices.

If the customer organization accepts the responsibility of knowledge maintenance, the best way to ensure a good know-how transference is appointing specific people for this task from the very beginning of the project. These people can participate in the project as apprentices in order to become familiar with both the programming environment and the knowledge representation.

9.2.6 External consultants.

External consultants may be useful for different aspects, such as project management and auditing, methodological aspects, third party validation, or to provide help in specialised topics of AI, communications, hardware, etc.

9.2.7 Training.

- **Knowledge engineers:** They should have an introductory training in the application area before starting expert system development. It is very difficult to obtain a broad view of the domain from domain experts because they tend to think in terms of highly specific cases. Moreover, initial interviews are the most important ones, because interpersonal communication will be established and the main concepts will appear. If the knowledge engineers are able to conduct them and detect the key points early, the whole process will be easier and faster. It is clear that a minimum basis about the application area can be very useful.

- **Domain experts:** A certain knowledge about expert systems may be useful, but it is dangerous to train domain experts in using development tools because this knowlege can influence how the expert expresses his or her knowledge about the domain, facilitating its elicitation but reducing its quality.

- **Development team:** If the tools or techniques that will be used in the development are new to the development team, an intensive training by vendors or specialists can avoid substantial delays or deficient implementations.

10 System specification.

As in any other information system, it is strongly recommended to develop a written functional and architectural specification before coding. The analysis required to obtain such specification constitutes the first stages of knowledge acquisition.

There are some differences between system specification in conventional projects and in expert system ones. Firstly, tools used to develop expert systems allow a very high level coding, so that there is not a clear frontier between specification and coding. Secondly, incremental development implies *per se* that coding begins before a full specification is developed.

If incremental development is used as life cycle model, designers must be very careful in determining the scope of the global system specification: it must cover the full breadth of the system, because increments will be defined based on it, it must be deep enough to select good increments, but not too deep to restrict increment development based on early decisions. If the right frontier is fixed, good increments will be selected and system specifications will be expanded without excessive modifications.

Once the specific objectives are clearly how they will be achieved, software and hardware can be properly chosen, assuming they were not part of the user requirements accepted in the previous phase.

10.1   Models in the development process.

In any project, two systems are always present:

- **The object system.** By object system we mean the real human and material environment where the final application will work. This object system is described as a collection of data: text books, information obtained from experts, reports, etc. This data is at a low level of abstraction and poorly structured.

- **The final system.** It is an implementation that performs the desired task, respecting all the requirements determined by the customer and by the technology (hardware and software) used. This system is described as a collection of programming code.

The purpose of the development team is to implement a good final system from the data obtained about the object system. These are some possible approaches to this task:

- **Map directly from object system to final system.** Rapid prototyping is a good example of this technique. In previous sections advantages and disadvantages of rapid prototyping were discussed, so we will only mention a few problems:

  - Except in very simple applications, this mapping can be very difficult to establish, because there is a large gap between the data about the object system and the primitives that the programming language or the expert system tool offer.

  - Lack of flexibility. Because there is not any intermediate step, any change affects the whole development. Thus, the final system may become difficult to enhance and maintain.

  - Often the final system is more tuned to the possibilities of the environment used than to the needs of the object system.

- **Establish more abstract intermediate models.** Many conventional methodologies establish different models between the initial object system and the target final application. Most of these ideas are applicable to expert systems. KADS [SBBW88] methodology distinguishes these levels:

  - **Object system data.** Obtained during knowledge acquisition.
  - **Conceptual model.** It is a model of the object system obtained from the abstraction and structuration of the acquired data. The conceptual model describes the domain expertise in a real-world oriented way, that is, with absolute independence of implementational considerations. Some conventional methodologies, such as ISAC, propose models with the same purpose, usually called object system model.
  - **Design model.** This model is still at a high level of abstraction, but it is implementation oriented. At this level, problem solving methods are chosen, user requirements (i.e. interface) and external requirements (i.e. performance) are considered, possibly limiting the conceptual model. The design model in expert system developments is very similar to the conceptual models proposed in some conventional system development methodologies.
  - **Detailed design.** At this level the general design decisions explicited in the design model are developed until the full system is specified.
  - **Final system.**

The conceptual model must be developed by the knowledge engineers, the design model needs both knowledge and conventional software engineering techniques. The detailed design and the final implementation should imply the application of well known techniques so that they may be developed by less experienced members of the team.

The distinction between conceptual and design model is useful because the way in which a problem is solved may change but the problem itself and how the expert solves it remain the same. For example, in a control problem the expert can use heuristic knowledge, but the final system may use it or exploit the computational power and perform simulations using a generate and test method.

The set of models commented above is easy to integrate in an incremental development life cycle. The design model represents the frontier between global specifications and splitted development.

10.2 Conceptual model.

The objective of the conceptual model is to model expertise. In many expert systems this model resides only in the developer's mind, or, at most, has a plain, inprecise structure. In order to allow design and detailed design models to be developed in a consistent manner, knowledge engineers must make an effort to formalize this model. The first step should be to structure the knowledge they obtain in different levels of abstraction and functionality. Again following KADS methodology, these levels may be:

- **Domain layer.** It represents the static knowledge of the domain and is composed by concepts and relationships between them. The knowledge engineer must try to

132

extract the underlying models that the expert uses. There are many formalisms that manage concepts and relations, the most extended are semantic networks and the Entity-Relationship model.

- **Inference layer.** This level is formed by the knowledge about *what* can be done with the domain knowledge. These operations, termed inferences, may be viewed as abstract functions that generate a set of outputs from a set of inputs. The operation should be expressable as a verb, and parameters may be concepts or generalizations of concepts, for example compute(data) – data, classify(instance concept) – generic concept, obtain(concept) – data, display(data). The inference layer must describe what the operation performs, but the method or paradigm applied to accomplish it belongs to the design model.

- **Task layer.** The tasks are ways in which inferences are combined order in to achieve a particular goal. Standard formalisms such as dataflow diagrams, pseudocode, etc may be useful to describe this essentially algorithmic knowledge.

- **Strategic layer.** It represents the problem solving knowledge of the experts, that is, how they drive the process of generating a solution. It is usually highly heuristic and it is the most difficult knowledge to formalize. It may be splitted into different sublevels of metaknowledge.

The general idea is that strategic knowledge lets the expert use the appropriate task knowledge to generate a solution of a problem defined on the domain knowledge. Task knowledge uses inference knowledge to operate on the domain knowledge.

10.3 Design model.

The design model must define the future system in its functional, behavioural and structural aspects. To cover them it should :

- **State future system functionality.** Specification must clearly define what the system will do and which requirements (i.e. response time or consistency constraints) must be met. Criteria should be established in order to evaluate if these goals are accomplished by the final system. This framework will be useful both as a guide in the development phase and as a valid reference during the system transfer and customer's evaluation.

- **Select paradigms to implement the desired behaviour respecting all the requirements.** Once the functions to be performed by the system have been defined, methods or combination of methods can be selected to achieve them. Conventional software engineering and AI provide a wide variety of techniques such as search algorithms, hierarchical classification, different kinds of planning (constraint satisfaction, opportunistic, etc), forward and backward chaining, ATN parsing, production (rule-based) systems, temporal reasoning, etc. There is a vast literature on these techniques and their use. These paradigms provide both an abstract description about how the problem will be solved and a set of design elements, which are the physical parts that support the method. For instance, if a production system will be used to perform some heuristic reasoning, production rules and a rule interpreter must be used. This design elements provide a lot of information about how the system can be organized architecturally.

- **Establish modularization.** In expert system development, modularity is greatly affected by the need to make system structure reflect expert knowledge structures and behavior. On the other side, modules without knowledge component may be as important as the others, and must be managed with the same care. The knowledge structures and, therefore, the final decomposition may not become rigid until several acquisition-implementation-validation cycles have been done. But once the major components of the system are known, each one can probably be treated as an independent module. When the main modules have been identified, the architectural specification must define:

  - The tentative decomposition, specifying the functionality of each module in a proper level of detail. Isolating configuration, control and dataflow structures and input-output management in specialized knowledge bases is very useful.
  - All the interfaces and control structures between modules. A basic interaction protocol may be a good solution. Nowadays, the most extended approach are blackboard-based architectures.

  A good decomposition and formalized interaction protocols between modules will improve maintainability and system evolution and development may be faster if several modules can be built in parallel.

- **Define interfaces.** All interfaces with other systems and humans must be specified. Interfacing an expert system with other information systems may be difficult because it usually manages data in a very flexible way, while the latter are generally very restrictive about types and formats. At least a full communication protocol should be established so that an interface level may be developed in parallel, possibly by another team. The expert system development team should build the final interface using this communication primitives. This approach makes future enhancement and migration easier. Man-machine interface may be considered as a subproject with its own particular incremental development, so a very deep specification at the first stages can be dangerous. Anyway, if some low level operations are defined (i.e. expose menu) or a standard interface, such as X11, is chosen, the same criteria can be used.

Programming and documentation conventions may be established. These standards should be an aid for actual and future developers of the system, so they must not be too restrictive and annoying.

10.4 Tool selection.

It is very common to select expert-system tools too early in the project. There are some general-purpose shells that are really powerful, so this may not have significant consequences. Moreover, some stable teams have their standard toolkit and use it in its full functionality with high productivity rate. In other cases, the target system determines the choice, or forces to fully develop a special-purpose shell (e.g. due to real-time requirements, special operating system or hardware, etc). Nevertheless, the general criterion recommends to postpone the choice until the application and the resources to implement it are well known and the most appropriate tool or tools can be selected.

Whether looking for a standard tool or for a specific tool to be used in a given project which has been properly specified, there are some criteria to keep in mind :

- **Integration.** The tool must allow efficient integration with all subsystems in the final system architecture and it should not be a handicap for the evolution of the expert system. The era of the open systems and architectures is beginning and expert systems are not an exception. If this integration is easy, clear and elegant, system maintenance and enhancement will be easier. Development tools based on the C language are the most promising ones.

- **Portability.** Ideally, the tool should operate on the same hardware platforms as the other subsystems in the target architecture. If the tool is supported on a wide range of platforms, additional options are retained to migrate the system at a later date, but one must be cautious in evaluating vendor's claims because there is a difference between making a tool run on a platform and being able to provide effective support and up-to-date releases on it. Nowadays there are some broadly proven tools that run from a PC to a multiuser mainframe, allowing (more or less) a good integration with the different environments. Again C-based tools are the most portable.

- **Representational and problem-solving power.** Good representational and problem-solving power, if well used, increase productivity and allow for easier maintenance and evolution. But integration and portability should be sacrificed in order to use sophisticated reasoning strategies only if there is not any other chance. In fact very few cases, if any, justify it.

- **Performance.** Most expert-system tools perform adequately if the knowledge base is well-designed. If performance troubles appear (in many cases due to insufficient physical memory) and the tool is really portable the system may migrate to a more powerful platform, of course if customer's budget allows it. In general, if a given good expert-system tool running in a conventional workstation, with the physical memory configuration recommended by the vendor, has performance limitations then the design should be revisited looking for inefficient coding, rather than puchasing a more powerful machine.

- **Vendor reliability and support.** As with any other software product, a vendor's support for the tool during the project development (i.e. training, hot-line, special customizations) and along system's life is very important. Vendor's financial viability and its commitment to its product should be assured before purchasing a tool because an application based on a tool that has become obsolete or even discontinued will hardly survive.

- **Easy learning and use.** This criterion is influenced by the use of the tool that the development team will make. If it is acquired for a single project, an easy-to-use shell can reduce costs if it is powerful enough for the desired application. But if future uses are considered, training delays may be accepted. Most of the current shells are easy to learn, but to use them in all their power requires a long time experience.

10.5   Development and final hardware selection.

Usually final hardware is determined by user requirements. If not, the choice must be based on the estimated requirements of the application and the environment where it will be used. If one of the main expected benefits of the expert system is knowledge distribution, then PC or low-range workstations would be the most adequate platform. If time performace is critical, then high-range workstations or special-purpose hardware (i.e. Lisp machines) should be chosen. Some expert-system shells allow full integration in corporative mainframes, with remote and multiuser operation.

It is desirable to develop the application in the same platform where it will finally run. But if target hardware selection is deferred - because the tools used for development allow a wide range of final platforms - or if it is not a good development environment (like most mainframes) the application may be developed in a different site and migrated when it has been tested and validated. In this case new tests and tuning should be carried out in order to ensure total acommodation of the expert system in the final environment.

10.6   First prototype.

The results of this phase must be formalized in written documentation, which will be probably refined several times along the life cycle of the system. But as relevant as this technical documentation is an initial prototype which reflects this information in such a way that people not directly involved in the project (i.e. customer's managers or users) can understand it and initiate the feedback inherent to an incremental development.

The first prototype(s) should focus on describing future system's functionality, structure and behaviour rather than formalizing first knowledge acquisition so they need not be operative, probably they must not. In order to avoid too detailed prototypes, which belong to the scope of the next phase, the first prototype should be released in a short term, say a few weeks from the start in a medium size project.

Again man-machine interface may be treated as a parallel project and have its own prototypes.

Some prototypes may be developed until the behaviour prototype has been accepted and the incremental development phase begins.

11   Incremental development.

If system specification was well performed, successive increments will be selected and developed covering the full breadth of the application in an easy-to-integrate way. But incremental development, especially when applied to expert systems, has a danger inherited from its prototyping component: if developers focus too deeply in a slice of knowledge they will generate increasingly deep and robust solutions to that slice, but in many cases they would not be iterating towards a balanced, deep and robust general solution. In expert systems there is always a slightly better prototype, which handles the problem a little better. Developers (especially the project leader) must be able to stop at the proper level, otherwise there will be problems with schedule and budget.

Depending on the knowledge component of each increment, this phase may be closer to a conventional detailed cycle (design, codification, testing, integration) or to an iterative, prototype-based one (knowledge acquisition, formalization, codification, validation, feedback, etc, until a final formalization and coding have been accepted).

11.1 Detailed design.

When a part of the system functionality is assigned to an increment, intensive knowledge acquisition over it must be performed. Knowledge and software engineers will extend the design model, formalizing specific knowledge and devising structures to handle this increment.

Detailed design should be as deep as possible, and the functionalities of the selected tool or tools should be considered. Decision tables, inference and decision trees, truth tables and more conventional formalisms like dataflow diagrams or pseudocode descriptions may be useful in this phase.

11.2 Codification.

If detailed design is properly performed, the gap between it and the tool functionalities should be very thin, and codification should not be very different from other conventional projects. Possibly, the only difference will be the power of the tools used.

11.3 Verification and validation.

For each increment two kinds of testing must be performed:

- **Verification:** Each generated item, which can be specification, documentation or code, must be tested to ensure that it meets the specifications and requirements of the previous level. In other words, verification ensures that every item does what it should do, and does it well. The final purpose is to generate a product of quality.

- **Validation:** This kind of testing ensures that the generated items meet the needs, wishes and requirements that led to the development of the product. The final purpose is to generate the right product.

Notice that both concepts refer to well done requirements analysis and specification phases. Even if a serious effort has been performed in these phases, expert system projects may encounter troubles in testing their results:

- Many requirements are expressed in terms of functionality or behavior, and this makes them difficult to evaluate. For example, a requirement might be "generate a good, feasible plan for restoration". It is not always evident how to measure "goodness" and "feasibility" of a generated plan.

- Because specifications are not completed when detailed design and codification start, a priori test plans can not be developed. This may be dangerous because misconceptions that affect design and/or code may also affect test plans.

- Many paradigms and tools used in expert system development are declarative rather than imperative, that is, their code describes what must be done, but an explicit statement about how it should be performed is not provided. The clearest examples of this are production systems, where a number or rules encapsulate knowledge and interact in a (more or less) non deterministic way. This behavior is very useful for problem solving, but creates additional problems for verification and validation, for example:
    - It is difficult to determine all possible execution paths, and therefore, test them.
    - It is not always clear that (quasi-)independent modules that were individually tested and approved will interact without problems.
- Uncertainty and truth management may be very difficult to test in complex systems.

11.3.1 Verification techniques.

There are some formal proof methods to check whether an item meets its specifications. In conventional programming, some tools, such as Annotated ADA, can manage the proof process, but designers must perform a significant part of the previous work (i.e. defining preconditions, postconditions, etc.). Due to this overhead, formal proof techniques are not much used. In knowledge-based systems, these techniques are not yet developed enough to be useful.

Some quality-assurance techniques are based on reviews of the generated specifications, documentation or code. These reviews may be informally performed by other members of the team (walk-throughs) or may be more formal and exhaustive, usually performed by external consultants (inspections). Walk-throughs and inspections are broadly used because not only are errors uncovered, but also style and quality can be evaluated, corrected and standardized.

Many expert-system shells provide a set of tools like syntactic error checkers, cross-reference generators, type checkers, consistency checkers, and different debugging and inference-tracing aids. These tools are useful in verifying the implementation, and some design errors may also be detected with them. If formal design and specification tools have been used (i.e. pseudocode, logic, petri nets, etc) *ad hoc* verification tools may also be developed. In medium-size and large-size projects the benefits obtained by the use of these tools will justify the effort of developing them. Moreover, this kind of tool may be useful for later projects, becoming part of the development resources or even a commercial product [Kah].

As mentioned in 7, example cases may be very useful in incremental verification and validation.

11.3.2 Validation techniques.

If requirements specifications are established, either initially or through incremental development, part of the validation testing can proceed like in conventional systems. User testing based on prototypes or on partial implementations may provide the required feedback information. The rest of the validation work refers to knowledge and it must be carried out by experts.

138

Expert validation may be performed by original domain experts and/or by other experts. Two techniques may be used:

- **Direct review.** In this case, experts validate the application by defining typical or pathological problems and evaluating how the system solves them. Usually original experts perform an extensive direct review and later the system is evaluated by external experts.

- **Turing Test.** It is a validation technique where an expert (or group of experts) is shown two solutions for the same problem (or set of problems). One solution is the output of the expert system and the other was generated by another expert in the domain. The evaluating expert doesn't know how the solutions were obtained and is asked to evaluate them. If the expert shows no preference for one of the two sources, it is assumed that the expert system adequately models the human expertise.

When using experts to evaluate or validate an expert system, some of the problems of knowledge acquisition may arise. For example, some experts may disagree with other experts. Additionally, experts will hardly differentiate the conclusions generated by the system from the whole execution and the interface. Thus, developers must know exactly what are they validating (the solution, how the solution is generated, the interface, etc). Again, all test cases must be added to the regression test library.

11.4   Integration.

All the lessons learned during the development of an increment must be incorporated to the system specification. The impact of this feedback must be carefully analyzed in order to guarantee full compatibility and integrability with previous and future increments.

12   Final system integration and testing.

Along this iterative process, system specifications will become more and more comprehensive and solid. When the last increment has been finished, the whole specifications must be as complete as they should be in a waterfall model, the only difference is that they are already implemented.

As commented before, this phase may imply from some esthetic work to a full revision of detailed design and recoding. System testing and tuning efforts will vary depending on the scope of this phase.

13   System transference.

Once the final system has been validated and accepted, it has to be installed in its definitive environment.

The application must be tunned, and probably unexpected troubles will appear when integrating it with customer's software and hardware.

Users must be trained to use the system effectively. The first contacts with the application are crucial, so they must receive all the support they need and the development team must be ready to make last-minute modifications in order to assure a good landing. User manuals and complementary documentation, such as tutorials or case examples, should be released at the very beginning of this phase.

Most troubles can be avoided if all the involved groups have actively participated in the previous phases and especially in this one.

If possible, a phased transfer should be done, perhaps overlapped with incremental development.

14   Maintenance, enhancement and support.

All information systems need some degree of maintenance and enhancement in order to evolve as their environment does. The more dynamic the environment is, the more maintenance effort will be required. Expert systems are especially sensitive to the environment, so very small changes in it may render a good expert system obsolete.

At this moment, automatic learning, that is, information systems that adapt themselves to environment evolution, is far to be practical, outside the research laboratories. Therefore, only two different approaches are possible:

- **Maintenance by specialists.** This is the most usual in conventional software. There are two possibilities, which are compatible:

  - **Maintenance by the developer organization.** The project does not finish with customer's acceptance, but the organization that developed the system must also provide technical support and maintenance along the full software life cycle. Many conventional information systems follow this model and probably this is the most appropriate. The only requirement is that the developer organization must be able to commit qualified people and resources to this not very pleasant task. Expert system technology has very recently left research environments and many developer organizations are not big and stable enough to accept this task.

  - **Maintenance by the customer organization.** In this case, data processing or R+D departments of the customer organization "inherit" the system. Customers may feel more confident about their own future than about developer's, or confidential data may be involved so they may prefer to keep control and responsibility of the system. Another situation may be that the customer organization has used developers to capture the *know how* of expert systems development in order to create its own development team. In any case, if the customer organization will perform maintenance, enhancement and user's support, a complete transference of the system and technology must be attained or else the system will degenerate.

- **Maintenance by users or experts.** A lot of research is devoted to it. Users and experts know exactly which enhancements the system needs and what is becoming obsolete, so if a proper set of tools is provided to them, they may be able to make the system evolve with them. This approach has many questions to be solved, for example:

- System design and documentation must be consistent with the system itself. It is not possible to require users to keep technical documentation updated. If automated tools perform this task, it should be closer to reverse engineering (from code to specifications, still a research theme) than to serious, formalized maintenance.

- Users *feel* how the system should evolve, but it is usually difficult for them to formalize what should be done. Consequently, they will hardly implement their *feelings*.

- Users can argue that they are users, not developers. And they would be right.

15   On the man-machine interface.

Ergonomy is a critical factor for success in almost every product. Information systems are no exception. A good interface can make a mediocre application successful. Conversely, a poor or awkward interface is influential enough to bring a technically perfect expert system to disaster.

The information flows and the requirements to be fulfilled by the interface are part of the requirements analysis and system specifications. It is not a good practice to split interface development into the different modules of the application. Man-machine interface should be considered as another increment, having its particular life-cycle and integration plan. During interface development requirements that do not refer to the system objectives but increase its usefulness may be discovered. For example, a graphical on-line representation of the key values of a power plant may be very useful to its operators. An expert system developed as an aid in failure diagnosis may access these data looking for symptomatic values and generate warnings only if strange behavior is detected. But if such representation - a graphical display of the data - is also provided, the operators will make a wide use of the expert system, and their attention will be captured when a warning is raised. The system will be useful in its task and a great extra value may be obtained with little additional effort.

Even if these oportunistic enhancements are not possible or desired, man-machine interface must be developed carefully. Designers must be very sensitive to the environment where the application will be used. For example, if the expert system will be used in emergency situations, it must provide only the required information and it must be presented in an easy-to-understand manner, such as graphically. If the user will have his/her hands busy, devices such as keyboards or mouses must be avoided.

The expert system must use the same language as the user, and if graphical constructions or mnemonics are used (i.e. icons, drawings, commands, etc) they must be as close as possible to the user's knowledge and skills.

If the man-machine interface is developed seriously, it will be the first increment started and the last one finished.

Further reading : [Sac91], [SBBW88], [Mag88], [Fab90], [CGJ89], [Efs89], [IZMG90],

[MHB89], [Kea90], [KD], [Pay91], [Ste87], [Enr91], [PGRA90], [KD], [OGMR90], [RAH91].

Part IV

Future trends.

16   From Expert Systems to Knowledge-based Modules.

Most of the references about successful expert systems refer to stand-alone systems (without interaction with other systems) or to on-line expert systems that interact with their environment but still maintaining their identity as somehow special systems. This is the approach of the previous chapter. More recently, conventional information systems with one or several components (modules) developed using expert systems technology (knowledge-based modules) are appearing in scene.

This phenomenon denotes that the software engineering community is beginning to incorporate knowledge-based technology as one more between many other resources they can use in the development of information systems. This dedramatization is very desirable because knowledge-based modules can be a good solution for scheduling, monitoring, some kinds of control and other tasks that are present in many conventional information systems.

The main condition for this normalization is a complete integration of knowledge enginering in software engineering.

16.1   Architectural integration.

At this moment, architectural integration of knowledge-based modules in conventional systems is a fact. Many expert system shells provide full integration with conventional programming languages and database managers. Moreover, some powerful and broadly used tools automatically generate C, FORTRAN ,ADA, or even COBOL or PL1 versions of the application. This computer-generated code is not very readable, but this is not a relevant requirement, because all the work may be performed using the shell, and the code is only generated to be compiled and linked with other hand-written modules.

16.2   Methodological integration.

Methodological integration is not merely a technological problem, but also a human one. Many analysts and organizations identify strongly with their development procedures and tools. These are part of their *corporate culture* and, for this reason, evolution in this field is very difficult.

Although some attempts have been carried out, classical methodologies and techniques seem not to be powerful or flexible enough to accommodate knowledge-based paradigms. There are some emergent paradigms that may be very useful. We think that the most promising one is object orientation (OO). The OO paradigm is interesting not only for itself and its potential integration with knowledge-based techniques, but also because it has attracted software engineers from the very beginning, thus providing an indirect way for integration.

16.2.1 Object orientation.

Combining OO and rule-based paradigms is not new. Some classical models of knowledge representation, like semantic networks and frames, are based on the same concepts as objects: entities, relationships, generalization/particularization, attached procedures and inheritance. Because of this, many expert-system shells provide structures very similar to the ones proposed by the OO paradigm.

The integration of OO with knowledge-based techniques could be deeper. Thus, the conceptual model described in 10.2 can be mapped into the OO paradigm:

- **Domain layer.** Object orientation provides a representational power similar to semantic networks or frames, the paradigms most commonly used to model the domain layer.

- **Inference and task layers.** OO paradigm includes the concept of services, to represent algorithmic knowledge attached to objects. The message-passing mechanism provides interaction between objects in a very flexible way. Inferences and tasks may be easily mapped into services. Encapsulation features are not as standardized as relations, services and message passing, but they may provide a mechanism to define the visibility constraints of the services, thus allowing to distinguish inference and task layers.

- **Strategic layer.** The mapping of this level into OO structures is not as clear as that in the previous layers. Some formulations of the OO paradigm include demons or condition-triggered services that may be very useful to formalize this highly heuristic layer. A lot of research must still be performed in this area, but the current state of the art provides mechanisms more powerful, flexible and natural than conventional paradigms.

Design and detailed design models can also be mapped into OO.

The interest in OO stems not only from its representational power but also from research that is being performed in order to define a methodological framework to developers. New approaches to modularity, maintainability, reusability and other significant concepts are appearing related with OO methodologies. Knowledge engineering should adapt some of these ideas to its special characteristics.

Object-oriented analysis and design may provide a common environment for both knowledge-based and more conventional modules and systems.

Further reading : [CY90], [War89], [Boo86]

143

References

[Anj87] A. Anjewierden. "Knowledge acquisition tools,". *AICOM*, pages 29–38, August 1987.

[Bec91] T.J. Beckman. "Selecting expert-system applications,". *AI Expert*, pages 42–48, February 1991.

[Boe81] B. W. Boehm. *Software Engineering Economics*. Prentice-Hall, Englewood Cliffs, New Jersey, 1981.

[Boo86] G. Booch. "Object-oriented development,". *IEEE Trans. on Software*, pages 211–221, February 1986.

[CGJ89] M.A. Carrico, J.E. Girald, and J.P. Jones. *Building Knowledge Systems*. McGraw-Hill, New York, 1989.

[CY90] P. Coad and E. Yourdon. *Object-Oriented Analysis*. Prentice Hall, Englewood Cliffs, New Jersey, 1990.

[Efs89] J. Efstathiou. *Expert Systems in Process Control*. Longman Group UK Limited, 1989.

[Enr91] P. Enrado. "Expert-system resource guide,". *AI Expert*, pages 52–59, May 1991.

[Fab90] J.M. Faba. "Metodologia de construccion de sistemas expertos,". *ALI*, pages 35–43, May 1990.

[Fox90] M.S. Fox. "AI and expert system myths, legends, and facts,". *IEEE Expert*, pages 8–20, February 1990.

[IZMG90] A. Irgon, J. Zolnowski, K.J. Murray, and M. Gersho. "Expert system development. a retrospective view of five systems,". *IEEE-Expert*, pages 25–40, June 1990.

[Kah] G. S. Kahn. "From application shell to knowledge acquisition system,".

[KD] P.J. Kline and S.B. Dolins. "Problem features that influence the design of expert systems,".

[Kea90] M. Kearney. "Making knowledge engineering productive,". *AI Expert*, pages 47–51, July 1990.

[Mag88] B. Maguire. "An incremental approach to expert systems development,". In *8th Avignon International Workshop*, 1988.

[MHB89] K.L. McGraw and K.H. Harbison-Briggs. *Knowledge Acquisition: Principles and Guidelines*. Prentice Hall, Englewood Cliffs, New Jersey, 1989.

[OGMR90] T.J. O'Leary, M. Goul, K.E. Moffitt, and A. E. Radwan. "Validating expert systems,". *IEEE-Expert*, pages 51–58, June 1990.

144

[Pay91]   E.C. Payne. "A modular knowledge-flow model,". *AI Expert*, pages 36–41, 1991.

[PGRA90]   D.S. Prerau, A.S. Gunderson, R.E. Reinke, and M.R. Adler. "Maintainability techniques in developing large expert systems,". *IEEE-Expert*, pages 71–80, June 1990.

[RAH91]   G. Ribar, F. Arcoleo, and D. Hollo. "Loan probe: Testing a BIG expert system,". *AI Expert*, pages 43–51, May 1991.

[Ros67]   P.E. Rosove. *Developing Computer-based Information Systems*. John Wiley & Sons, New York, 1967.

[RU91]   W. Remmele and B. Ueberreiter. "Systematic knowledge acquisition in expert systems,". *Siemens Review. R&D Special*, pages 9–14, Spring 1991.

[Sac91]   E.D. Sacerdoti. "Expert-systems development,". *AI Expert*, pages 26–33, May 1991.

[SBBW88]   G. Schreiber, J. Breuker, J. Bredeweg, and B. Wielinga. "Modelling in KBS development,". In *8th Avignon International Workshop*, 1988.

[Ste87]   L. Steels. "The deepening of expert systems,". *AICOM*, pages 9–16, August 1987.

[War89]   P. Ward. "How to integrate object orientation with structured analysis and design,". *IEEE Software*, pages 74–82, March 1989.

INTRODUCTION TO FUZZY LOGIC AND FUZZY SETS

Dr. Ranjan Vepa

Department of Aeronautical Engineering, Queen Mary and Westfield College,
University of London, London, E1 4NS, U.K.

1. Introduction

Rule based expert systems are based on facts. It must be recognised at the outset that the truth or falsehood of a fact is entirely based on human perception. Every fact is true only because it is believed to be so. However as we know more about the subjects pertaining to the fact, that is as we gather further evidence, the fact may become less and less true and eventually it is quite possible that 'fact' is in fact false! In most logic processing a fact is assumed to be true and is added to a list facts. The fact then stays on this list permanently unless it is explicitly deleted from it. What is essential to the ability to move the fact onto and off the "true" list as the evidence changes. In many practical cases a fact undergoes a gradual metamorphosis as the evidence changes. The problem with logic systems is that it is not possible to incorporate time or "dynamics" into logic processing as in the case of differential systems. On the other hand given the evidence at a particular instant of time one may not be willing to assert that something was true but just that it is likely or unlikely with a certain degree of certainty. In this sense the status of facts are rather more vague than just being either true or false.

Logic may be extended in several ways to account for this feature of the real world. One approach is to admit that the status of a fact can change from false to true and vice-versa. This approach generally gives rise to what is known as non-monotonic logic. Monotonicity refers to the feature of classical logic where the pool of facts may be either steadily increasing or steadily decreasing. At this point it would be essential to distinguish between logic and reasoning. Classically reasoning has been restricted to the study of formal philosophical and mathematical logic. The term formal imples that logic is concerned more about the form of logical statements and less about their meaning. Thus logic is concerned primarily with the syntax of statements and less about their meaning. Non monotonic reasoning therefore implies that the pools of known facts can both increase or decrease during a certain period of time as the reasoning progresses. Thus non-monotonic logic allows for a certain vagueness as it permits facts to change there status from true to false with the availability of further evidence.

An alternative approach is to associate a degree of certainty or truthwith each fact, thus allowing for patially true facts with different shades of meaning between true and false. This is the basis of Fuzzy Logic and is a form of continuously multivalued logic as opposed to classical two valued logic. Finally by allowing for and defining the dynamics of change of the shades of meaning during the metamorphosis of a fact in non-monotonic logic we could bridge the gap between non-monotonic and multi-valued logic. This would involve not only maintaining a fact list but also the complete justification for each fact. When there is change in the status of any fact in the fact list, the effects of this change must be propogated speedily through the entire fact list. The fact list must include also all facts that are yet to be proven. Thus only justifiable facts are used for the puposes of reasoning. Facts can be added to or deleted from the list of facts that have been justified and it is this character that gives it the non-monotonic feature.

2. Uncertain versus Unknown Knowledge
At first is is of some interest to consider how one can deal with a situation when the degree of truth associated with a fact is unknown.

A very good example is the case of expert control systems applied to an aircraft showing symptoms of flying too low. In case of aircraft it is essential that in the cockpit there is an indication to the pilot of the aircraft's altitude with reference to the sea level but also the altitude of the terrain with reference to the same frame. The knowledge base may have to rule which has the form:
IF (TERRAIN-HEIGHT IS HIGH) AND (AIRCRAFT-ALTITUDE IS HIGH) THEN (ACQUIRE ALTITUDE)

147

This implies of course that the pilot must know the terrain height as well as the aircraft's altitude in order to implement such a rule. If either one of the two indications are not available due to a failure he must make an assumption of some sort and proceed to draw a conclusion. Thus the assumption or hypothesis may take the form,

IF (TERRAIN-HEIGHT IS UNKNOWN) THEN (SET TERRAIN-HEIGHT IS MAXIMUM).

At some later instant however this may lead to a contradiction and so the hypothesis must be retracted. Logics where such a retraction is possible naturally are non-monotonic in nature since the fact has now to be withdrawn from the pool of facts to which one had been steadily adding facts.

Most conventional expert systems do not allow for any form of non-monotonicity. Although it appears deceptively simple to be able to modify a conventional logic based expert system to perform non-monotonic reasoning, it requires that one has an opinion knowledge of all the facts that are yet to be proven and the basis or justification for these facts so one can propagate the effect of any fact changing its status from true to false and vice-versa. This can be an impractical approach especially since the justification in many cases the justification may involve an extensive knowledge base by itself. Thus non-monotonic logic processing may require machines that are far larger than ones that are actually used at the present time.

3. Boolean Logic and its quantification

Before actually describing fuzzy logic in detail it is worth looking at Boolean logic and see how one can quantify the degree of truth in the logic and interpret the logical connectives AND, OR, XOR (Exclusive OR) and NOT by suitable mathematical operations.

Consider two propositions A and B which can be true (T) or false (F). The logical operations (A AND B), (A OR B), (A XOR B), (NOT A) and (NOT B) can be represented by a truth table as shown below.

We now introduce the degree of truth or membership value of each of the preparations A and B. The membership value $\mu(A)$ is defined to be equal 1 if the proposition A belongs to a set consisting of all propositions that are true. i.e. if A is true $\mu(A)$ is defined to be equal to zero if A is false. Similarly the membership value $\mu(B)$ is defined to be 1 if B is true

A	B	A AND B	A OR B	A XOR B	NOT A	NOT B
T	T	T	T	F	F	F
T	F	F	T	T	F	T
F	T	F	T	T	T	F
F	F	F	F	F	T	T

Table 1. Truth table for logical operators AND, OR, XOR and NOT.

and 0 if B is false. We also assume at this stage that the membership value can be either 1 or 0 only. We may now define the membership values associated with the propositions A AND B, A OR B, A XOR B, NOT A and NOT B. Examining the truth table we note that we may associate the functions min $(\mu(A),\ \mu(B))$, max $(\mu(A),\ \mu(B))$, $1 - \mu(A)$ and $1 - \mu(B)$ to define the membership values of the propositions A AND B, A OR B, NOT A and NOT B respectively. The exclusive OR operator (XOR) is defined as

A XOR B = (A AND NOT B) OR (NOT A AND B)

$$= \max\left[\ \min\ (\mu(A),\ 1-\mu(B)),\ \min\ (1-\mu(A),\mu(B))\right]$$

Thus we may rewrite the truth table in Table 1 as shown below in Table 2.

A $\mu(A)$	B $\mu(B)$	A AND B $\min[\mu(A),\mu(B)]$	A OR B $\max[\mu(A),\mu(B)]$	A XOR B	NOT A $1-\mu(A)$	NOT B $1-\mu(B)$
1	1	1	1	0	0	0
1	0	0	1	1	0	1
0	1	0	1	1	1	0
0	0	0	0	0	1	1

Table 2. Truth table for logical operators in terms of the degree of truth

The truth Table 2 is completely equivalent to the Table 1 if we interpret 0 as F and 1 as T.

4. Fuzzy Logic and Fuzzy Sets

One may now contemplate that a fact can be somehow intermediate between true and false and this seems a natural extension to two-valued logic. Consider the proposition "The Oven is hot". On the basis of traditional propositional logic the proposition "The Oven is hot" may be true or false. While on theory definitely be prepared to say that the proposition is true

149

if the temperature is greater than 100°C and that the proposition is false
if the temperature is less than 0°. However if the temperature is 35°C and
the truth of the proposition depends on human perception heat in the oven
it is quite impossible to make a definite statement about the truth or
falsehood of the proposition in fact ideally the proposition can be said to
be equally true and false if the temperature is 35°C. Thus we may
associate a membership value or degree of truth of 0.5 and 1 when the
temperature is between 35°C and 100°C and a membership value between 0 and
0.5 when the temperature is between 0°C and 35°C. This is the basis of
fuzzy logic where there are an infinite number of truth values. Fuzzy
logic is also connected closely to the use of symbolic linguistic terms,
such as 'hot' and 'cold', 'small', 'medium' and 'large' etc.

In fact considering the earlier example one may associate a degree of truth
with each value of temperature of a 'hot' body. Thus there is another
aspect to the application of fuzzy logic and this is the notion of a fuzzy
set. As an example we again consider the earlier proposition 'The Oven is
hot'. We have just seen that the notion of 'hot' indicates a range of
temperatures with various levels or grades of membership values. We
therefore cannot associate a specific membership value with this
proposition. Thus we introduce the notion of fuzzy set. The proposition
"The oven is hot" implies that the temperature of the oven belongs to a
fuzzy set, "HOT", which is defined by a continuous membership function
defined according to Fig. (1). Where $\mu(T) = 0$ at 0°C and below 0°C, $\mu(T) =$
0.5 at 35°C and $\mu(T) = 1$ at 100°C and above 100°C.

When connectives AND, OR and NOT are used with fuzzy propositions which
have membership values between 0 and 1 we may continue to define the
associated membership values using the min( ), max( ) and 1-( )
functions respectively. Thus the above connectives can be used more or
less as they stand as the basis of Fuzzy logic, to deal with uncertainty in
expert systems. In fact Fuzzy logic has been sufficiently extended to
support particular ways of interpreting, true, very true, slightly true,
not very true etc.

If the membership value of a proposition A that is true is given by $\mu(A)$,
the connective "very" is assumed to suggest the membership value, $\mu^2(A)$
while the connective "slightly" or "more or less" is assumed to suggest the

membership value $\sqrt{\mu(A)}$. Although this part of theory is often ignored in many practical applications, the term fuzzy logic is more often used to

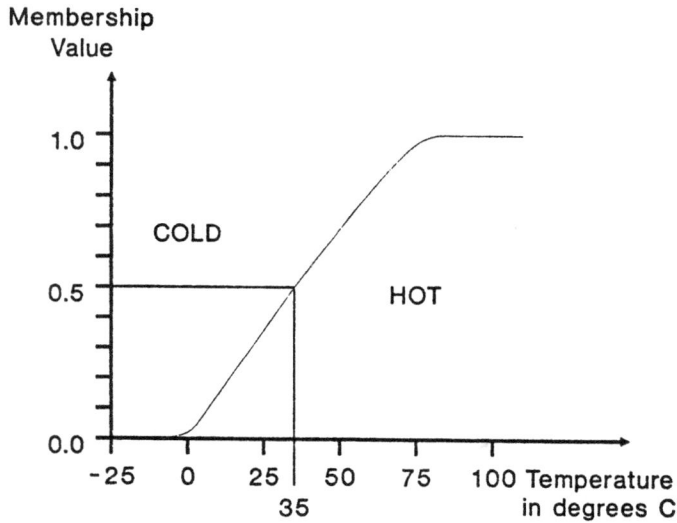

Fig 1.   Example of a membership function

signify the application of the max and min functions to combine estimates of uncertainty.

In fuzzy logic we have rules that take the form:

    R : IF X is A THEN Y is B

which implies that if X belongs to a fuzzy set A then Y belongs to a fuzzy set B. Given a particular fuzzy set X and a premise that X is A^* one may apply the rule to this premise and conclude that Y belongs to a fuzzy set B^* where the Fuzzy set B^* is characterized by a membership function $\mu_B{}^*$ given by,

$$\mu_B{}^*(v) = \max_{u \in U} \left[\min\{\mu_A{}^*(u); \mu_R(u,v)\} \right]$$

where $\mu_R(\ )$ is the membership function associated with the fuzzy relational implication R. The membership function,

$$\mu_R(u,v) = \mu_{Y/X} = \max \ [\min(\mu_A(u), \ \mu_B(v)), \ 1 - \mu_A(u)]$$

and is known as a relational distribution. The notions that are essential to computing inferences in practical application will be introduced and

151

described in detail in a latter chapter.

4.1. Comparison with probabilistic methods :

Before introducing some of the basic concepts related to fuzzy sets it is
worthwhile comparing fuzzy logic and probabilistic methods as such a
comparison provides a thorough understanding of fuzzy logic, its merits as
well as its drawbacks.

The main feature of fuzzy logic and the associated methods is that the
techniques of interpreting the levels of truth or belief depend a lot on
the perceiver. For example if one is considering a coke oven in a steel
plant the concept of hot may refer to temperatures well over $1000^{o}C$ and a
cold oven may refer to a condition when the temperature is less than $200^{o}C$.
There is therefore no absolute concept of 'HOT' and 'COLD'. Thus in many
practical situations requiring an absolute or objective definition
designers turn to probability theory for an answer.

Fuzzy logic deals with a different kind of uncertainty than probability
theory. Probability predicts the likelihood that a random sample will
belong to a set while fuzzy logic evaluates the degree to which the
particular sample belongs to the set. For example if an oven has been
switched on for some time an evaluation can be made as to how well the
temperature in the oven fits the condition of being hot. This is a fuzzy
uncertainty. Probability would predict the chance of the oven being hot
before the temperature in the oven was measured or the chance that a random
observer would classify the oven as being hot. On the other hand if the
oven was already labelled as 'hot' or 'not hot' neither method would be
needed. Fuzzy logic deals with the evaluation of degrees to which certain
conditions have occurred as opposed to prediction whether or not they will
occur. Fuzzy logic may therefore be viewed as an approximate logic for
action while probability theory is an approximate logic for prediction.

Although there are many theories of probability, in the subjective theory
or probability all probabilities are related to physical events. Like
fuzzy logic probability is a measure that can take values between 0 and 1.
Thus probabilities are objective and measurable. This feature is of
fundamental importance in certain applications. As an example if one
considers the tossing of a coin where the probability of a coin landing
heads or tails is ideally 0.5. This fact can also be checked

experimentally by throwing the coin a very large number of times and counting the number of times we come up with heads. However in many situations probability is equated with belief. For example when we say that the probability of survival of a spacecraft for a period of seven years is 0.98 we cannot really verify this by experiment. Thus fuzzy logic has as much a claim to be correct and valid in practical situations as probabilistic logic.

Probabilistic logic is particularly useful in the propagation of belief when additional evidence becomes available. In artificial intelligence parlance this is known as "Truth maintenance" and is of fundamental importance. It is often customary in expert systems to assign "certainty factors" to rules which are interpreted as fuzzy membership functions in certain parts of the system and as probabilities for purposes of propagation of belief.

For example a particular set of symptoms may appear after the occurrence of a certain failure with a probability of 0.8. However if the failure itself is due to the presence of a certain fault with a probability of 0.9, the probability that could be assigned to the symptom depends on both these values. Most expert systems multiply 0.8 and 0.9 and assign a probability 0.72 to the symptom. On the other hand if a rule such as

IF (A AND B) THEN C

is given, the proposition C is assigned a probability by taking the minimum of $P(A)$ and $P(B)$ based on Fuzzy logic rather than on probability theory.

5. Basic Fuzzy Set Theory

Fuzzy sets can be manipulated in the same way as non-fuzzy sets with the operations of union, intersection and complementation being defined by the simple maximum and minimum operations on the corresponding membership functions.

If A_1 = (temperature of oven is greater than 100°C) and

A_2 = (oven is hot)

where the corresponding membership functions are represented in Fig. 2 and Fig. 3 then, the following may be defined.

UNION: The Union of the fuzzy sets A_1 and A_2 which are both members of a generalized ordinary set which will be referred to the 'universe of discourse', U, is denoted by $A_1 \cup A_2$ which belongs to the Universe of

Membership
Value

1.0

0.5

0.0

0   100   200   Temperature, T
                                      in degrees C

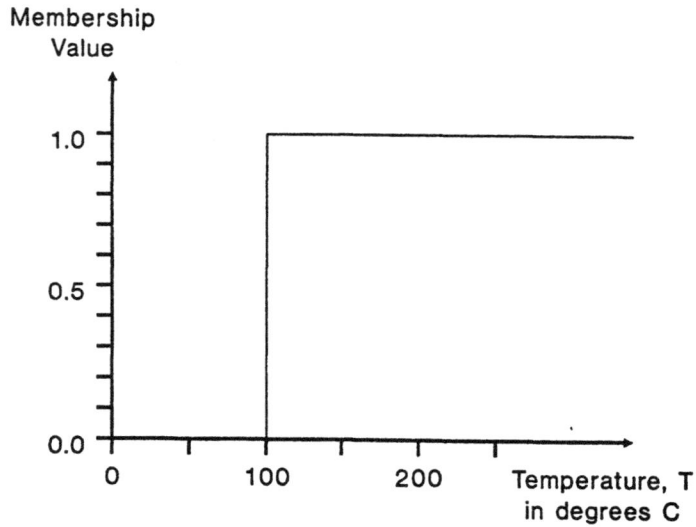

Fig. 2   Membership function of A_1

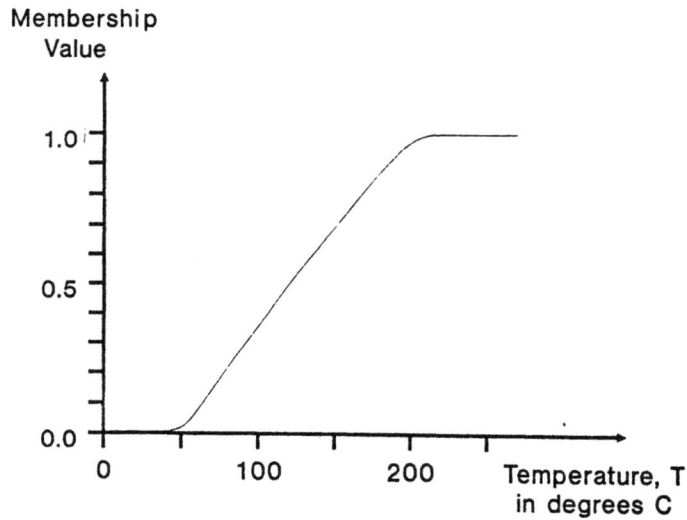

Membership
Value

1.0

0.5

0.0

0   100   200   Temperature, T
                                      in degrees C

Fig. 3   Membership function of A_2

discourse and is characterised by a membership function given by

$$\mu_{A1 \cup A2}(u) = \max [\mu_{A1}(u); \ \mu_{A2}(u)], \ u \in U$$

This corresponds to the connective 'OR' and is shown in Fig. 4.

INTERSECTION: The intersection of fuzzy sets A_1 and A_2 is denoted by

154

$A_1 \cap A_2$ which belongs to the Universe of discourse and is characterised by a membership function given by

$$\mu_{A1 \cap A2}(u) = \min [\mu_{A1}(u); \mu_{A2}(u)], \quad u \in U$$

This corresponds to the connective 'AND' and is shown in Fig. 5.

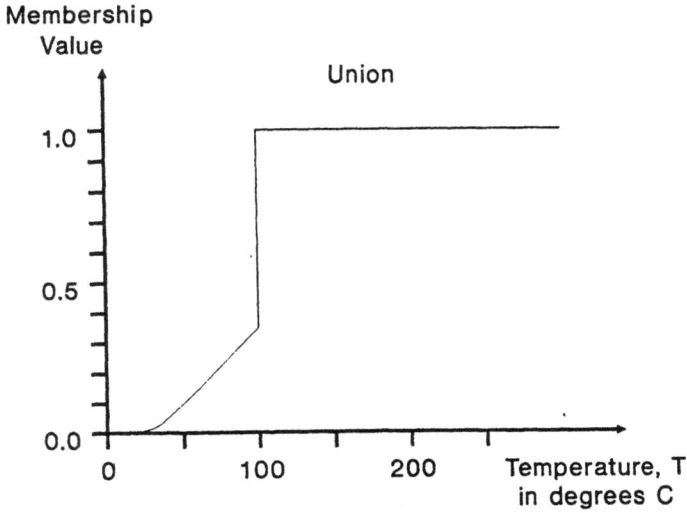

Fig. 4.   Membership function $\mu_{A1 \cup A2}(u)$

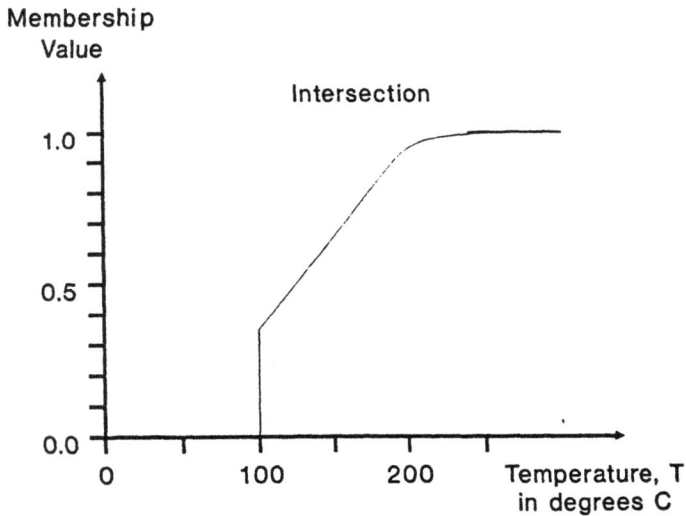

Fig. 5 Membership function $\mu_{A1 \cap A2}(u)$

COMPLEMENT :   The complement of a fuzzy set A_1 is denoted by $\overline{A}_1$ with a

155

membership function defined by

$$\mu_{\bar{A1}}(u) = 1 - \mu_{A1}(u)$$

This corresponds to the negation 'NOT' and is shown in Fig. 6.

Considering two fuzzy sets defined as two different universes of discourse $A \equiv A_1$ and the fuzzy set B = (time spent in oven is about twelve minutes), then we may define the Cartesian Product as follows:

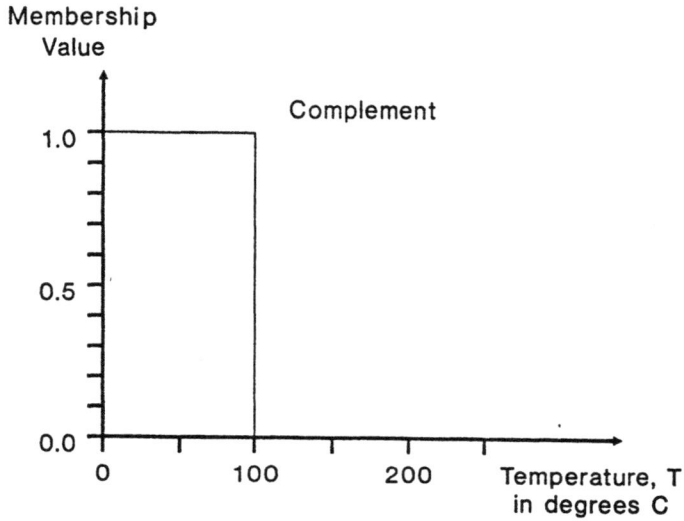

Fig. 6. Membership function $\mu_{\bar{A1}}(u)$

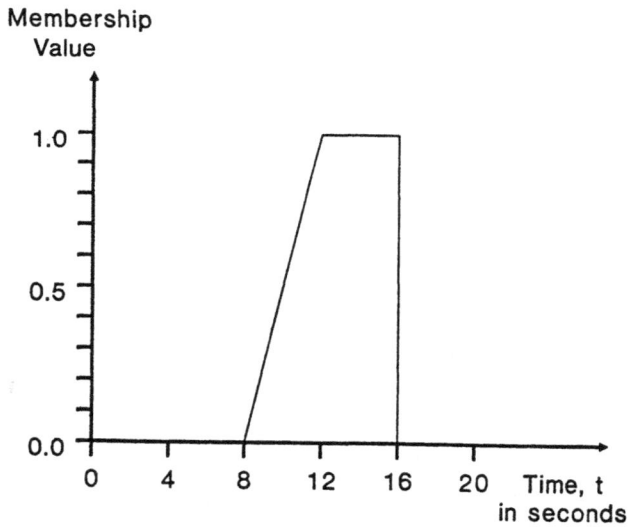

Fig. 7 Example of the membership function $\mu_B(v)$

156

CARTESIAN PRODUCT: The Cartesian Product of fuzzy sets A and B in the universes of discourse U and V, is a fuzzy set denoted by A x B in the Cartesian product space U x V with its membership function defined by,

$$\mu_{A \times B} (u, v) = op (\mu_A(u), \mu_B(v))$$

where 'op' is some operation defined a priori and $\mu_A(u)$ and $\mu_B(v)$ are membership functions. For the example considered, $\mu_B(v)$ is shown in Fig. 7.

Fuzzy Relations. The fuzzy relation in U and V is a fuzzy set denoted by R with a membership function denoted by $\mu_R(u, v)$.

Examples of Fuzzy Relations

If both A and B are propositions then A x B denotes a proposition that is a composition of A and B. The three most commonly used composition rules are logical AND, OR and logical implication.

a) Conjunction or logical AND: Proposition and related by means of an 'AND'. If the proposition is,

X is A AND Y is B

then $\mu_{A*B} (u, v) = max (\mu_A(u), \mu_B(v))$

b) Disjunction or logical OR: Propositions are related by means of an 'OR'. If the proposition is

X is A OR Y is B

then

$$\mu_{A+B} (u, v) = min (\mu_A(u), \mu_B(v))$$

c) Implication Propositions are related by means of "IF ....THEN...". For example : IF x is A THEN Y is B. The membership function for this case is discussed separately.

Other examples of fuzzy composition are:

i) Disjoint sum : Propositions are related by the connective XOR. If the proposition is X is A XOR Y is B then

$$\mu_{A \ XOR \ B} = max \left[min(\mu_A, \ 1-\mu_B), \ min(1-\mu_A, \ \mu_B) \right]$$

where the arguments of the membership functions have not been indicated explicitly.

157

Apart from fuzzy composition there are other examples of fuzzy relations such as:

TRUTH QUANTIFICATION : For example if one has the proposition :

IT is 'r' THAT x is A

where 'r' may be VERY TRUE, QUITE TRUE or MORE OR LESS TRUE etc. The corresponding membership function is

$$\mu_{rA}(u) = \mu_r[\mu_A(u)]$$

If r is VERY TRUE, $\mu_{rA}(u) = \mu_A^2(u)$

and r is MORE OR LESS TRUE, $\mu_{rA}(u) = \sqrt{\mu_A(u)}$.

THE COMPOSITION OF RELATIONS

The composition of two relations R and S in U x V and V x W respectively is a fuzzy relation in U x V x W denoted by R ∘ S with its membership function defined by :

$$R \circ S := \max \left[\mu_R(u,v), \mu_s(v,w)\right]$$

Fuzzy Implication we are now in a position to consider fuzzy implications and logical inferences. If we consider the example of a relation expressed as,

R : IF X is A THEN Y is B

Fuzzy implication is expressed as a fuzzy relation R using the cartesian product space,

R = A x B

where R denotes a fuzzy set. This uses statements of the form:
IF X is A THEN Y is B.
That is if we know that Y is B. In classical logic the equivalent form is

IF X THEN Y

That is to say that if X is true Y is true. However if X is untrue we can make no inference about Y. Whether it is true or false the statement IF A THEN Y will be interpreted to be correct. The only way we can show that the inference statement is itself false would be to produce a case where Y is known to be false when A is known to be true.

Thus we are in a position to construct a truth table (Table 3.) for the

158

implication 'IF X THEN Y' which can be interpreted as (X AND Y) OR NOT X or alternatively as (X AND Y) X OR NOT X.

X	Y	X AND Y	(X AND Y) OR NOT X	[(X AND Y) XOR NOT X]
T	T	T	T	T
T	F	F	F	F
F	T	F	T	T
F	F	F	T	T

Table 3. Truth Table representing 'IF X THEN Y'.

The IF-THEN implication is usually symbolized in fuzzy logic as:

    IF X is A THEN Y is B

or alternatively as

    IF X is A→ Y is B.

In fuzzy logic there are several ways of interpreting the implication. The most popular approach is one proposed by MAMDANI [1974] and in this case the implication is interpreted as follows. If we know that 'X is A' we can deduce that 'Y is B'. However if it is not known that 'X is A' either partially or fully we can make no deductions and the statement 'IF X is A THEN Y is B' is not interpreted to be correct. This is based on the argument that there are other rules in the rule base to cater to case when 'X is A' is not known either partially or fully. Thus fuzzy implication is interpreted to be equivalent to the AND connective and the membership function is defined by

$$\mu_{Y/X} = \min \ (\mu_X(x); \ \mu_Y(y))$$

The classical solution would be

$$\mu_{Y/X} = \max \ [\min\{\mu_X(x); \ \mu_Y(y)\}; \ 1 - \mu_X(x)]$$

An alternative proposed by Zadeh is based on the arithmetic sum

$$\mu_{Y/X} = \min \ (\mu_X(x) + 1 - \mu_X(x); \ \mu_Y(y) + 1 - \mu_X(x))$$

$$= \min \ (1 \ ; \ \mu_y(y) + 1 - \mu_x(x))$$

5.1. COMPOSITION RULE OF INFERENCE

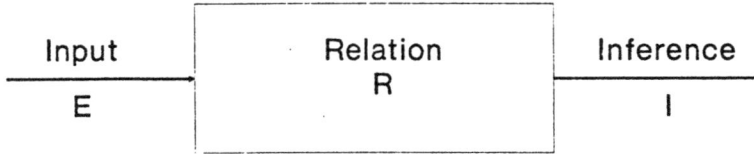

Fig. 8　　　　　　　　A Relational System

The composition rule of inference is used to draw an inference using a fuzzy composition,

$$I = E \circ R$$

where E is the fuzzy set of actual inputs to the relational system consisting of a single relation R and I is the inference.

The membership function for I is defined by

$$\mu_I(i) = \max_{e \in E} \left[\min\{ \mu_E(e); \ \mu_R(e,i) \} \right]$$

The maximisation here is with respect to e for each i hypothesized.　This is known as the max - min convolution.

If E is a crisp valuable that is a fuzzy singleton with a membership

$$\mu(e) = 1 \text{ for } e = e_1 \text{ and } \mu(e) = 0 \text{ for all other } e, \text{ then}$$

$$\mu_I(i) = \mu_R(e,i) \text{ for } e = e_1 \text{ and } \mu(e) = 0 \text{ for all other } e, \text{ then}$$

$$\mu_I(i) = \mu_R(e)|_{e=e_1}$$

In this case when the fuzzy relation R is a fuzzy implication of the type:

$$\text{IF } e \text{ is } E_k \text{ THEN } i \text{ is } I_k$$

μ_{R_k} has the form,

$$\mu_{R_k}(e,i) = \min (\mu_{E_k}(e); \ \mu_{I_k}(i))$$

In practice the relational system may be a collection of rules R_k, $k = 1$,

2, 3.... k........ The rules may be interpreted as:

IF.......THEN ....(for k = 1)

ELSE

IF...... THEN ....(for k = 2)

ELSE

IF...... THEN..........(for k = 3)

ELSE etc.

In this case the 'ELSE' connective is interpreted as the 'OR' connective and,

$$\mu_I (i) = \max_k (\mu_{R_k} (e,i)|_{e=e_1})$$

$$= \max_k \min (\mu_{E_k} (e_1); \mu_{I_k} (i))$$

This is known as the max - min rule of inferences, and is illustrated in Fig. 9.

rule 1 : **IF** INPUT IS POSITIVE SMALL (PS) **AND** CHANGE IN INPUT IS ZERO (ZR) **THEN** OUTPUT, U IS ZERO (ZR)

rule 2 : **IF** INPUT IS ZERO (ZR) **AND** CHANGE IN INPUT IS POSITIVE SMALL (PS) **THEN** OUTPUT, U IS NEGATIVE SMALL (NS)

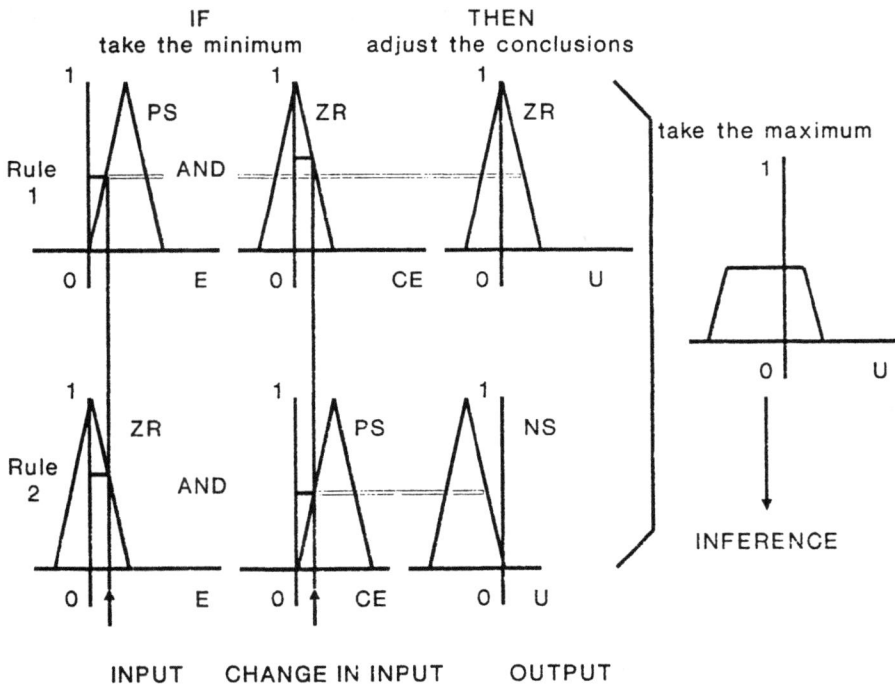

Fig. 9 Illustration of the max-min rule of inference

161

In standard logic, Modus Ponens is a natural deduction rule stating that if the hypothesis 'A → B' and 'A' are both true then 'B' is true. In fuzzy logic this idea is generalised. However Generalised Modus Ponens (GMP) differs from the classical version in two respects : matching is not required to be exact and predicates A, B are not required to be crisp (they can be fuzzy).

For example if A, A*, B and B* are fuzzy statements then GMP will come up with 'Y is B*' from the given premise and implication as follows:

Premise :   X is A*
Implication: IF X is A THEN Y is B
Conclusion: Y is B*

For instance given the premise and the implication,

Premise: This Banana's skin is dark.
Implication: IF a Banana's skin is dark THEN the Banana is over ripe.

From the definition of the composition rules $\mu_{Y/X}(U,V)$ is the membership function of the fuzzy relation. Using generalised modus ponens we conclude that 'Y is B*'. Hence the grade of membership of Y can be obtained from μ_A and μ_B as follows:

$$\mu_Y(v) = \max_{u \in U} (\min [\mu_A(u); \; \mu_{Y/X}(u,v)])$$

Using Zadeh's definition of an implication

$$\mu_{Y/X}(u,v) = \min [1, \; 1- \mu_A(u) + \mu_B(v)]$$

Hence $\mu_Y(v) = \max_{u \in U} [\min[\mu_A(u), \; 1 - \mu_A(u) + \mu_B(v)]$

Classical Modes Tollens can also be extended to fuzzy logic. For example from,

IF X is A THEN Y is B

and given that, Y is NOT B*,

we can conclude that, X is NOT A*.

162

This is called Generalised Modes Tollens.

5.2 DISCRETE FUZZY SETS:

In practice one deals with discrete sets. A finite discrete subset fuzzy subject of k elements is represented in standard fuzzy notation as the union of fuzzy singletons, μ_i/x_i, where μ_i is the membership value corresponding x_i, using the + sign as the notation for union.

i.e. $F = \mu_1/x_1 + \mu_2/x_2 + \ldots + \mu_K/\mu_K$

$$= \sum_{k=1}^{k} \mu_k/x_k = \bigcup_{k=1}^{k} \mu_k/x_k$$

Thus the fuzzy set tall is represented as:

   TALL = (0/5', 0.12/5'4", 0.4/5'8" , 0.6/6', 0.9/6'.5", 1/7',
         1/7'6", 1/8')

The reader is referred to Giarratano and Riley(1989), Zimmerman(1985) and Ng and Abramson(1990) for a detailed explanation on fuzzy sets and there applications in expert systems.

6. REFERENCES:

Giarratano, J. and Riley, G.(1989), *"Expert Systems: Principles and Programming"*, PWs-KENT Publishing Co., BOSTON.

Mamdani, E. H.(1974), "Application of Fuzzy Algorithms for Control of a simple dynamic plant," *Proc. IEEE*, 121, 1585 - 1588.

Ng, Keung-Chi and Abramson, Bruce (1990), "Uncertainty Management in expert Systems", *IEEE Expert*, 4 , 30-48.

Zadeh, Lotfi A.(1965), "Fuzzy Sets", *Information and Control*, 338 - 353.

Zimmerman, H. J.(1985), *"Fuzzy Set Theory and its Applications"*. Kluever-Nijhoff Publishing Co..

AN INTRODUCTION TO NEURAL NETWORKS

Gordon Wells

Instituto de Cibernética, Univ. Politécnica de Cataluña
Diagonal, 647, 08028 Barcelona, Spain.

1 Overview

In recent years, the subject of neural networks has become increasingly popular, both within the university research environment and in the commercial marketplace. It has become quite common to hear references to neural networks in relation to a wide range of subjects, including computer science, neurophysiology, robotics, automatic control, communications, physics, mathematics, psychology, business, and many other fields where new applications are being found and investigated.

Neural networks, or "connectionist techniques", are a novel data-processing paradigm with exciting prospects for handling many of the traditionally most challenging computational problems. While advanced techniques such as artificial intelligence have aimed to expand the capabilities of existing computers by automating human reasoning processes at the *software* level, neural networks are an attempt to mimic the processing capabilities of the brain at the *hardware* level –at the level of individual neurons and their overall behavior when combined into large, highly interconnected networks. In doing so, it is hoped that processors can be achieved with many of the most desirable abilities and behaviors of biological systems: highly parallel computation for data-intensive tasks such as vision and speech processing, adaptive learning and self-organization of data, generalization, robustness towards noisy or imprecise data, and fault-tolerance through distributed or redundant information coding. Such capabilities and performance have been difficult to achieve with traditional computers based on the Von Neumann serial processing paradigm.

Despite their sudden wave of popularity and the present boom in neural networks research, development and commercialization, the field itself is not new. Within a broad context, attempts to develop artificial neural systems may be seen as one of the newest among countless approaches to understanding and modelling our own brain processes, a human endeavor as old as civilization itself, motivated by the need to extend and enhance our own

164

capabilities with increasingly powerful tools. Formally speaking, work in the neural network field began in the early 1940s, when the first simulations of neural processes were performed in order to validate current theories regarding neurophysiology. Although early research was often hindered by limitations encountered with existing neural models, recent new insights into neurophysiology and successful results with advanced connectionist techniques have spurred a resurgence of interest and financial support for the field. Currently, departments dealing in one way or another with neural networks may be found in almost any university, several companies have begun to market neural network technologies, and information on the subject abounds in a vast number of publications.

Neural networks are being applied in numerous areas involving signal processing and complex pattern recognition and classification tasks, such as image processing, voice and character recognition, noise filtering, data compression, time-series prediction and optimization. Another important research area is automatic control, where neural networks are being investigated for the adaptive control of robots and autonomous vehicles, and the identification and control of complex or unknown processes. The capacity of neural networks for adaptive learning, non-linear signal mapping and fault-tolerance are being exploited to fill many of the persistent gaps present in classical control techniques, and several large-scale research and development projects are currently underway to apply neural networks in complex space, industrial, and commercial control applications.

The following sections provide an introduction to neural networks which includes a discussion of their underlying concepts, neurophysiological basis, advantages and limitations, as well as a description of several of the most important neural networks and their applications. Special emphasis is placed on multilayer feedforward networks and backpropagation since, on the one hand, they are currently the most well-known and widely used networks for most applications and therefore deserve special attention in an introductory work, and secondly, because they are the networks most commonly being applied in the neural control applications described in other chapters of this book. Notwithstanding, other neural architectures and algorithms are also briefly described.

Part I

Origins and Underlying Concepts

2 Neurophysiological Basis: Modelling Biological Hardware

To date, many different computational techniques have been successfully developed to handle a wide variety of tasks, many in order to supplement processing capabilities which humans lack, such as complex and extensive mathematical calculation, and many in an attempt to mimic and extend certain human capabilities. Some of the most sophisticated of these, such as artificial intelligence, have aimed to model human knowledge representation and reasoning capabilities on a very abstract level.

Although such methods represent a significant advance over other programming techniques insofar as they permit the automation of human computational processes which were previously very difficult or impossible to model due to being unquantifiable, inexact, or changing over time, they all continue to rely on *symbolic programming* used in conjuction with traditional computer hardware implementing the Von Neuman *serial processing* paradigm. Hence, artificial intelligence techniques have enhanced existing computing power only on the software level, as they typically involve the use of highly abstract symbolic schemes such as frames, objects, rules, or logical predicates which more accurately model human knowledge representation and the algorithms used to manipulate them.

The novelty of the neural network approach is that it is attempt to mimic biological processing capabilities at a lower level than ever before: at the *hardware* level. Artificial neural network models aim to mimic the brain's processing in terms of the functionality of the individual neurons themselves, and its overall capabilities when many neurons are combined into highly interconnected networks. At this level, computation no longer takes place in a serial fashion, but involves *sub-symbolic processing* over a *parallel, distributed architecture* of simple processing units. Input data no longer consists of symbolic representations of knowledge, but of *parameterized representations* of data such as patterns, signals, process variables, or state vectors which are processed in a *parallel, distributed* manner over the network. Figure 1 illustrates both of these computational schemes. The motivation and advantages of using the biological model of parallel distributed processing are described below.

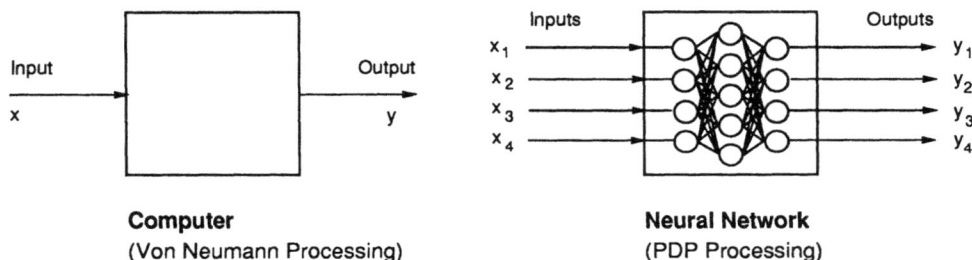

Figure 1: Traditional vs. Neural-network processing. Traditional computers process a single data stream sequentially. Neural networks implement parallel, distributed processing of many data streams using many simple, highly interconnected processing units.

3   Motivation

There are several good reasons for using brain neurophysiology as a computational model. First of all, it is quite easy to see that the brain is a system which works, therefore it must be possible to build artificial systems based on the same principles. The brain is capable of handling complex tasks such as sensory processing, recognition, adaptive motor control, discrimination, classification, generalization and adaptive learning with relative ease, all of which have so far been very difficult or impossible to accomplish with existing computational methods. Although amazing power and speed have been achieved with current computers for complex mathematical calculations and memory operations, the surprising performance with which the brain handles data-intensive tasks such as walking and talking has been so far unattainable with these same computers. In addition, biological systems are able to learn adaptively from experience and form their own internal representations of knowledge. This is due to their massively parallel processing architecture, a characteristic which also provides them with a high degree of fault tolerance even in cases where considerable damage is sustained. It is hoped that artificial neural network models may be developed with similar properties, and which can mimic, and hopefully improve upon the brain's particularly appealing processing capabilities. Such artificial neural systems could be exploited to solve challenging problems in many application areas.

Another benefit to be gained from developing simulations of biological neural systems is that it helps to confirm existing theories regarding neurophysiology, and to gain new insight into how the brain works. Naturally, neurophysiologists are interested in making sure their understanding of the brain is correct, and developing simulations of this behaviour is a good way to accomplish this. The fact that artificial neural networks implementing parallel, distributed architectures and adaptive learning rules can be made to learn, recognize complex patterns, discriminate and generalize similar to humans and animals in behavioral experiments is good reason to believe that the brain also accomplishes these tasks by the same mechanisms [Maren 90].

The fields of neurophysiology and neural networks have always involved much interaction between researchers in both areas. The first neural network simulations, performed by McCulloch and Pitts in 1943, were inspired by the work of neurophysiologists such as Donald Hebb and Karl Lashley, who studied the mechanisms of learning and memory in the brain. Since then, as new knowledge of neurophysiology has become available, it has been readily incorporated into existing simulations, often with greatly improved results. Major discoveries have often given rise to entirely new neural network designs. Successful results with neural network simulations have in turn helped to validate assumptions and theories about how the brain works. Two specific areas of neurophysiology which have been the focus of intense research are the processing of visual and auditory sensory information. In each of these cases, the combined efforts of neurophysiologists and neural network researchers have resulted in the development of very powerful working models of the retina and the cochlea, both of which offer promising possibilities for applications in computer vision and automated speech processing.

4 Contributions from Other Fields

Despite their essentially biological basis, many developments in artificial neural networks have stemmed from ideas in fields such as engineering, mathematics, and psychology. As so little is still known about neurophysiology, current artificial neural networks continue to be mere oversimplifications of biological systems. Many of the original, neurologically based network designs have been extended, modified and enhanced in a pragmatic way to provide them with specific functionalities needed for particular applications.

Mathematics has been used to produce several very important theorems regarding the capabilities and limitations of certain network architectures. Learning rules such as Least Mean Squares and the Generalized Delta Rule, used to train multilayer feedforward networks, were developed using differential calculus. Other learning algorithms, such as reinforcement learning, are based on principles from behavioral psychology, and the learning rule used in the Boltzman machine is the simulated annealing optimization method borrowed from quantum physics.

In addition, it must be kept in mind that the biological model is one of several other approaches to parallel distributed processing. As a result, neural networks research has become a highly multidisciplinary field which many prefer to call simply "connectionism".

5   Advantages

It is clear that neural networks represent a radically different approach to information processing than any other existing computational technique, even to the extent of seeming unfeasible since they borrow very little from tried and tested methods. Thus it is appropriate to ask the question: *What are the advantages to be gained from modelling biological processing at the neurological level ?*

The major processing advantages of biological neural systems which neural networks research aims to exploit include:

- Adaptive Learning. This is the ability of neural networks to learn how to perform certain tasks by being presented with illustrative examples.

  For example, a network can be trained to recognize and distinguish between various handwritten characters simply by showing it several different versions of each and the actual characters they represent. Another network may be trained to model a complex chemical reaction using only sets of typical input and output data. A robot arm with a neural controller may "teach itself" to approach and grasp objects within a predefined space by making repeated attempts and correcting its mistakes based on error measurements.

  In all of the above cases, no a-priori models or explicit representations of the systems are necessary for the networks to learn their respective tasks. No complex geometric algorithms are needed for the character recognition, no kinetic equations of the chemical process are required, and no model of the robot arm kinematics are necessary. This is similar to the way humans learn to distinguish between sparrows and bluejays by seeing examples of each, to play a melody by ear after hearing it one or more times, to walk and talk by repeated attempts and corrections, or to throw a basketball by trial and error.

  If necessary, a neural network which is already trained and in operation can be made to undergo continued learning by constantly presenting it with new examples. In this way, the network may improve its ability to discriminate between similar inputs or adapt to changing environments.

  The sole concerns when designing neural networks to perform such tasks are choosing the *appropriate architecture* for the network, developing an effective *learning algorithm*, and selecting a good set of *training examples*. These are often not easy tasks, as will be discussed below in more detail.

- Self-organization. During training and operation, a neural network organizes and creates its own representations of the information it receives. This is quite different from traditional programming methodology, which requires all data to be explicitly specified by the programmer. Consequently, neural networks are well suited for problems for which it is either difficult or impossible to define an explicit model, program or rules for obtaining the solution. This is one clear advantage, since a vast number of such problems exist.

When a neural network is trained for character recognition, for example, it will learn to recognize certain features in written characters and will create its own internal representations of these features. The robot arm described above will form its own internal mapping of its environment and of the inverse kinematics required to transform the desired trajectories into the appropriate control commands.

One very useful result of this capability for self-organization is that the neural network gains the ability to *generalize*. A network trained to recognize certain patterns will create its own *categories* of patterns, so that when presented with a slightly different or unfamiliar pattern it will attempt to associate it with one of the categories of patterns it has already learned. This is analogous to how we are able to understand a slightly mispronounced word, read messy handwriting, or recognize a melody sung out of tune.

Another way to interpret this property of neural networks is that they are able to process *noisy, distorted, incomplete or imprecise data*. Thus, in a properly trained network, a very sketchy "A" is still recognized as an "A", and a very noisy control signal is treated as if it were smooth and noise-free. This feature of neural networks makes them particularly suited for classification, pattern recognition, and signal filtering applications.

- Fault tolerance via redundant information coding. There are two types of fault tolerance exhibited by neural networks. The first is fault tolerance with respect to *data* which is noisy, distorted or incomplete, which results from the manner in which data is organized and represented in the network. This aspect was described above.

The second type of fault tolerance is with respect to the physical degradation of the network itself. Up to a certain point, if individual neurons, or even entire portions of the network are destroyed, it will continue to function properly. When damage becomes so extensive that network behavior starts to be affected, the effect is a *gradual degradation* in performance, as opposed to a completely disabilitating failure.

This same phenomenon is observed in biological neural networks. Every day, many of our brain cells die and are never replaced. Nevertheless, the brain continues to function properly with no apparant loss of ability. Even in cases where a significant degree of brain damage occurs, the remaining neurons are often capable of fully assuming the functions of those which were lost.

The reason neural networks are capable of this type of fault tolerance is that they store information in a distributed (or redundant) manner. A single "piece" of knowledge or information is not encoded in a single neuron, but over many. This contrasts with traditional information processing technology, in which each element of data is stored in a specific memory location. In this case, even a minor hardware malfunction can cause catastrophic data loss and program failure.

- Fast processing speed. One of the most critical issues in data processing technology has always been that of processing speed. As attempts are made to automate processes involving greater and greater complexity and volumes of information, the need for highly efficient data processing software and hardware becomes increasingly urgent.

Traditionally, research in computer efficiency has dealt primarily with improving *serial* software and hardware to make them run faster, by either increasing algorithm efficiency, designing processors with improved architectures, or optimizing memory capacity and management.

Recent efforts, however, have attempted to *parallelize* data processing by dividing computing tasks into smaller subtasks and spreading them over several processors operating simultaneously. This has proven to be a complicated problem and often less than optimal results are obtained, since ways must always be found to restructure and distribute essentially serial software, written with serial programming techniques, into separate tasks which may be processed in parallel. In many cases it is difficult or impossible to decompose serial processes into parallel ones and correctly control their processing sequence.

Biological neural networks are massively parallel processors by design. The brain is capable of handling complex tasks involving immense volumes of data in real time, such as vision, speech processing and motor control, by spreading processing over millions of neurons acting as individual, highly interconnected processors. Artificial neural networks are modelled on this architectural paradigm in an attempt to make real-time processing of data-intensive tasks feasible. Clearly however, the parallel processing capabilities of neural networks will never be properly exploited until they are implemented in suitable parallel hardware.

- Compatibility with existing technology. As with any new idea or technology, a novel processing method like neural networks would have little chance of success and acceptance if it were incompatible and difficult to integrate with other established technologies. Neural networks are practical in this respect since they may be easily implemented using existing techniques and easily integrated in almost any type of system.

Currently, powerful neural simulators may be written in common programming languages such as C or Fortran using a minimal amount of code, since they often consist of nothing more than a few matrix operations. Numerous commercial development tools are also available which allow the simple construction of networks which may be readily integrated into other programs.

6 Biological Neural Networks

In order to better understand artificial neural networks, it is helpful to take a brief look at the elements of biological neural systems which originally inspired them. The first neural simulations were performed as an aid to understanding and demonstrating the basic principles of biological nervous systems being studied by neurophysiologists. Early research focused on the primary electrophysiological functions of individual neurons and how they communicate by passing electrical signals between each other. Many of these functions are now fairly well understood and have been successfully incorporated into neural network models, which as a result have been shown to exhibit several of the fundamental behaviors of their biological counterparts.

The following sections provide a brief overview of the fundamental elements of biological neural systems.

6.1 The Neuron

The neuron is the fundamental building block of the nervous system. Neurons are living cells, and the cell body of the neuron contains the usual mitochondria, Golgi apparatus, rough endoplasmic reticulum and free ribosomes.

Just as many different types of cells exist to fulfill specific purposes in the body, there are many different types of neurons, each having a particular shape, size and length depending on its function and utility in the nervous system. The shape of the cell body can be round, triangular, drop-like, pointed on two ends or stellate depending on whether it is an olfactory, auditory, muscular, tactile or other type of neuron.

While each type of neuron has its own unique features needed for specific purposes, all neurons have two important structural components in common. These may be seen in the typical motor neuron shown in Figure 2. At one end of the neuron are a multitude of tiny, filament-like appendages called *dendrites* which come together to form larger branches and trunks where they attach to the cell body. At the other end of the neuron is a single filament leading out of the cell body called an *axon*, which has extensive branching on its far end. These two structures have special electrophysiological properties which are basic to the functionality of neurons as *information processors*, as we shall see below.

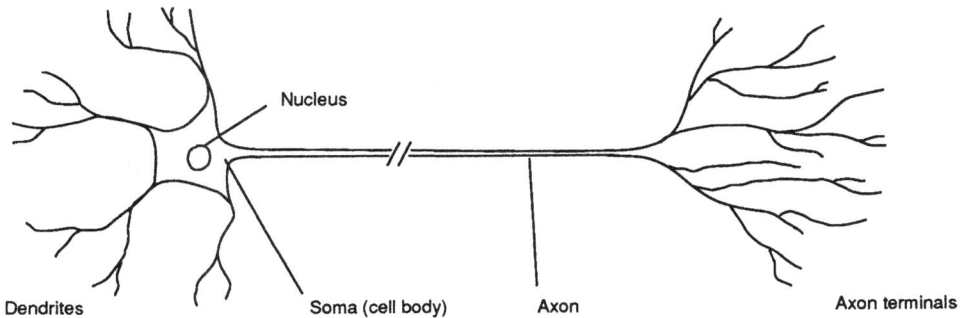

Figure 2: A typical motor neuron. (Adapted from [Jutten 90].)

6.2 Communication between Neurons

In essence, neurons are tiny electrophysiological information processing units which communicate with each other by means of electrical signals. Neurons are connected to each other via their axons and dendrites. Signals are sent through the axon of one neuron to the dendrites of other neurons, hence dendrites may be represented as the inputs to the neuron, and the axon is the neuron's output. Note that each neuron has many inputs through its multiple dendrites, whereas it has only one output through its single axon. The axon of each neuron forms connections with the dendrites of many other neurons, with each branch of the axon meeting exactly one dendrite of another cell at what is called a *synapse*. Actually, the axon terminals do not quite touch the dendrites of the other neurons, but

171

are separated by a very small distance of between 50 and 200 angstroms. This separation is called the *synaptic gap*, and is illustrated in Figure 3.

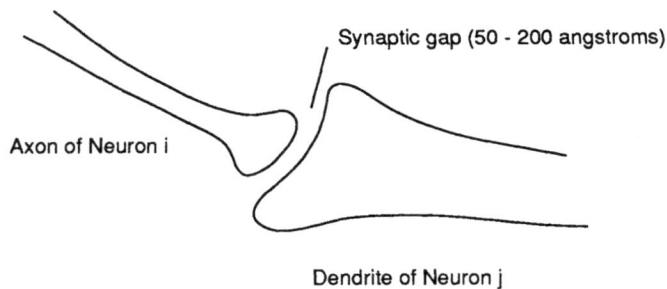

Figure 3: The synapse. (Adapted from [Jutten 90].)

Neurons produce electrical signals by varying the voltage potential across their cell wall. They accomplish this by controlling the distribution of charged ions inside and outside the cell. The chemicals which contain these ions are called *neurotransmitters*, and include ions such as sodium, potassium, chlorine and calcium. One neuron communicates to another by transferring electrical energy along the axon to the axon terminals, and from here to the next cell by releasing neurotransmitters into the synaptic gap. In this way, the conductivity in this region is increased and the signal can pass from the axon of one cell to a dendrite of the target cell. Some ions enhance the membrane potential by polarizing the membrane or increasing the ionic difference, while others depolarize the nerve by opening channels in the membrane so the ions can pass through. It is this depolarization which increases the conductivity of the cell and allows the signal to be transmitted through the nerve.

Obviously, the distance travelled by a given nerve signal may vary. Whether or not a nerve signal is passed on from one cell to the next is determined by the *amount of stimulus* a neuron receives at its incoming dendrites. This is also called the degree of neuron *activation*. Neuron activation, in turn, depends on three main factors: 1) the *number of incoming synapses* receiving input signals, 2) the *concentration of neurotransmitters* released into these synapses, and 3) the *synaptic strength* of each of these connections. The greater the overall depolarization of the incoming synapses, the larger the portion of the cell which will be depolarized. Depolarization generally decreases or degrades over time and distance within the cell. However, if there are enough incoming transmissions occurring nearly simultaneously, and if their overall connection strengths are great enough, the depolarization will spread to the base of the axon (called the axon hillock) and the axon itself will then depolarize and fire an electrical signal.

Unlike in the rest of the neuron, the depolarization is transmitted down the axon without degradation. This sudden burst of energy along the axon is called *axon potential* and has the form of a short, quick electrical impulse, or "spike" (see Figure 4). The magnitude of the axon potential is constant and not related to the amount of stimulus. However, neurons typically respond to a stimulus by firing not just one, but a barrage of successive axon potentials resembling the string of pulses in a digital signal. What varies is the *frequency* of axonal activity, which means that nerve signals are not entirely digital in nature, but contain analog information as well. Neurons can fire between 0 and 1500 times per second, as illustrated in Figure 5. Thus, information is encoded in the nerve signals in two ways: 1) as the *instantaneous frequency* of axon potentials, and 2) as the *mean frequency* of the signal.

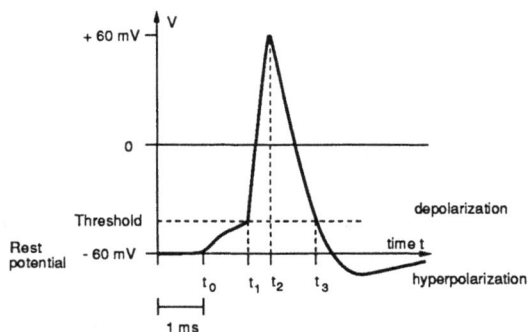

Figure 4: Axon potential. (Adapted from [Jutten 90].)

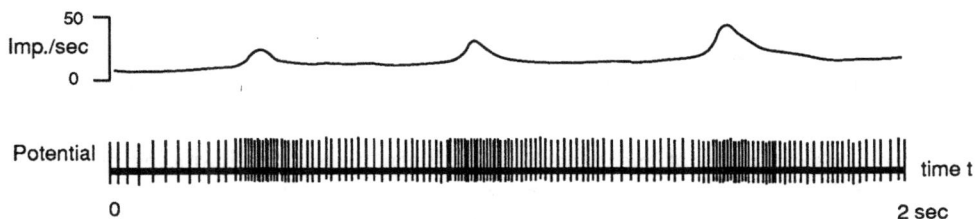

Figure 5: A sequence of axon potentials. (Adapted from [Jutten 90].)

The communication between individual neurons may be then summarized as follows: Neurons are simple processing units with many inputs from other neurons, and a single output which is distributed to many other neurons. Each neuron acts as a sort of threshold saturation function which, when it receives enough overall electical stimulus from its incoming connections, thus activating the neuron beyond a certain threshold level, produces an output signal of constant magnitude and variable frequency. This signal is then passed on as input stimulus to other neurons, whose activation may then be affected.

It is evident that these basic features of the dynamic behavior of biological neurons can be easily simulated with mathematical models. Later sections will describe how this has been accomplished.

6.3 Networks of Neurons

It is easy to see that, although the basic function of individual neurons is quite simple, they do not promise any spectacular processing capabilities by themselves. It is the interaction of many neurons combined into large networks which give the brain its ability for learning, recognition, reasoning, discrimination, generalization and motor control.

The human brain is comprised of millions of neurons, estimated at some 10 to the 12th power. This is the largest and most complex biological neural network, and hence has some impressive capabilities. However, even the minute brain of the fly, with comparatively few neurons, is all that is needed to provide flight control, picnic detection, and sensory processing from several hundred eyes and two antennae for this pesky little autonomous vehicle. Although even the smallest biological networks are large compared with any of the models tested so far, modest artificial networks with only a few neurons have been made to reproduce many of the most fundamental behaviors of biological systems.

173

Obviously, neurons might be connected and combined in a variety of ways to achieve many different architectural configurations. In fact, a rich diversity of neuronal architectures have been observed in biological nerve tissue. The most challenging problem for neurophysiologists is discovering the functions provided by particular neuronal structures. Many such structural mechanisms have been found.

First of all, the possibilities of interconnecting the multiple inputs and outputs of neurons are conceivably infinite. In the human brain, it appears that nerve cells typically have between 1000 and 10,000 synapses with other neurons. A few small subsystems of interconnected neurons have been traced in humans and animals to discover the basic relationships between their architecture and function. For example, the pain reflex is mediated by a sensory and a few motor neurons. One neuron senses a painful stimulus and carries the message to the spinal cord. The information is passed to motor neurons via a single set of synapses. The motor neurons transmit signals to muscles which in turn move the body away fro painful stimulus [Maren, p. 32]. This particular subsystem is simple in nature, but the brain consists of many more subsystems, and systems within systems, which are all generally much more complex to analyze.

A few other basic structural mechanisms have been discovered in biological neurvous systems. *Inhibitory connections* are one example. An inhibitory connection between neurons has the effect of one neuron blocking, or reducing the activation, of another. Inhibitory stimulus or learning in the nervous system has the function of reinforcing or focusing in on particular neuronal activities, while penalizing inappropriate or conflicting ones. An example of the role of inhibition in human learning, on a large scale, is that of learning a particular task or behavior by not only learning what to do, but at the same time learning what not to do. Appropriate actions are reinforced, while inappropriate ones are inhibited.

Other provisions have been observed in the brain for processing *dynamic information*. Much of the information processed in the brain is dynamic in nature. Speech is one example, in which words and sentences are understood based on previously heard syllables and words. One dynamic mechanism is the *variation in output firing frequency* exhibited by individual neurons. Time-varying information is encoded in these frequency modulations and processed accordingly by neurons and neuronal structures sensitive to them. At the level of communication between individual neurons, a high signal frequency is the equivalent of a strong signal, while on a larger scale, certain firing frequencies by particular neurons or groups of neurons can also have the effect of activating other regions of the brain which are "tuned" to these frequencies.

Another way in which the brain processes dynamic information is by means of *feedback processing loops*. These are networks in which information is passed forward from one set of neurons to another, and then back to the previous one for further processing. In some cases this mechanism is used to retain information needed for later processing, while in others it serves to focus or refine information in an iterative manner.

Short term memory is one example of a reverboratory system in the brain. Here, memory is kept active for several seconds or minutes by circulating patterns of neural activity repeatedly. By maintaining the memory active, enough time is allowed for recalling information, or for storing information in long term memory by making the necessary neural modifications.

6.4 Interacting Network Systems

The brain is capable of sophisticated processing tasks by virtue of equally sophisticated neural architectures. Many of the brain's most high-level processing capabilities have been linked to large scale structural features. The most prominent of these are *layers*, *columns* and *regions* of neural tissue.

Information reaching the brain from sensory input passes through many successive layers of neural tissue as it is processed. Some layers may perform a preliminary filtering of data, while others are responsible for feature extraction, association and distribution to other parts of the brain. The visual cortex is multi-layered network in which each succesive layer extracts and refines certain features from the visual scene received, and finally distributes this information to the brain to perform abstract recognition functions. The main cortex is also organized into six main processing layers, within these, several columns of neural tissue have been discerned which act as individual functional units. The brain is further subdivided into specific regions responsible for functions such as motor control, sensory processing, and abstract reasoning.

6.5 Learning in Biological Neural Networks

The ability to learn is clearly one of the most remarkable features of biological neural systems. Humans and animals learn continually from interaction with their environment, whereupon the learned information is automatically assimilated into the brain and affects their behavior thereafter.

Many researchers have investigated biological learning in an attempt to discover exactly how and where learned information is stored in the brain. Several of the most important results include the following:

- Synaptic activity facilitates neural communication. Donald Hebb suggested that synaptic activity facilites the ability of neurons to communicate. Thus a high degree of activity between two neurons or groups of neurons at one time could facilitate their ability to communicate at a later time. By the same token, if two events are active at the same time in the brain, the synapses involved would be facilitated such that later activity in either set of neurons would reproduce the event it previously activated in the other.

  These ideas established the basis for learning or association by the brain and have guided researchers for many years in their efforts to uncover the possible mechanisms of learned facilitation [Maren 90]. Many artificial neural models have also incorporated Hebbian-type principles in their learning rules, examples of which will be described in the following sections.

- Learning strengthens synaptic connections. One mechanism which has been found that partially explains Hebb's theory is the strengthening of synaptic connections as a result of their activity. Recent work suggests that the concentration of cell-wall proteins near the synapse may be increased by synaptic activity to provide a permanent basis for learning.

- Learning stimulates neuronal growth. Other researchers have observed that long-term learning causes a significant increase in the number of connections between

175

neurons in certain regions of the brain. While some "hard-wired" structure is already present in the brain from birth, neuronal growth increases dramatically over time.

7 Artificial Neural Networks

As discussed earlier, "artificial" neural networks were first conceived as models of biological networks in an attempt to verify, explore, and hopefully mimic their unique processing capabilities. For this reason, they are often called *artificial neural networks, neural network models*, or *neural network simulations*. Nevertheless, artificial networks are far from exact models of their biological counterparts. Most of the network models devised so far are gross oversimplifications of biological networks which implement only a few of their most basic and well understood structural and functional characteristics, while borrowing a much greater portion of their design from engineering and other techniques more readily available for achieving desired network behaviors. Since, from this standpoint, the use of biological terminology can be somewhat misleading, many researchers advocate a more conservative classification of artificial networks simply as a novel computational scheme based on the parallel distributed processing paradigm involving many highly interconnected processing units. Those who prefer this idea often use the terms *connectionism* and *connectionist techniques* to describe artificial neural networks.

Regardless of their name, artificial neural networks may be thought of as biologically inspired mathematical models for parallel distributed processing. The fundamental elements of neural network models are the *neuron*, the *architecture* into which many neurons are arranged and interconnected, and a *learning rule* for modifying the connection strengths.

In the following sections, the main elements of neural network structure and operation will be described.

7.1 The Artificial Neuron

The typical artificial neuron used in practically all neural network models is shown in Figure 6. In this diagram, the neuron has been represented in such a way that the correspondence of each element with its biological counterpart may be easily seen.

Like biological neurons, artificial neurons are simple processing units which may receive many inputs from other neurons, but produce a single output. Just as biological neurons may receive incoming signals from many other neurons through their multiple dendrites, artificial neurons may have many inputs in the form of incoming connections from other neurons. Similar to the single nerve axon which branches to other nerves, artificial neurons produce a single output value which may be propagated to several other neurons.

The artificial neuron is really nothing more than a simple mathematical equation for calculating an output value from a set of input values. A neuron's output is equal to its *activation*, which is calculated as the weighted sum of all incoming activations from connected neurons, modified by a certain transfer function. The functions by which the total input and activation are calculated are described below.

Elements of Artificial Neuron

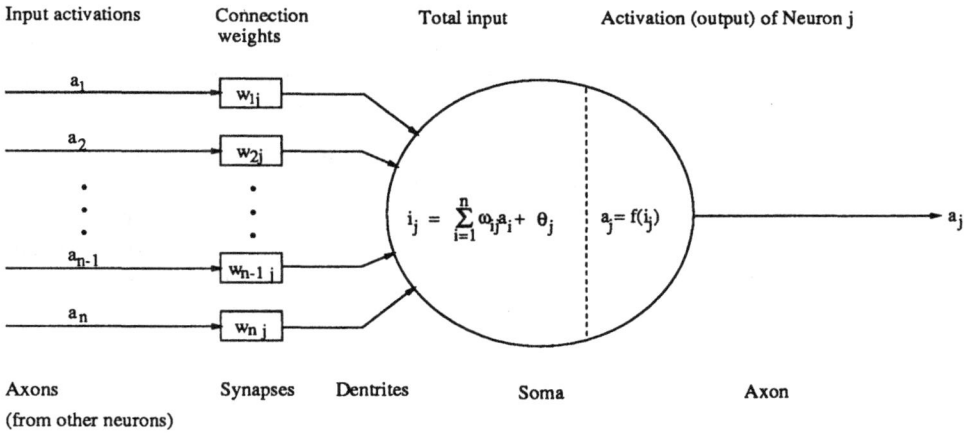

Figure 6: The neuron model.

7.1.1 Static vs. Dynamic Neurons

Regardless of the form of the activation function itself, neuron models may be first classified according to their temporal behavior, which depends on the manner in which the activation function is applied. Neurons may be given either *static* or *dynamic* behaviour by computing their output as a function only of the current input, or as a function of the current input and the current activation value, respectively. Thus, a static neuron j applies its transfer function f only to its current total input i_j to produce an activation value:

$$a_j = f[i_j] \tag{1}$$

while a dynamic neuron j applies its transfer function f to its current input $i_j(t)$ and its current activation $a_j(t)$ to produce a new activation value $a_j(t+1)$:

$$a_j(t+1) = f[i_j(t), a_j(t)] \tag{2}$$

Static neurons are typically used for tasks consisting of essentially static input/output mappings, such as association and classification.

By making a neuron's activation dependent on its previous activation state, information with temporal (dynamic) content may be appropriately processed. Dynamic neurons are useful for applications in which sequential or time-varying input patterns are encountered, such as speech processing, signal filtering, and robot motion control.

One method for implementing dynamic activation in neurons is to provide neurons with a memory function, which allows it to sum its activation value over a given time interval. Another is to use recurrent (feedback) connections, by including the neuron's own output as one of its inputs. Networks with recurrent architectures will be described in later sections.

7.1.2 Connections Between Neurons

As in the nervous system, the effectiveness, or strength of the connections between neurons is a variable parameter. In fact, this adaptability of synaptic (connection) strength is precisely what provides neural networks with their ability to learn and store information, and consequently is an essential element of connectionist models.

In order to simulate the variable synaptic strength found in biological nerve connections, each of the incoming connections from other neurons has an assigned strength called a *connection weight*. Thus each incoming activation is multiplied by its corresponding connection weight, and the result is input to the neuron. The total input i_j to a neuron j is simply the weighted sum of the outputs (activations) from each of the connected neurons a_i multiplied by their corresponding connection weights w_{ij}, plus a *bias* or *offset* term θ_j (often called a "threshold term"):

$$i_j = \sum_{i=1}^{n} w_{ij}a_j + \theta_j \tag{3}$$

The bias term serves as a sort of zero adjustment for the overall input stimulus to the neuron.

Excitatory and *inhibitory* connections are represented simply as positive or negative connection weights, respectively.

7.1.3 Neural Activation Functions

Although most neuron models sum their input signals in basically the same manner, as described above, they are not all identical in terms of how they produce an output response from this input.

Artificial neurons use an *activation function*, often called a *transfer function*, to compute their activation as a function of total input stimulus. Several different functions may be used as activation functions, and in fact the most distinguishing feature between existing neuron models is precisely which transfer function they employ.

In general, it is desired that the transfer function have a threshold behavior, since this is the type of response exhibited by biological neurons: when the total stimulus exceeds a certain threshold value, a constant output is produced, while no output is generated for input levels below the threshold. Early neural network simulations, in which activation was a constantly increasing linear function of input, produced very unstable network behaviors since neural activations tended to increase without bounds. Much improved performance was obtained when limits were imposed on neuron activation.

Four of the most common activation functions used in neuron models are shown in Figure 7. Although other functions have been used, these are the most widely used activation functions:

- Threshold logic function, also called a *binary* or *sgn* function. This function is used to implement simple binary nodes whose activation is 1 if the summed input exceeds the neuron's threshold, and 0 otherwise. Note that there is a *bipolar* variation of this and the other activation functions shown which has limits between -1 and 1. Threshold

178

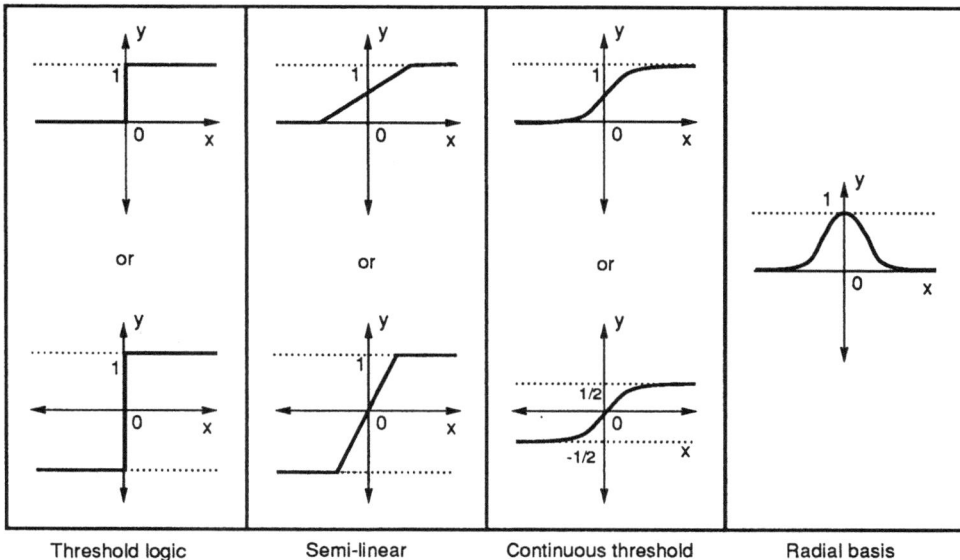

| Threshold logic | Semi-linear | Continuous threshold | Radial basis |

Figure 7: Common neuron activation functions. (Adapted from [Maren 90].)

logic functions have been used in early versions of the Hopfield, Perceptron, and Adaline networks. Networks with binary nodes are easy to implement in hardware (which partly accounts for the early widespread interest in the Hopfield network), but are often very limited in their learning and performance capabilities.

- Semi-linear function. This function may be viewed either as an increasing linear function with upper and lower limit values, or as a threshold logic function with a linear transition region. There is both an upper and a lower threshold value in this function. For total input values below the lower threshold, the activation is 0 (or -1), and for input levels above the upper threshold, the activation equals 1. Between the two thresholds, output is a linear function of input. Like the threshold logic function, these functions are discontinuous and therefore not differentiable at their transition points, which limits the type of learning rules which may be applied to them and consequently their applications.

- Continuous threshold function (sigmoid). The introduction of continuous, or smoothly limiting threshold functions as neural activation rules was part of a major breakthrough in the neural network field. During the 1960s and early 70s, researchers began to discover serious limitations to the capabilities of two-layer feedforward networks with hard-limiting activation functions, such as the Perceptron, and neural network research virtually came to a halt for many years. Part of the reason was that, in order to extend their capabilities to more complex problems, more layers of neurons were needed, which required a more sophisticated learning rule allowing the adjustment of the connection weights in the intermediate layers.

In 1974, Paul Werbos presented a new learning rule for Perceptron-type networks called *backpropagation*. With this method, it is possible to efficiently calculate the connection weight adjustments at every layer of a feedforward network having any number of intermediate layers. The weight increments are obtained using the derivative of the network's outputs with respect to its inputs, which is calculated

179.

by computing the derivative at each intermediate layer and propagating these values back through successive network layers using the chain rule.

Consequently, in order to calculate the derivatives of each node's activation, a continuous, derivable transfer function is necessary. Both the threshold logic function and the semi-linear function are discontinuous and have undefined derivatives at their transition points. Werbos suggested the *sigmoid* function, which is very similar in shape to the threshold logic and semi-linear function but is continuous and smoothly increasing at its transition points, and therefore is continuously derivable. The sigmoid has the following equation:

$$a_j = \frac{1}{1 + e^{-\alpha i_j}} \tag{4}$$

where

a_j is the output activation of neuron j

i_j is the total input to neuron j

α is a constant.

Like the hard-limiting functions, the sigmoid is bounded and nearly constant for input values either below or above the lower and upper thresholds. It is typically defined to be bounded between 0 and 1, -1 and 1, or $-1/2$ and $1/2$. The width of the sigmoid's central transition region may be modified by varying the constant α, which is useful for affecting the learning behaviour of sigmoid nodes.

The great success achieved with backpropagation networks in numerous application areas since their introduction has made the sigmoid the most widely used continuous threshold function. Nevertheless, other functions with similar properties may also be employed. The *hyperbolic tangent* has frequently been used, for example, which yields output values between -1 and 1.

- Radial basis function. This is still a fairly uncommon transfer function, although recent theoretical results may contribute to its popularity in many application areas. The *gaussian* function is typically used, which has the equation:

$$a_j = e^{-\frac{1}{2\sigma_j^2} \sum_{i=1}^{n}(x_{ij}-c_i^j)^2} \tag{5}$$

where

a_j is the output activation of neuron j

x_{ij} is the input from node i to node j

c_i^j is the *center* of node j's transfer function for input i

σ_j is the *width* or *spread* of node j's transfer function.

Unlike the preceeding functions, the radial basis function yields its maximum value only for inputs very near its center, and is nearly zero for all other inputs. The center and width of the radial basis function may be varied to achieve different network behaviours.

180

Radial basis nodes are useful for applications involving *continuous function mappings*, such as encountered in system identification and control. It has recently been demonstrated [Girosi 89] that any continuous function can be uniformly approximated by linear combinations of gaussian functions. Researchers have exploited this result [Sanner 91], in conjunction with sampling theory, to develop a constructive method for designing stable adaptive neural controllers using a three layer feedforward network with a single hidden layer of gaussian nodes.

In some cases radial basis functions may also be used to improve performance in classification tasks, since they provide smooth interpolation between classification regions as opposed to the sharp discontinuities often observed with sigmoid function nodes.

7.2 Network Topologies

Topological structure is the most important characteristic of neural network models, since the functional behavior, and hence the applications for which a particular network is best suited, are mostly determined by its structure at the topological level. Consequently, networks are often classified according to their topological features.

The topology of neural networks refers to the architecture into which multiple neurons are arranged and interconnected. The features which characterize network topology, and the various ways in which these features may be varied, are illustrated in Figure 8.

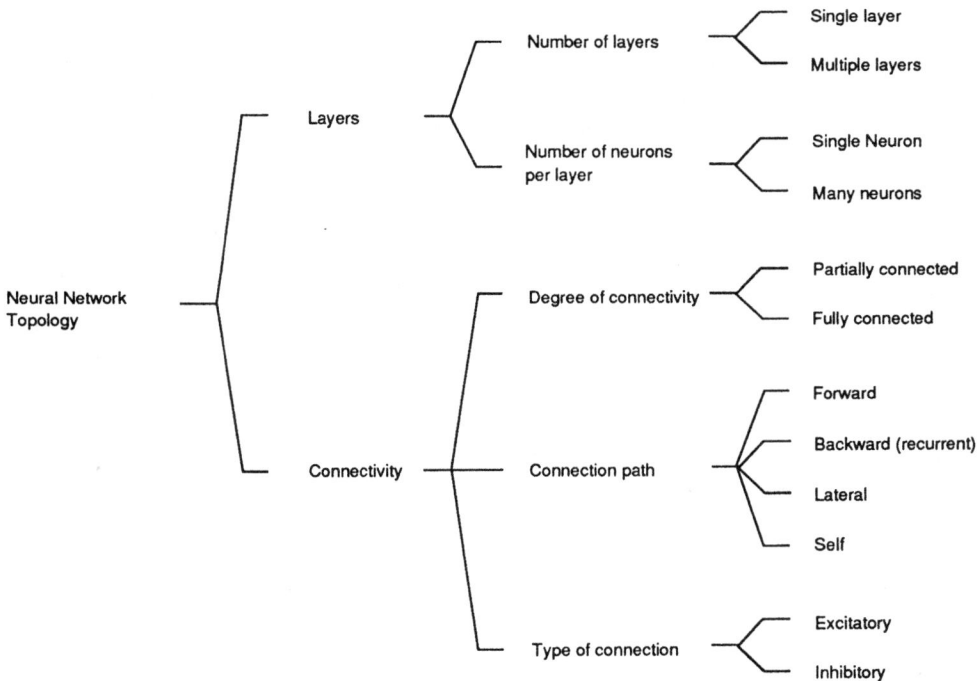

Figure 8: Characterization of neural network topology.

The two main elements which determine a network's architecture are the *layers* of neurons it contains and the *connections* made between neurons. Both the number of layers and the

181

size of each layer may be varied. Existing network models are usually grouped as single-layered, bilayered or multilayered. The number of neurons per layer is a highly variable parameter which is difficult to specify when designing networks, as it depends on the form and type of input data to be processed, the output desired, and the type and complexity of the processing task. Values typically range between one and several thousand nodes per layer.

Neurons may be connected in numerous ways to produce different information flow behaviors within a network. Each node may be connected with only a few other nodes, or with all nodes. In practice, most networks are generally fully connected, since "unimportant" connections will often acquire a connection weight of zero after training, which is equivalent to no connection.

Two major classes of networks are *feedforward* and *recurrent* networks. Networks with forward connections pass information forward to successive layers, but not back to preceeding ones. If a network has backward connections as well, it is termed a feedback, or recurrent network. Nodes can also have recurrent connections with themselves (self connections). Other networks have lateral connections between nodes of the same layer. Finally, connections may be either excitatory or inhibitory, depending on whether their connection weights are positive or negative, respectively. One important class of networks with both excitatory and inhibitory connections are called *competitive-learning* networks.

7.3 Learning in Neural Networks

Learning in neural networks refers to the process of acquiring a desired behavior by changing the connecting weights. The main rules to produce these weight changes may be classified into three important categories [Torras 89], each of which borrow concepts from behavioral theories of learning in biological systems.

- *Error-Minimization Rules* These rules compute the necessary change in the connection weights by presenting the network given input pattern, comparing the obtained response with a desired rsponse known a-priori and then changing the weights in the direction of decreasing error. This type of learning is also referred to in literature as *supervised learning*. Important examples of this category are the perceptron learning rule, the Widrow-Hoff (or Delta) rule and the backpropagation algorithm, which are dealt with later in this chapter.

  This type of learning requires the knowledge of sets of input patterns and desired responses to be "taught" to the network, usually called training sets. Furthermore, the network dynamics consists of two clearly different periods: the *training period* when the weights are adapted to learn some desired behaviour and the *operation period* when the network works with fixed weights to produce outputs in response to new patterns. Error-minimization rules are used extensively in training feedforward and some recurrent networks.

- *Correlational Rules* These rules adjust each connection weight according to the correlation between the activations of the two neurons involved in that particular connection. In this case, no explicit examples of the desired behavior are provided, a fact which has caused this type of learning to be known as *unsupervised*. Furthermore, no separation between the training and operation periods exists in these networks.

The learning rules in this category may be considered variants of the Hebbian learning rule, closely related to the first observations of learning in biological systems, which updates the value of any weight in a network by adding it a term which is proportional to the product of the activation values of the two neurons involved in the connection.

This type of learning has been extensively used for associative memory applications and also in the so-called competitive-learning models. In these models the layers in a network are organized hierarchically and excitatory connections exist between groups of neurons in successive layers; neurons within layers are organized in clusters with inhibitory connections within each cluster.

The name "competitive" stems from the fact that each neuron in a cluster competes with the others to react to the pattern coming from the previous layers, in a behavior also known as "winner-takes-all". This type of model may learn to extract features and classify stimuli successfully.

- *Reinforcement Rules*

  In this type of learning rule, the network is also presented with sample input patterns, but they do not require being supplied with the desired responses to input patterns. An overall measure of the adequacy of the obtained response is enough to drive the network to the desired behavior. This measure, usually called *reinforcement* signal is fed back to the network in order to reward correct behaviours and penalize incorrect ones, just like an animal may be trained to behave in a particular way by repeatedly rewarding good performances and penalizing poor ones.

  The learning proceeds through the following steps: First, the network computes the outputs produced by the presented input with the current values of the weights. Then, the output is evaluated and the reinforcement signal is injected into the network. Subsequently, the weights are adapted according to the reinforcement signal by increasing the weights which contributed to good performance, or decreasing those which caused a poor performance. The network then seeks a set of weights which tends to prevent negative reinforcement in the future.

  Reinforcement rules, usually seen as an intermediate approach between *supervised* and unsupervised training, are also considered as random search procedures guided to maximize a performance index or reward. Feedforward and recurrent networks may also be trained using this type of rule.

Part II

Some Major Networks and their Applications

8   Multilayer Feedforward Networks

The earliest neural networks were kept simple in structure in an attempt to get a first idea of the basic processing behavior of interconnected artificial neurons. Several binary nodes were arranged into architectures of two or more layers and linked by strictly feedforward connections –that is, connections which pass signals forward through successive network layers, but never back to preceeding ones. Lateral and self connections were also excluded. These models gave rise to an important class of networks called *multilayer feedforward networks*.

Two early models, the Perceptron and Adaline networks, proved capable of performing simple linear classification tasks. Both of these networks employed a similar learning procedure which lacked the sophistication needed to overcome this limitation. The later development of an improved learning rule for the perceptron called *backpropagation* allowed training networks with additional layers, and thus greatly extended its classification abilities to more complex, non-linear problems. The success achieved with multilayer backpropagation networks in many application areas has made it by far the most widely used network, and has made multilayer feedforward networks practically synonymous with backpropagation.

8.1   The Perceptron Network

The Perceptron, developed in 1958 by F. Rosenblatt, is considered to be the first artificial neural network. It consists of simple bipolar threshold-logic nodes arranged into two or more layers with feedforward connections, as shown in Figure 9. In order to be useful as a classifier, the input layer has at least two nodes, and the output layer contains at least one. The original perceptron investigated by Rosenblatt had two layers, consisting of a variable number of input nodes connected to a single output node.

The Perceptron Network

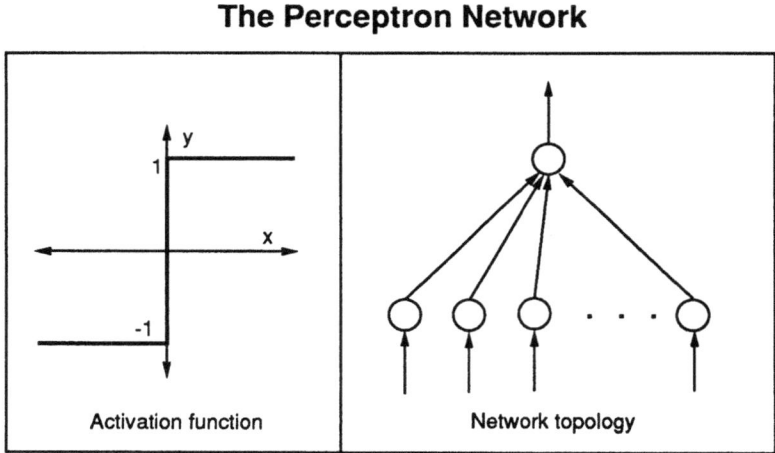

Figure 9: The Perceptron network: transfer function and architecture. (Adapted from [Maren 90].)

With the exception of the input layer, the nodes in the Perceptron implement the functions of the basic neuron model: The incoming activations a_{ij} to each node j are multiplied by their corresponding connection weights w_{ij}, summed, and a threshold value θ is subtracted (assuming a positive value is used). The bipolar (-1, 1) threshold function is then applied to the result to produce the neuron's output. The nodes in the input layer are not true neurons in the sense that they do not process their inputs with a transfer function, but

184

merely act as "distribution units" which feed each input to every node in the next layer. (In fact, this is a widely, but not universally used convention for all multilayer feedforward networks [Kröse 90].)

8.1.1   Training the Perceptron

The goal of the perceptron network is to learn simple associations between given input vectors, $\vec{x}$, and desired output (activation) values a^d. The network may be taught to associate any number of input vectors with their corresponding outputs. Often, these input-output pairs may represent examples of two or more categories or classes of patterns between which the network must learn to distinguish. In these cases it may be said that the network's objective is to learn to *classify* or to form *decision boundaries*.

If the network is to be taught a classification task, the training data, called the *training set*, must be chosen carefully so as to include good examples of each of the categories the network is to learn. If the patterns from two or more categories resemble each other too closely, it may be difficult for the network to learn to distinguish between them. Thus the first step in training the perceptron (and any other multilayer feedforward network) is to choose a good training set.

To train the perceptron, the connection weights and the threshold θ are first set to small random values. An input vector $\vec{x}$ is selected from the training set and input to the network. The resulting output a_j is then compared with the desired output a_j^d, by computing the difference δ_j :

$$\delta_j = a_j^d - a_j \tag{6}$$

If $\delta_j = 0$ (node j produced the desired output), the weight w_{ij} is left unchanged. If $\delta_j \neq 0$ (node j gave an incorrect response), connection weight w_{ij} is modified according to the equation:

$$\Delta w_{ij} = \alpha \delta_j x_{ij} \tag{7}$$

where

Δw_{ij} is the change to be made to the connection weight between neurons i and j

α is a constant

x_{ij} is the input from neuron i to neuron j .

The thresholds θ are adjusted by a similar rule.

This criteria is applied to each of the connection weights w in the output layer. This process is repeated for each of the input vectors $\vec{x}$ in the training set. One cycle through the training set represents a single training iteration. Training is continued through several iterations until all the input vectors are correctly classified.

There are several features of the perceptron training algorithm worth noting. First of all, the Perceptron undergoes *supervised* training, since the desired output a^d is used to

185

compute the modifications to the connection weights. Secondly, the perceptron's training rule closely resembles Hebb's learning law in that the connection weights are updated by an amount proportional to the incoming activation from the connected neuron. The only difference is that when the correct output is achieved, the connection weight is not modified. Finally, it must be pointed out that the learning procedure may only be used to adjust the connection weights of the output units. No similar method was found for adjusting the connection weights in networks with intermediate layers. As will be explained below, this fact proved to be a serious limitation on the capabilities of the original perceptron model.

8.1.2 Capabilities and Limitations of the Perceptron

The perceptron's ability to perform classification tasks stirred considerable interest in the computational possibilities of neural networks in the early 1960's. This optimism was largely supported by the *convergence theorem* which Rosenblatt proved with regard to the perceptron, which stated that if a set of connection weights exists which will allow a perceptron to perform a given classification task, then the network will converge to a solution in a finite number of training steps. With this, the first neural computational device had been developed, which was able to learn entirely from the iterative presentation of examples, and whose convergence to a solution was guaranteed. Figure 10 illustrates how a perceptron can be made to learn a classification problem.

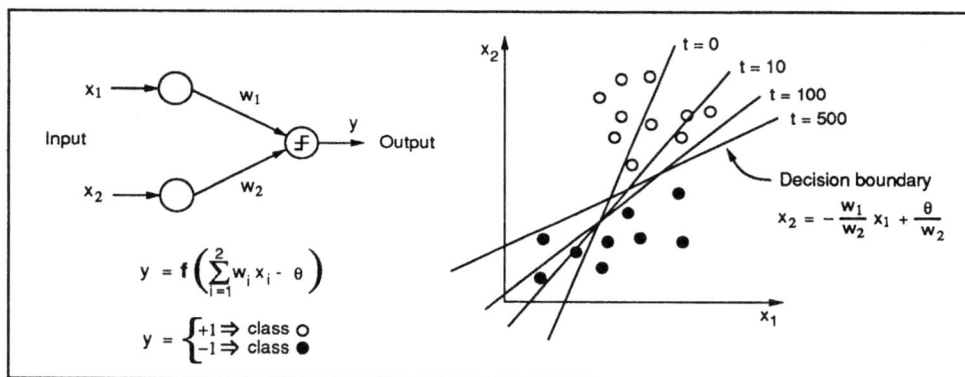

Figure 10: An example of how a two-layer perceptron with two inputs learns to solve a linear classification problem. The network has formed the correct decision boundary between the two input classes after 500 training iterations. (Adapted from [Lippmann 87]).

While the perceptron's initial results appeared promising, they were later shown to be very limited in the type of problems they could solve. In 1969, Minsky and Papert analyzed perceptrons thoroughly and concluded that they were incapable of performing many of the most basic computations, such as parity and connectedness. One of their most discouraging results was that a two-layer perceptron cannot represent even a simple exclusive-or (XOR) calculation: given two binary inputs, output 1 if exactly one of the inputs is on, and output 0 otherwise. The XOR problem may be thought of as a pattern classification problem in which there are four input patterns and two possible output classes. Figure 11 shows the problem represented in this way, and illustrates that no linear decision boundary may be drawn which separates the two output classes. This is because the XOR is *not linearly separable.* The two-layer perceptron was shown to be capable of classifying only linearly separable decision spaces, since the decision boundary it forms is merely a hyperplane (a line, in two dimensions). Unfortunately, the vast majority of computations are much more complex than this.

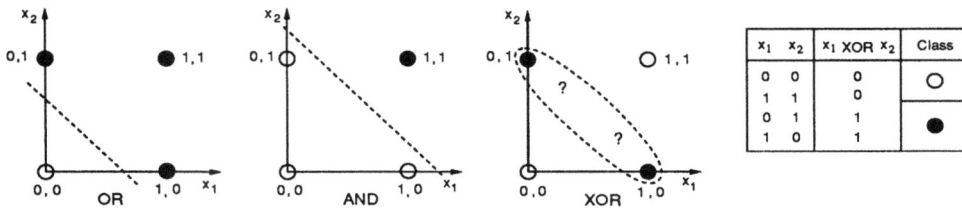

Figure 11: The XOR problem: a classification problem which is not linearly separable. The truth table shows the four possible input patterns and their corresponding outputs (classes), which are represented by the two types of circles. No linear decision boundary (line) may be drawn which separates the two output classes into two distinct regions. (Adapted from [Kröse 91].)

In order to be able to form more complex decision boundaries, an additional layer of intermediate, or "hidden" units was necessary. Figure 12 illustrates the types of decision boundaries that perceptrons can form when hidden units are added. Despite having established the need for hidden units, Rosenblatt's training algorithm was incapable of calculating the weight adjustments for layers preceeding the output layer. A more powerful training method was needed.

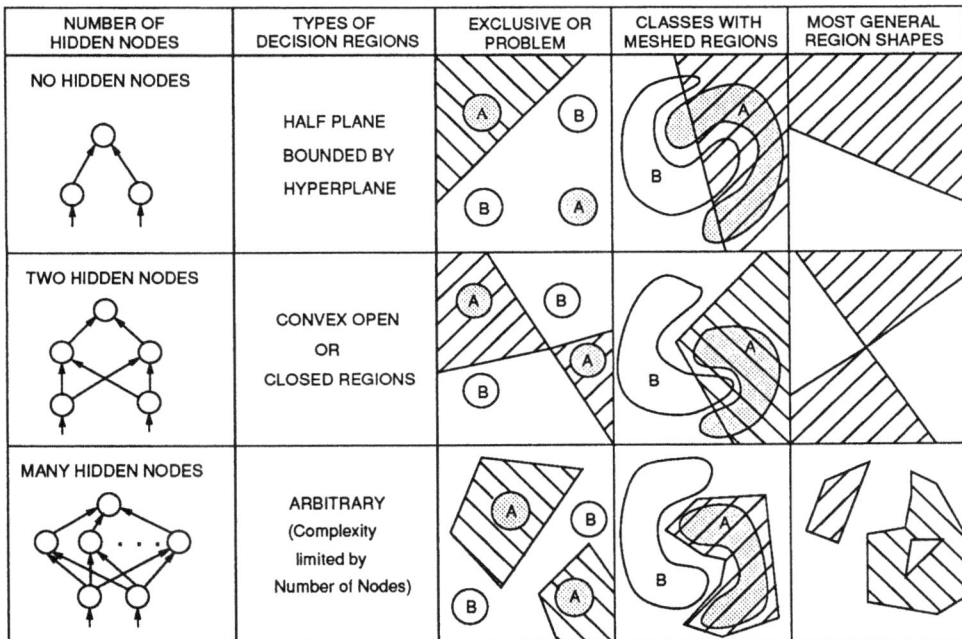

Figure 12: The types of decision regions that can be formed by single and multi-layer perceptrons with binary units, depending on the number of hidden nodes. (Adapted from [Lippmann 87].)

187

Minsky and Papert summarized these arguments in their book *Perceptrons*, which many researchers claim was largely responsible for creating widespread disillusionment in the neural network field and diverting major government funding away from neural network research for several years following its publication. In fact, they had merely pointed out the current limitations of existing networks, and emphasized there additional research should be focused. Nevertheless, during the 1970s neural networks lost much of their credibility and many researchers were forced to migrate to other fields. Not until 1975, when Paul Werbos introduced the backpropagation training algorithm for perceptrons with hidden units, did neural networks begin to show renewed possibilities, and even then the recovery of the field was slow.

Backpropagation will be discussed below, following a brief look at Adaline, a network very similar to the perceptron and whose learning rule served as a basis for backpropagation.

8.2 The Adaline Network

Another early feedforward network is Adaline, meaning ADaptive LINear Element, which was introduced by Widrow and Hoff (Stanford University) very shortly after Rosenblatt's perceptron. The name derives from the linear sum of inputs computed by the neurons, and the learning rule used to adapt the connection weights. While Adaline is practically equivalent to the perceptron, its learning rule was later used in the development of the backpropagation algorithm.

8.2.1 Adaline Architecture

Architecturally, the Adaline network is identical to the perceptron. It consists of an input layer of two or more neurons with feedforward connections to a single output neuron, as shown in Figure 13. There is a version with multiple output nodes called *Madaline*, for Many Adalines. The output node computes the linear sum i of the inputs x_i from each input neuron connected to output neuron j, multiplied by their corresponding connection weights w_{ij}, plus a threshold θ:

The ADALINE and MADALINE Networks

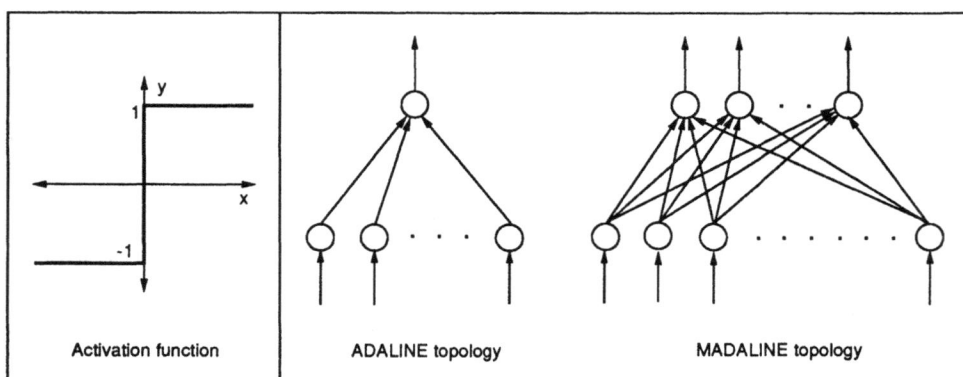

Figure 13: The ADALINE and MADALINE networks: activation function and architecture. (Adapted from [Maren 90].)

188

$$i = \sum_{i=1}^{n} w_{ij} x_i + \theta \qquad (8)$$

The bipolar threshold function (-1, +1) is applied to the result to produce an output.

Widrow and Hoff developed a simple hardware implementation of this network consisting of a set of controllable resostors connected to a circuit which summed the currents produced by the input voltage signals. This summation block was followed by a thresholding circuit which output either -1 or +1 depending on the polarity of the sum.

8.2.2 Training Adaline: The Delta Rule

The main difference between Adaline and the perceptron is the way their learning rules are applied. Each has a slightly different basis, but yield identical formulas for adjusting the connection weights. The objective of Adaline's training algorithm is to minimize the *mean summed square error* between the desired and achieved outputs of the network for the entire training set. For this reason, it is often called the *LMS (least mean squares) learning procedure.*

This error function E is given by

$$E = \sum_p E^p = \frac{1}{2} \sum_p (a^d - a)_p^2 \qquad (9)$$

where the index p ranges over the set of input patterns and E^p represents the error for pattern p. The variables a^d and a are the desired and actual outputs when pattern p is presented. However, unlike for the perceptron, Adaline's training rule is applied to the activation value *before the threshold activation function is applied.* Thus, the activation value used for making weight adjustments in Adaline is simply the weighted sum of inputs given by eq. (8):

$$a = \sum_{i=1}^{n} w_{ij} x_i + \theta \qquad (10)$$

where w_{ij} and x_i are the connection weight and input from neuron i to neuron j, respectively. This is the important distinction between Adaline and the perceptron: by using the activation value which has not yet been forced to a threshold value, Adaline's weights are adapted in a way which is much more sensitive to the *real difference* between the desired and achieved activations. An additional advantage to using the linear activation value before applying the threshold function is that the delta rule may be used with both continuous and binary neurons.

The LMS procedure finds the values of all the weights which minimize the error function E by a method called *gradient descent.* Ths idea is to adjust each weight by an amount proportional to the negative of the derivative of the error when a particular input pattern is presented. The weight adjustment is thus given by:

$$\Delta w_{ij} = -\gamma \frac{\partial E^p}{\partial w_{ij}} \qquad (11)$$

189

where γ is a proportionality constant. By the chain rule, the derivative is

$$\frac{\partial E^p}{\partial w_{ij}} = \frac{\partial E^p}{\partial a} \frac{\partial a}{\partial w_{ij}} \qquad (12)$$

Since the activation value used is simply the linear sum of the inputs (eq. (10)),

$$\frac{\partial a}{\partial w_{ij}} = x_i \qquad (13)$$

and

$$\frac{\partial E^p}{\partial a} = -(a^d - a) \qquad (14)$$

and thus eq. (11) becomes

$$\Delta w_{ij} = \gamma \delta x_i \qquad (15)$$

where $\delta = a^d - a$. this difference δ is called "delta", which is why the LMS procedure is often called the *delta rule*.

Clearly, this formula is identical to the one used to adapt the perceptron's connection weights. It is used in a similar iterative training procedure as for the perceptron. But again, the difference lies in the activation values used to calculate the difference δ, and in the fact that the LMS procedure is based on the gradient descent technique, involving derivatives, for minimizing the overall error. In the next section it will be shown how the technique used in deriving the delta rule was extended to obtain a generalized form applicable to networks with hidden units.

8.2.3 Applications of Adaline Networks

Adaline networks belong to a general class of algorithms called *adaptive linear filters*. They have been and are still being used in many application areas where adaptive linear filters have traditionally been employed. These include adaptive equalization filters for high-speed modems, adaptive echo cancelers for long-distance telephone and satellite communications, and signal prediction. Another example of a noise-canceling application is to cancel maternal heartbeat in fetal elecrocardiograph recordings [Maren 90].

8.3 The Backpropagation Network

As discussed in the previous sections, the types of problems which may be solved by two-layer feedforward networks is very limited. Although Minsky and Papert pointed out in 1969 that including additional layers of "hidden" units would overcome many restrictions (see Figure 12), the solution of how to adjust the weights from input to hidden nodes remained unsolved.

The solution came in 1975 when Paul Werbos introduced the *backpropagation training algorithm*. This method is a generalization of the delta rule with which, by employing a

190

continuously-derivable activation function, the errors for the hidden nodes are determined by back-propagating the errors of the output nodes using the chain rule.

Despite having a few limitations of its own, the backpropagation algorithm greatly expanded the problem-solving capabilities of multilayer feedforward networks and opened up a wealth of new application areas. The many successful results obtained since its introduction, especially for complex classification tasks, renewed interest in the neural network field and began a new wave of research worldwide.

8.3.1 Network Architecture

The structure of a backpropagation network is that of a perceptron with one or more output nodes and an additional intermediate, or *hidden* layer between the input and output layers, as shown in Figure 14. Several hidden layers may be included in the networks but, as will be explained below, a single layer of hidden nodes can perform exactly the same functions as many. Hidden units are often called *associative*, *predicate* or *feature representation* units by virtue of their ability to extract and represent features from the network inputs.

The Backpropagation Network

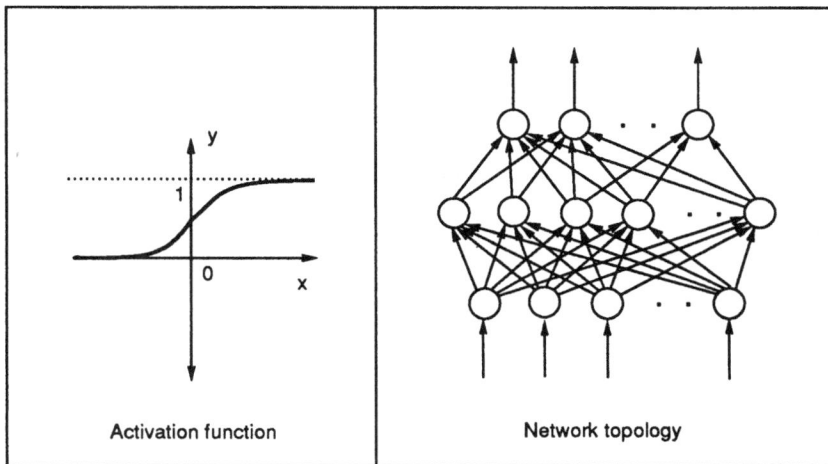

Activation function   Network topology

Figure 14: The Backpropagation network: activation function and architecture. (Adapted from [Maren 90].)

All connections are feedforward between adjacent layers, and networks are generally fully connected, that is, each node is connnected with all nodes in the following layer. It must be kept in mind that the term "backpropagation" refers not to the direction of the connections within the network, but to the way in which errors are propagated backwards through the network layers during training.

The nodes of the backpropagation network compute the same linear sum of weighted inputs plus threshold as those of Adaline and the perceptron. The key difference, which made the development of backpropagation possible, is the use of a smoothly limiting activation function. Most commonly, this function is the sigmoid, as shown in Figure 14, which has the desired thresholding characteristics of binary neurons but with smooth transition points. Since this function is continuously derivable, the chain rule may be applied to calculate the error in the hidden layer, and hence the correct weight adjustments for the intermediate connections.

8.3.2 The Generalized Delta Rule

Backpropagation training is an extension of the delta rule for Adaline networks, and therefore is based on the same principles: it is a supervised training method in which the connection weights are adapted using a formula which minimizes the mean square error between the desired and achieved network outputs when each input pattern is presented. The extension consists of using the actual neuron output after applying the transfer function, which, by being continuously derivable, allows these errors to be propagated across the hidden layer in order to calculate their weight adjustments in the same way as for the output layer.

The output activation of a node j in the network is given by

$$a_j = f(i_j) \tag{16}$$

where the total input i_j from all nodes i is simply

$$i_j = \sum_i w_{ij} a_i + \theta_j \tag{17}$$

As for Adaline, the objective is to minimize the overall network error $E = \sum_p E^p$ for all input patterns p in the training set. In practice, this is often performed in an iterative fashion by minimizing the error over the n output units after presenting each pattern p:

$$E_p = \frac{1}{2} \sum_j (d_j - a_j)_p^2 \tag{18}$$

Using the same criterion as for the delta rule, the weight adjustment is defined as proportional to the negative derivative of the error with respect to the weight:

$$\Delta w_{ij} = -\gamma \frac{\partial E^p}{\partial w_{ij}} \tag{19}$$

In the same manner as before, the derivative is

$$\frac{\partial E^p}{w_{ij}} = \frac{\partial E^p}{\partial i_j} \frac{\partial i_j}{\partial w_{ij}} \tag{20}$$

From eq. 17, the second factor is

$$\frac{\partial i_j}{\partial w_{ij}} = a_i \tag{21}$$

and the first factor is defined as "delta":

192

$$\delta_j = -\frac{\partial E^p}{\partial i_j} \tag{22}$$

Substituting the latter two equations into eq. (19), a weight updating rule is obtained which is equivalent to the delta rule:

$$\Delta w_{ij} = \gamma \delta_j a_i \tag{23}$$

The difference is that for the network with hidden layers, δ_j must be found for each successive network layer, since their inputs are different. As will be shown, for a network with any number of hidden layers (assuming all neurons are equivalent), there will be *two deltas*: one for the output layer, δ_j, and another for the hidden layers, δ_h.

To derive both δ, the chain rule is recursively applied to eq. (22):

$$\delta_j = -\frac{\partial E^p}{\partial i_j} = -\frac{\partial E^p}{\partial a_j}\frac{\partial a_j}{\partial i_j} \tag{24}$$

By eq. (16), the second factor is

$$\frac{\partial a_j}{\partial i_j} = f'(i_j) \tag{25}$$

which is simply the derivative of the activation function, in this case the sigmoid.

In order to calculate the first factor of eq. (24), the cases for both output and hidden layers must be considered:

- **Output layer**

  For any output unit j, the derivative may be obtained directly from the error definition of eq. (18):

  $$\frac{\partial E^p}{\partial a_j} = -(d_j - a_j) \tag{26}$$

  which is the same result as obtained for the standard delta rule. Substituting this and eq. (25) in eq. (24) yields

  $$\delta_j = (d_j - a_j)f'(i_j) \tag{27}$$

  The activation function f is the sigmoid, whose equation is

  $$a_j = f(i_j) = \frac{1}{1 + e^{i_j}} \tag{28}$$

  and its derivative is easily obtained:

$$f'(i_j) = \frac{\partial}{\partial i_j} \frac{1}{1 + e^{i_j}}$$

$$= \frac{1}{(1 + e^{i_j})^2}(-e^{i_j})$$

$$= \frac{1}{(1 + e^{i_j})^2} \frac{-e^{i_j}}{(1 + e^{i_j})}$$

$$= a_j(1 - a_j) \tag{29}$$

The error signal δ_j for an output unit can now be written as

$$\delta_j = (a_j^d - a_j)a_j(1 - a_j) \tag{30}$$

This expression, substituted in the delta rule given by eq. (23), is used to update the weight w_{hj} of any output unit j after presenting each pattern p to the network:

$$\boxed{\Delta w_{hj} = \gamma(a_j^d - a_j)a_j(1 - a_j)a_h} \tag{31}$$

where the indices h and j refer to nodes in the last hidden layer and the output layer, respectively.

- **Hidden layers**

  Unlike for the output nodes, the desired outputs of the hidden nodes are unknown. However, they may be computed by expanding the error derivative using the chain rule as follows:

$$\frac{\partial E^p}{\partial a_j} = \sum_j \frac{\partial E^p}{\partial i_j} \frac{\partial i_j}{\partial a_h} = \sum_j \frac{\partial E^p}{\partial i_j} \frac{\partial}{\partial a_h} \sum_h w_{hj} a_h = \sum_j \frac{\partial E^p}{\partial i_j} w_{hj} = -\sum_j \delta_j w_{hj} \tag{32}$$

Substituting in eq. (24), and using the index h for hidden units:

$$\delta_h = f'(i_h) \sum_j \delta_j w_{hj} \tag{33}$$

Using the derivative of the sigmoid obtained in eq. (29) yields the delta for hidden units:

$$\delta_h = a_h(1 - a_h) \sum_j \delta_j w_{hj} \tag{34}$$

Thus, the equation for updating the weights in the hidden layer is:

$$\boxed{\Delta w_{ih} = \gamma a_i a_h(1 - a_h) \sum_j \delta_j w_{hj}} \tag{35}$$

where the index i refers to nodes in the layer preceeding the hidden layer.

8.3.3 Training with Backpropagation

Backpropagation networks are trained in much the same way as perceptron and Adaline networks. All weights are first initialized to small random values. The input patterns in the training set may then be presented in one of two ways: either all the inputs may be presented to obtain their corresponding error signals and the overall error used to perform a single weight adjustment, or the inputs may be presented individually and the weights adjusted after each one. In both cases, eq. (31) is used to adjust the weights of the output layer, and eq. (35) is used to modify the weights for all hidden nodes. Likewise, either procedure is repeated until the network converges to the correct solutions. Several empirical results indicate that making weight adjustments after presenting each input pattern results in faster convergence [Kröse 90].

8.3.4 Limitations of Backpropagation

While the backpropagation algorithm has proven very useful for training multilayer feedforward networks, it is not without its shortcomings. Training times are typically long, and the incorrect selection or presentation of training examples can often result in "network paralysis", a condition in which training comes to a standstill due to the network becoming too focused on a particular subset of patterns. In addition, there is no theorem which guarantees that the backpropagation procedure based on gradient descent will converge to a global minimum, as in the case of the perceptron, and networks may often become trapped in local minima.

Several advanced algorithms and enhancements to backpropagation have been developed in an attempt to overcome some of these limitations. However, there is still much theoretical ground to be covered before reliable, constructive methods are available for efficiently designing and training multilayer feedforward networks.

8.3.5 Applications of Backpropagation Networks

Until quite recently, it was generally believed that the problem complexity which a backpropagation network could handle was not only dependent on the number of hidden nodes, but on the number of hidden *layers* as well [Lippman 87]. It was later proven, however, that a single layer of hidden nodes is sufficient to perform virtually *any* non-linear function mapping to any desired degree of accuracy [Girosi 89]. The implications of this fact are quite significant, since all processing tasks consist of nothing more than some sort of mapping between inputs and outputs. As mentioned above, however, the fact that multilayer networks can perform such mappings says nothing about what architectures are necessary, nor about how they may be trained for a particular task. Nevertheless, backpropagation and its variations are currently the best algorithms available for training these networks, and have proven adequate for applying them to a variety of practical tasks.

Backpropagation networks have shown great success in wide range of applications involving pattern classification and signal processing. They have been used extensively for signal filtering and signal interpretation tasks, as well as for image compression, handwritten character recognition, time-series prediction and text reading.

Impressive results have been demonstrated with a system called NETtalk, which is capable

195

of converting printed English text into highly intelligible speech.

There has been much recent interest in applying backpropagation networks in control applications, particularly those in which adaptive controllers are required for systems involving nonlinearities and unknown parameters. The ability of networks with hidden units to adaptively learn complex nonlinear function mappings is especially appealing for such systems, for which classical control techniques have often not been applicable. Important applications include the adaptive control of robot arms for industrial and space applications, adaptive controllers for satellite attitude control, and autonomous vehicle navigation.

9   Recurrent Networks

All of the netowrks described in the previous sections had strictly feedforward connections, meaning that information flowed in a single direction between successive layers of neurons. Recurrent networks, in contrast, are a broad class of networks having feedback loops which circulate information back through the same or previous layers. Consequently, when an input pattern is presented to a recurrent network, it does not produce an output pattern in a finite number of steps, but instead causes a circular process, or "resonance" of neural activity within the network wherein the same layers of neurons are activated repeatedly. If the network is inherently stable, it will resonate for a time before settling to a stable state, at which time the neural activations stop changing and a constant output is produced. Otherwise, it may continue resonating indefinitely. When training recurrent networks, it is generally desired to obtain the set of connection weights which allow the network to stabilize to the desired output values.

Many different recurrent architectures have been investigated. The Hopfield network has a single-layer architecture and functions as a content-addressable, or *autoassociative memory*. The ART (Adaptive Resonance Theory) and BAM (Bidirectional Associative Memory) networks are two-layered networks with feedforward and feedback connections whose capabilities as *heteroassociative memories* have been studied. More complex architectures include the Jordan and Elman networks, whose multilayer recurent architectures were developed in an attempt to more adequately deal with input signals containing dynamic or temporal information, such as speech recognition and control of dynamic systems.

9.1   The Hopfield Network

In 1982, Hopfield presented a network consisting of a single layer of fully connected nodes. Thus its connections are recurrent in the sense that they are symmetric between any two neurons (see Fig. 15). The original Hopfield model employed binary valued nodes, whereas continuous activation functions were used in later models.

The Hopfield network was shown to have some interesting properties as an *autoassociative memory*, in which several patterns can be stored by adjusting the connection weights such that each pattern corresponds to one of the network's various stable states. When the network is presented with a noisy or incomplete input pattern, it will retrieve one of its stored patterns which most closely resembles it. Such a network may be thought of as a "content-addressable" memory, since information is searched for by content, rather than

196

The Hopfield Network

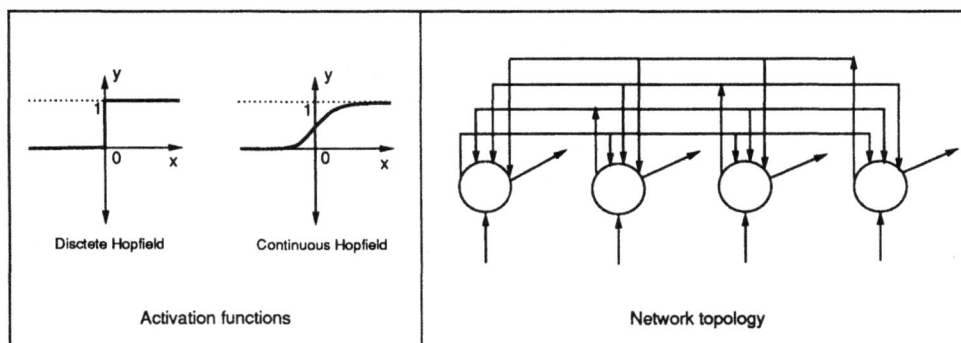

Figure 15: The Hopfield network: activation functions and architecture. (Adapted from [Maren 90].)

by its address in a particular location. To retrieve a stored pattern, only a portion of it need be specified, and the network will find the best match.

9.1.1 Training

The training procedure for the Hopfield network consists of a single step in which all of the examplars in the training set are used simultaneously to calculate the connection weights. These remain fixed for the remainder of its operation. To determine the weights, a Hebbian-type formula is used which basically computes the weight matrix as the sum of all the pattern vectors multiplied by their transpose. Thus, for a set of P input pattern vectors $\vec{x}$, each weight w_{ij} is given by [Kröse 91]:

$$w_{ij} = \begin{cases} \sum_{p=1}^{P} x_i^p x_j^p & \text{if } i \neq j \\ 0 & \text{otherwise} \end{cases} \tag{36}$$

9.1.2 Performance

The potential applications of Hopfield networks as content-addressable memories stirred considerable interest at one time. This was partly due to the fact that several researchers showed that they could be implemented in existing hardware with relative ease, since their wieghts are non-adaptive. Recent emphasis in many application areas on the need for networks capable of continued learning through adaptive weight adjustments has diminished this interest to some extent [Maren 90].

Other problems exist with regard to the performance and storage capabilities of Hopfield networks. When patterns resemble each other too closely (i.e. are not very orthogonal), often the network will not converge to a stable state or will produce an incorrect response. Extremely noisy or unfamiliar patterns can also cause the network to produce spurious stable states. Additionally, their memory capacity has been shown to be limited to only about 0.15 times the total number of nodes. Various advanced algorithms have been proposed in an attempt to improve the performance of Hopfield networks. Nevertheless, their use continues to be somewhat limited.

197

9.2   Boltzmann Machines

The Boltzmann machine is considered to be an extention of the Hopfield network to include hidden units. Like the latter, its nodes are fully connected with each other, such that pairs of nodes are symmetrically linked.

Boltzmann machines were conceived for use in solving optimization problems. Previous attempts had been made to use Hopfield networks for optimization by exploiting their capability of settling to minima (stable states) on an energy surface. The problem with Hopfield networks, however, is that they typically have many local minima, which, although desirable for storing multiple patterns in autoassociation tasks, is a serious limitation if the goal is to find a globally optimum stable state.

Convergence to a global energy minimum is possible with the Boltzmann machine by using a learning procedure called *simulated annealing*. The training algorithm it employs is somewhat more complicated than the others seen so far and will not be described in detail here. Further information may be obtained in [Ackley 85].

What is important are the underlying concepts of the simulated annealing algorithm. This method is based on the annealing principle of quantum physics which describes how heated materials, when their temperature is lowered very slowly in a controlled manner, can cool to very low energy states such as that of a highly ordered crystal lattice. In this way, local minima (stable states of higher energy) are avoided. The learning procedure used by the Boltzmann machine simulates this process by determining the *probability* p that a particular neuron becomes active (i.e. are in a particular "energy state"), according to the formula:

$$p = \frac{1}{1 + e^{-\Delta E/T}} \tag{37}$$

where ΔE is the sum of the unit's active inputs (its "energy state"), and T is the simulated "temperature" of the network. The network is trained by lowering its temperature using a suitable time-dependent function, called a "cooling schedule", and then making weight adjustments using these probabilities. The Boltzmann machine receives its name from the fact that, like a physical system, the network will eventually reach "thermal equilibrium", at which time the relative probability of finding it at either of two global energy states α or β will follow the Boltzmann distribution:

$$\frac{P_\alpha}{P_\beta} = e^{-(\varepsilon_\alpha - \varepsilon_\beta)/T} \tag{38}$$

where P_α is the proability of global state α, and ε_α is the energy of that state. Note that at thermal equilibrium the units still change state, but the probability of finding the network in any global state remains constant [Kröse 91].

As in the annealing process of physical systems, at high temperatures units display random behavior and are able to settle into the larger-scale features of the energy surface. As the network "cools", units have a lower energy and become more sensitive to smaller features. In this way the network is able to escape local minima as it first begins to cool and finally reach the overall minimum energy state.

In addition to their suitability for optimization problems, the architectural similarity of

Boltzmann machines to multilayer feedforward networks with hidden units makes them applicable to many of the same types of tasks as the latter, with the added advantage that their convergence to a globally optimal set of connection weights is more likely.

10 Concluding Remarks

In the previous sections, many of the underlying principles, capabilities and important examples of neural networks have been described. Clearly, neural networks are a promising new technology with great potential for a number of application areas in which traditional processing paradigms have proven inadequate. However, they are certainly not the solution to every type of problem, and existing methods should always be used when most applicable.

Research in neural networks is still in a very early stage of development. Most of the architectures and algorithms investigated so far are little more than oversimplified models of a few of the basic structural and functional features found in biological networks. Much theoretical groundwork remains to be laid before artificial neural networks begin to approach the complexity and computational power of biological systems, and in order that they may be designed and trained in a constructive manner to solve large-scale problems.

It is also evident that the power inherent in parallel computational paradigms such as neural networks is of little use if they are merely made to run as programs on existing serial computers. This power will only be tapped when they are implemented in suitable parallel hardware. Much investigation is currently underway in the application of VLSI technology to the design of neural chips.

The networks described here represent a small fraction of those which have been developed and investigated. No attempt was made to be comprehensive, but rather to introduce several of the most important classes of networks and illustrate their development with examples of each. While the multilayer feedforward and recurrent networks presented are among the most well known and used networks, many other architectures and learning paradigms exist, and new ones are continually being investigated. Other recurrent networks such as ART and Jordan networks, and competitive learning networks such as Kohonen's Self-organizing Feature Maps and the Neocognitron were considered beyond the scope of this introductory text. Further information regarding these networks may be obtained from the references listed below.

Acknowledgements

Research in neural networks and their applications to process control at the Instituto de Cibernética is partially supported by the research grant CICYT-TIC91-0423 of the Spanish Science and Technology Council.

The contribution of Mr. Joan Alex Ferré in the preparation of this text and the helpful advice and guidance of Prof. Carme Torras are gratefully acknowledged.

References

[**Ackley 85**] Ackley, D. H., Hinton G. E., and Sejnowski, T. J., "A learning algorithm for Boltzmann machines", *Cognitive Science*, Vol. 9, 1985

[**Girosi 89**] Girosi, F., and Poggio, T., "Networks and the best approximation property", *Artificial Intelligence Lab. Memo*, No. 1164, MIT, Cambridge, MA, October 1989.

[**Jordan 91**] Jordan, M. L., and Rumelhart, D. E., "Forward Models: Supervised Learning with a Distal Teacher", Occasional Paper 40, MIT Center for Cognitive Science, 1991.

[**Jutten 90**] Jutten, C., and Blayo, F., "Reseaux de Neurones: Principes, Paradigmes et Applications", Laboratoire de Traitement d'Image et de Reconnaissance de Formes, Institut National Polytechnique de Grenoble, 1990.

[**Kröse 91**] Kröse, B. J. A., and van der Smagt, P., "An Introduction to Neural Networks", Faculty of Mathematics and Computer Science, University of Amsterdamk, 1991.

[**Lippmann 87**] Lippmann, R., "An Introduction to Computing with Neural Nets", *IEEE ASSP Magazine*, April 1987.

[**Maren 90**] Maren, A., Harston, C. and Pap, R., *Handbook of Neural Computing Applications*, Academic Press, 1990.

[**Millán 90a**] Millán, J. del R., and Torras, C., "Reinforcement learning: discovering stable solutions in the robot path finding domain", *Proc. 9th European Conference on Artificial Intelligence (ECAI-90)*, pp. 219-221, Stockholm, 1990.

[**Sanner 91a**] Sanner, R. M., and Slotine, J.-J. E., "Direct Adaptive Control with Gaussian Networks," *NSL Report*, No. 910303, March 1991.

[**Sanner 91b**] Sanner, R. M., and Slotine, J.-J. E., "Gaussian Networks for Direct Adaptive Control", *NSL Report*, No. 910503, May 1991.

[**Sanner 91c**] Sanner, R. M., and Slotine, J.-J. E., "Stable Adaptive Control and Recursive Identification Using Radial Gaussian Networks", *NSL Report*, No. 910901, Sept. 1991.

[**Torras 89**] Torras, C., "Relaxation and neural learning: points of convergence and divergence", *Journal of Parallel and Distributed Computing*, Vol. 6, pp. 217-244, 1989.

PART II

CONTROL ENGINEERING

Road Map.

The second part of the course deals with *Control Engineering*. Basic knowledge on control engineering, like state-space representations, bode diagrams, stability, etc. is expected from the student and will not be discussed in this book. The text shows in what way control systems have evolved over the years into the current state-of-the-art. Then the dead-ends in control will be shown, including a first insight in where and how artificial intelligence can help solving the problems encountered.

There are two chapters:

Computer Control Systems: An Introduction. Here a survey is given of control and it is shown how control engineering has evolved over the years. From this, the step is made to distributed control systems. The chapters concludes with an indication on new directions for control.

Towards Intelligent Control: Integration of AI in Control. This chapter sets the scope for the main topic of this book Intelligent Control. Here an introduction is given, starting from an analysis of the needs for future control systems. The main subjects expert systems, neural networks and fuzzy control are already briefly introduced.

This road map should enable the reader to make up his own choice in this basic course, depending on his skills and future use.

COMPUTER CONTROL SYSTEMS: AN INTRODUCTION

M.G. RODD* and G.J. SUSKI**

*University of Swansea, Wales, Singleton Park, Swansea, Wales, SA2 8PP
**Lawrence Livermore National Laboratories, Livermore, California, USA

1. ORIGINS OF FEEDBACK CONTROL

As suggested by Otto Mayr (Mayr, 1970) "Every animal is a self-regulating system owing its existence, its stability and most of its behaviour to feedback controls. Considering the universality of this process and the fact that the operation of feedback can be seen in a great variety of phenomena, from the population cycles of predatory animals to the ups and downs of the stock market, it seems curious that theoretical study of the concept of feedback control came so late in the development of science and technology. The term "feed-back" itself is a recent invention, coined by pioneers in radio around the beginning of this century. And the exploration of the implications of this principle is still younger: it received its main impetus from the work of the late Norbert Wiener and his colleagues in the 1940s."

As defined by the American Institute of Electrical Engineers in 1951 "A Feedback Control System is a control system which tends to maintain a prescribed relationship of one system variable to another by comparing functions of these variables and using the difference as a means of control".

The purpose of such a system is to carry out a command automatically, and it functions by maintaining the *controlled variable* (the output signal) at the same level as the *command variable* (the input) in spite of interference by any unpredictable disturbance. The command signal may be either constant, as it is in the case of the temperature setting on a thermostat, or continuously variable, as it is in the case of the steering wheel positions in the power-steering system of an automobile. In all cases, if the feedback control system is to function effectively, it must be so designed that the controlled variable follows the command signal with the utmost fidelity (Mayr, 1970).

One of the first examples of a feedback control system was a water-clock in the 3rd century, produced by a Greek inventor, Ktesibios, working in Alexandria. The simplified block diagram for this is shown in figure 1(a), and this embodies the fundamental components of any typical control system. Figure 1(b) shows the classic, generic block diagram of such a system.

205

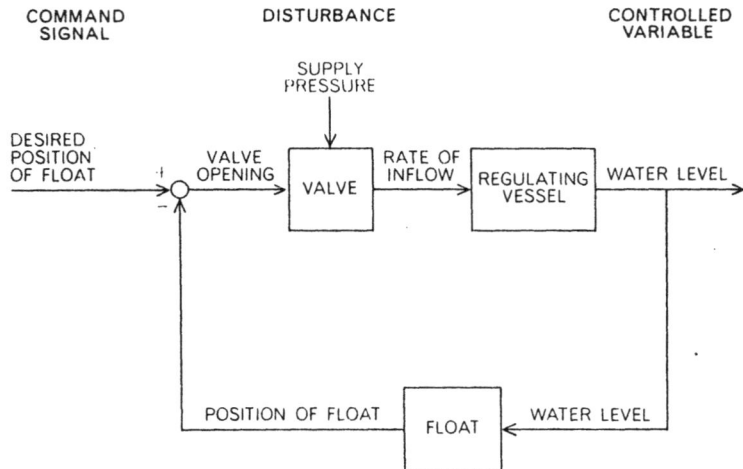

Figure 1(a): The feedback control loop of a 3rd century water-clock

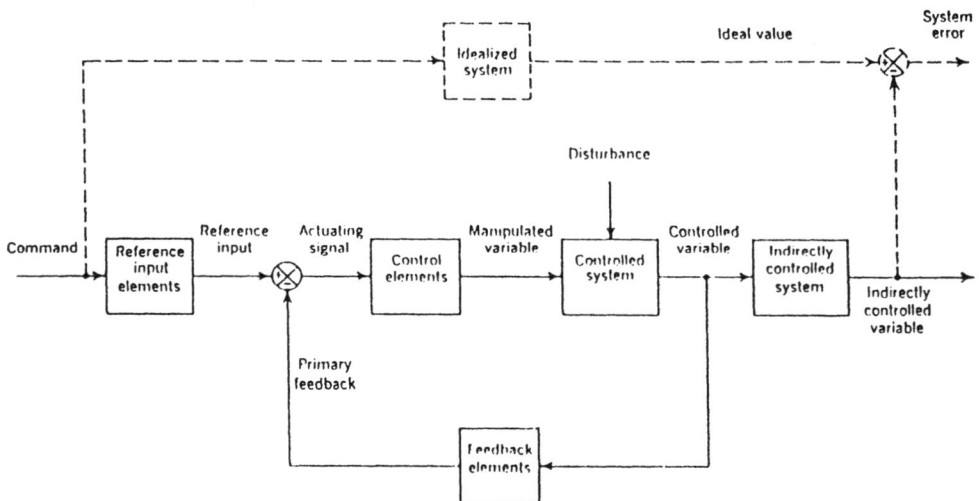

Figure 1(b): The generic block diagram of a control system

From such a system flows the vast array of control theory - the bulk of which is aimed at producing control strategies which provide optimal functioning of the system in any given situation. In reality, though, two prime practical characteristics must be considered.

(i)   Most processes being controlled are not well-defined mathematically - and, in many cases, have operations which are non-linear.

206

(ii)   Almost all but the most simplest processes involve the control of more than one variable. And again, almost without fail, the process variables interact. Change one variable, pressure say, and another variable, temperature say, also is affected.

As a result of these, and other problems to be discussed in this chapter, moving from the control of a single, simple, well-behaved loop, to a real-process or plant, involves a major reconsideration of theory and practical implementation. Of importance, too, is to remember that, if we take Mayr's thesis that control and regulation has a *universal objective*, then it has to be seen that control of a single variable, as important as it is, must be viewed in the context of the overall goal of the application. In a chemical process this is not just to keep a single vessel at a given temperature, but to produce, say, a chemical product as optimal as possible - not just optimal in terms of being the correct product, but as efficiently and as safely as possible.

2. TRENDS IN CONTROL

Few can doubt that we have come a tremendous way in the last fifty years - from the earliest beginnings of control theory to the point where we are now able to automate a wide range of real, industrial processes (Larson, 1980; Williams, 1990). Somewhat unfortunately, we have tended to split our automation activities into two sectors. The first group includes those relating to the so-called 'discrete' control world (as typically met in manufacturing), where control systems have largely tended to be based on relay-switching models - of course, now implemented by Programmable Logic Controllers (PLCs). Here we are essentially stopping, starting, waiting for, or monitoring a sequence of discrete events. In the second case, the continuous control world, we deal with processes which are time-varying and in which we have seen a shift from analog-based control loops to computer-implemented, but largely equivalent, forms of controllers.

If we step back from these two forms of control, we realise fairly quickly that not only do most practical situations involve a mixture of the two (e.g in many cases PLCs are used to start batch processes which, once initiated, fall under the control of PID controllers) but, in fact, the two technologies have much in common and we can argue that it is simply a question of different sampling rates!

Whichever view one takes, it is clear that we are now developing very complex systems, and indeed can justifiably claim that we have pushed automation technology to a point where we are extremely successful in cutting production costs and putting our processes under secure, robust control, which financially justifies the investments made. On this basis one can argue that we have gone, maybe, 90% of the way down the road towards achieving very satisfactory control.

However, the competitive world in which we now find ourselves demands that we go even further. The point here is simply that when one company has access to advanced control technology, all its competitors probably have the same access! Indeed, in most cases, having developed the automation technology for our own companies, we quite happily sell (or, in some cases, give) that technology away so that our competitors can use it! It is important, though, to remember that to really exploit the advantages of a new technology, users must have a sound, deep, hands-on understanding of it. Many of the claimed advantages are never balanced against the disadvantages, and as a result, many installations of the latest, bought-in technologies suffer from a lack of actual expertise by the on-site engineers. However, it is clear that a wide range of new automation techniques is now available to us, and the rush is on to use them.

As a result, the thrust in control has to be towards squeezing out that last 10% of possible maximum performance. Doing this, however, is not just a case of finding a slightly more efficient control algorithm - we have already done that. We need to take a complete overview of our overall manufacturing processes and our economic objectives in order to see how we can find these last few elusive performance improvements. Of course, this strategy will lie not only in the control philosophy adopted, but also in how we design the actual process and choose the components to implement it. It might well be that the introduction of a new, 'controllable' unit process could provide significant performance gains, as compared to the installation of a totally new control system!

Whether we are looking at the manufacturing or the process control situation, a few characteristics of the directions in which we should go are clear. Pointers to these come from those 'gurus' of early production engineering who were simply stating the obvious - that there is no point in manufacturing anything unless we have a market for it. But, once a market is there, we must manufacture 'it' as soon as possible (or else someone else will). In achieving these goals we have to produce the article in the most efficient way - not incurring any unnecessary loss during any stage of the production process - whether by sheer loss of return on investment in capital equipment, loss through inventory standing around idle, energy loss by a piece of metal cooling down and subsequently having to be reheated, or loss through unnecessary physical effort in moving a component from one part of the factory to another. A simple Just-in-Time philosophy must underlie all of automation. We need to go one stage further and say that before we even begin to produce anything, we need to *make certain* that we can do it - as efficiently as possible and so that we make a profit. Likewise, when we take on a new order, we must ensure that it does not adversely impact our current operations.

This all implies two things - firstly, that we cannot view any automation only at the localised level; automation requires total plant-wide control (or supervision) - not only of machines but also of human activities. Secondly, in order to predict *how* our system will perform under certain circumstances (since it is vital for us to be able to assess whether we should undertake a certain task), we must be in a position to model our production and control processes as accurately as possible.

We should realise, however, that desirable as it may seem to *totally* model a complete, complex process, this will be virtually impossible. Indeed, in many cases, it might not even be necessary. The degree of completeness required of the model will depend on a variety of factors - most of which relate to the ultimately desired control strategy. Also, we can be horribly misled in trying to model a process which is designed in a fundamentally wrong way - in terms either of its controllability, or considerations of reliability.

The point here is that now, when we are automating any process, we are no longer concerned only with the local situation, but have to track the total costs (resources, effort and energy) - right from the earliest stage through to the point at which the final product reaches the customer.

This is leading us inevitably towards highly integrated, enterprise-wide control - with the stress on *integration*. Just as we learnt, some twenty years ago, that controlling a series of single, isolated loops was inadequate because of the interaction between loops (and hence moved rapidly towards multi-variable control), we now find that, regardless of what the plant is, we must move towards integrating control of all components which make up the total process.

From this we can distil two important trends - firstly, a shift towards the total integration required for complete automation and secondly, in association with this aim, looking at each local controller (or group of controllers) and attempting to squeeze maximum performance as part of an integrated, co-operative system.

There are other influences which are now at work, and which must be considered. Not only are we driven by the final price of an article which we produce but, as our countries advance sociologically, there are escalating demands for improved safety and environmental considerations and, from our workers, a demand for better working conditions.

We are undoubtedly moving towards the next level of control - control which completely integrates the process - horizontally from the earliest raw-materials handling stage through to delivering a product to a customer, and vertically down from management right to the lowliest worker. At each level we require better and more integrated control.

In examining the nature of the various component parts which require to be integrated, we quickly discover that we have a range of elementary considerations which no one unifying, mathematically-provable, theory can cope with. Some relevant properties which the processes exhibit may include (Rodd, 1991) the following:

- *Inherent instability*
  The total process, or possibly many components of the process, might be dynamically unstable, in that not changing controllable conditions can result in deteriorating performance. In other processes, though, such instability is actually desirable - say, for example, in the open-loop control of a military aircraft!

- *Mixed continuous/batch operations*
  Whilst many processes have to operate continuously for many months, other processes have to operate in a batch mode. When the batch is changed, the operating parameters could well also have to be changed to meet the demands of a new product.

- *Incomplete/excessive data*
  Increasingly we have to deal with the fact that much observed data is unreliable, noisy and incomplete. In other cases, though, we find that too much data is captured and we have to select which data we are going to use.

- *Unidentifiable processes*
  Most processes which we have chosen to automate, (and have done so very successfully), have tended to be those which we can mathematically identify fairly accurately. There still remains a whole host of processes which we simply cannot identify.

- *Temporal problems*

  When we were dealing with simple, single-loop control systems, we could guarantee that the data we sensed from, and subsequently sent back to, our processes would be transmitted in time to meet the characteristics of the processes they were designed to support. Now, however, we are moving data over communication systems that are inherently non-deterministic, together with algorithms running on computers which are, again, non-deterministic. As we move towards bus-based systems (for example, at the lowest shop-floor, or fieldbus, level), these problems will be exacerbated. Also, as we strive towards integrated control, we need to synchronise activities, not just at a local level but right across the plant. Thus, for example, a hot ingot emerging from a furnace should not have to wait around before being processed - unless, during that time, the amount of cooling can be controlled so that the temperature, on entering the next stage, is correct, and no forced reheating (or cooling) is required. On a manufacturing line, robots need to be synchronised to guided vehicles and conveyor belts. Whilst this can be done at the local level, global control offers greater flexibility.

- *Operator interaction*

  Despite all the 'hype' of the 'workerless factory', and despite a few isolated examples of this, we have learnt (to our cost) that operators are, in most cases, essential ingredients in our automation systems. At the same time we have recognised that, given the complexity of our processes, operators must be supported by tools which allow them to act as efficiently and as appropriately as possible. We have seen, only too often, accidents which could have been prevented had operators taken the correct action.

It is useful also to dwell on the question raised above relating to the problem we have in many cases, of not being able to describe processes mathematically. Perhaps because of most control engineers' mathematical background, and an over-emphasis on theory, we feel we should always model a process mathematically before we can design a controller. We often go further and feel that unless we can derive a linearised model, from which we can minimise, say, a quadratic performance index, we cannot achieve any reasonable level of control! In practice, we must rid ourselves of these ideas, and look for a blend of mathematically-based tools (where appropriate) and some form of reasoning which is based, somehow, on common-sense - the common-sense which allows pilots, as unskilled aerodynamicists, to fly 747s with 500 people on board. After all, we do not require a degree in mechanical engineering in order to drive a car!

In viewing these characteristics against the goal of squeezing out the last possible percentage of possible maximum performance, it is evident that the control engineer must necessarily search for new integrating tools. The bulk of this text is aimed at providing the theoretical and implementational aspects of a range of such tools. The rest of this chapter, however, will investigate the ways in which such tools are integrated in practice. Chapter C.1 will continue with these issues, specifically looking at communication issues.

3. TOWARD DISTRIBUTED COMPUTER CONTROL SYSTEMS

Historically, the most prominent early uses of computer control were in the chemical process control industry beginning in the late 1950s (Astrom, 1985 and Lane, 1962). These early applications were at installations operated by Texaco (aided by TRW), American Oil, and Standard Oil (Stout, 1959).

In these first uses of digital computer control, a single computer was used in a supervisory role to calculate, and in some cases control, set points on analog controllers. The analog controllers, however, remained responsible for the primary control functions. The data available from sensors, oriented towards predetermined control schemes, limited the information available for calculations. Although digital computers were typically not in direct on-line control of processes (due in part to the poor reliability of the computers), a major reason for introducing computers in such systems was the need to improve the yield of processes through more timely adjustment of control valves and switches in response to changing process conditions and product specifications. Early computer control systems, however, suffered from high cost and poor reliability: their cost could be justified only by allocating to them many sensors, actuators and functions. (Astrom, 1985, Suski, 1988). While such factors limited the number of applications areas, the presence of a degree of machine "intelligence" in the control system justified their use. These early successes provided the impetus for the later development of lower-cost systems, better suited to process control.

Early 1962 brought the first uses of digital computers for direct process control in which analog control electronics were increasingly bypassed. One can define such systems as 1st generation computer control systems, see figure 2. A key demonstration of the technology was accomplished by Imperial Chemical Industries, Ltd. in England, using a Ferranti Argus computer. It was now easier to add interaction between control loops and, during this period, the first digital-control-oriented programming languages were developed. Progress remained limited, however, by the unavailability of inexpensive, reliable computers, until the development of the low-cost minicomputer changed the situation.

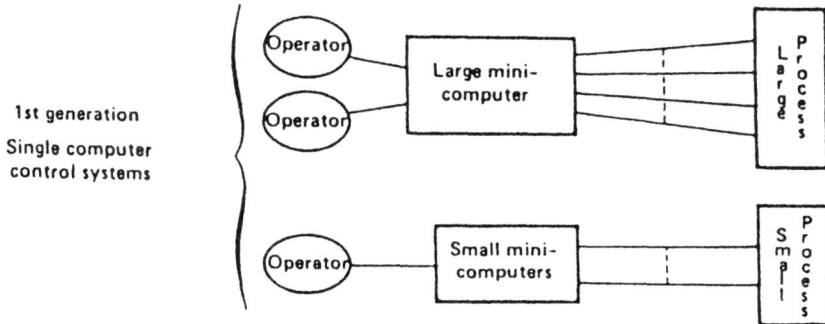

Figure 2: 1st generation, single computer control systems

3.1. Multicomputer control systems

As computers became smaller and less expensive during the late 1960s, applications expanded to include progressively smaller processes. Inexpensive minicomputer-based control systems were made possible through the development of low-cost solid-state electronics, particularly integrated circuit technology. This led to the economic feasibility of using digital computer control in simpler processes.

More importantly, this development made it possible to move the computer into educational institutions and their laboratories. By the early 1970s, a growing pool of graduating engineers was entering the industrial and scientific communities, with basic skills applicable to digital-computer-based process control.

It is estimated that in the period from 1970 to 1975, there was an order-of-magnitude increase (to approximately 50,000) in the number of process computers in use. As microcomputer technology became established, the cost and size of process control computers decreased, while their reliability improved substantially. The development of more-sophisticated languages and operating systems for minicomputers hastened their acceptance, particularly by the end-users who used the flexibility introduced by higher-level software to perform more sophisticated analysis and control of processes. The development of minicomputer networks in the latter 1960s and early 1970s provided the final impetus needed for the emergence of distributed computer control systems (DCCS).

This transition from first- to second-generation distributed control systems was characterised by their hierarchical nature (Fig. 3). Such systems were first marketed commercially in 1975, by Honeywell. In the second generation, the concept of local versus central computer control emerged. Smaller computers were located near the devices they controlled or monitored, often providing a local control panel with basic functionality for maintenance purposes. They were interconnected, usually, via serial RS-232C communication links at speeds up to 9600 bits/second, to one or more central computers which generally had greater processing capability, and more memory and disc storage, as well as peripherals such as printers, and a high-functionality operator interfaces.

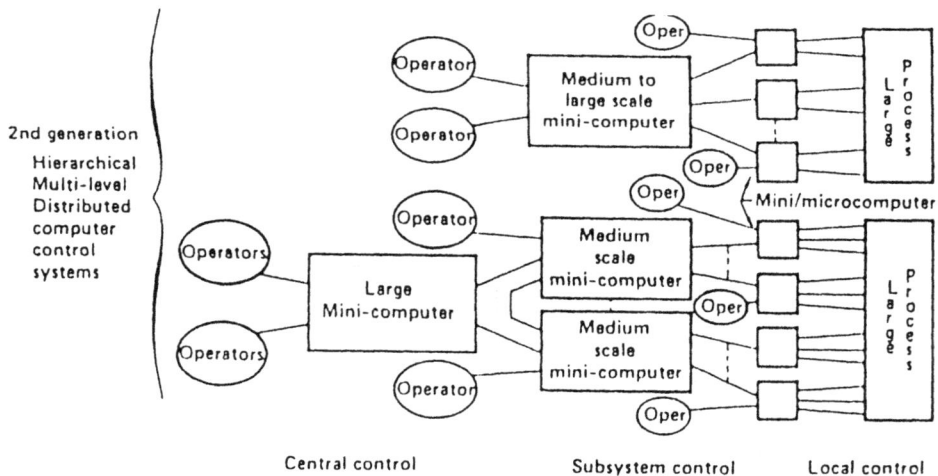

2nd generation

Hierarchical Multi-level Distributed computer control systems

Medium to large scale mini-computer

Large Mini-computer

Medium scale mini-computer

Medium scale mini-computer

Large Process

Mini/microcomputer

Central control   Subsystem control   Local control

Figure 3: 2nd generation hierarchical distributed computer control systems

These systems used graphics to display system status, supported operator commands entered through keyboards, and often included graphical input devices such as light pens. The

212

central computers performed analysis in addition to control, and also implemented feedback loops that spanned across interfaces connected to different local computers. The size of the process, or requirements of functionality, frequently led to additional levels of more-powerful computers. These were in turn connected by point-to-point network connections to the middle-level computers. With multiple levels of computers in such a system, a hierarchy of computers and functions emerged. Accordingly, these became known as *hierarchical distributed computer control systems.*

The majority of the distributed computer control systems in use today employ second-generation architectures. Figure 4 illustrates a detailed model of these systems and the environment in which they must operate. Figure 5 provides a simplified model of such a system, viewed primarily from the intercomputer communication point-of-view.

At the top of this model appears the overall organisational structure. It is simplified here to show a layer of top management which is responsible for the overall strategy of operations. If this is a manufacturing organisation, this level may be responsible for decisions as to the type and quantity of items to produce. It deals with long-term decision-making and can employ computer-based information, resource allocation, and decision systems of varying complexity.

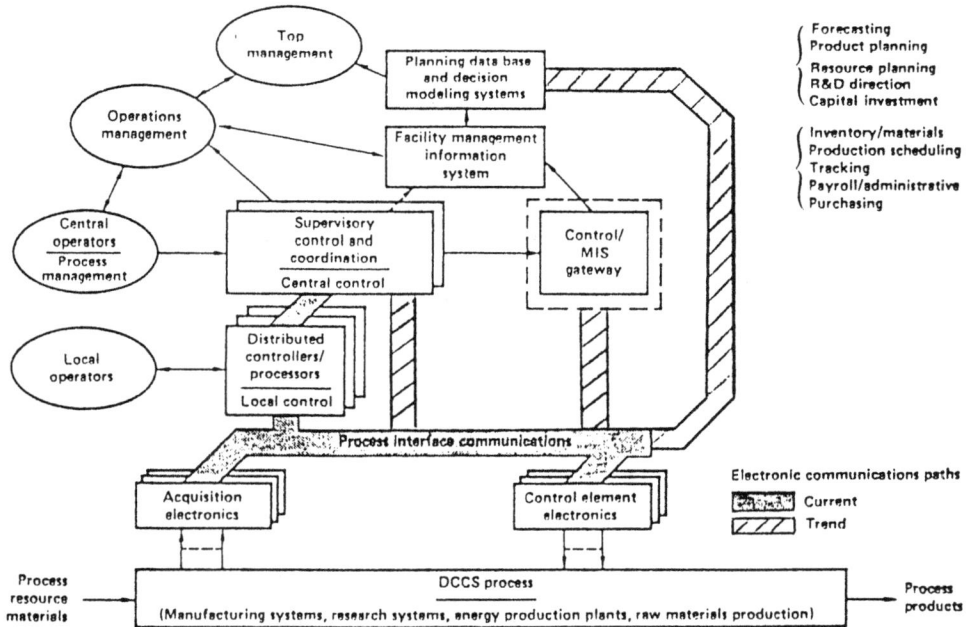

Figure 4: A model of the functioning of a distributed computer control system

213

The next level manages the daily operations of the facility which may have more than one process active. Management information system (MIS) aids may also be used at this level. Process status information in the MIS may require frequent updates to allow effective process supervision and reporting. The process under control appears in Fig. 4 at the lowest level. It is characterised by a set of inputs (resources) and outputs (products) and is monitored and controlled (via a set of sensors, actuators, and control elements) by one or more operators according to requirements established at the level of daily operations management.

Requirements for decision responsiveness increase from the top management level down to the process level. At the upper levels, decision time frames of days, weeks, and even years are typical. Conversely, the process may require response times varying from minutes to submilliseconds. The connection between management and the process is, of course, primarily through the communication mechanisms.

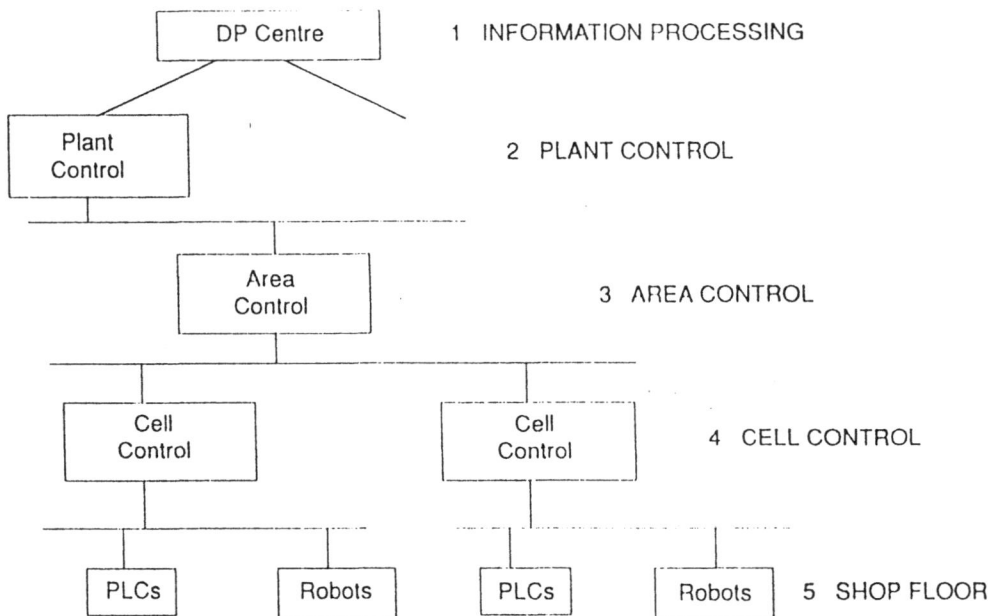

Figure 5: A simplified hierarchical model of a 2nd generation system

To accommodate a range of requirements the DCCS is, generally, hierarchical in its physical distribution of functionality. Electronic interfaces are used to allow the connection of small distributed computers (or controllers) to sensors and control elements. The distributed control computers provide local control functions and, often, local control stations. They may also process local feedback loops. They must be capable of handling information

214

transfer, not only with the process devices but also with one or more other computer(s) in the DCCS. Computer-to-computer information transfers include command, status, and often synchronisation information.

Other computers in the DCCS are responsible for higher-level functions including supervision, co-ordination of processes, feedback loops spanning across distributed processors, and computationally-intensive tasks. Computers at this level are also responsible for performing centralised control tasks and for providing the primary operator interfaces.

The characteristics of the communications media between devices and computers, and the degree to which process data can be transferred directly between management information systems and the process control system, are important issues. The latter is particularly important in the manufacturing community. These issues will be introduced later during the discussion on characteristics of third-generation architectures, and covered in detail in chapter C.1.

There are several reasons why distributed computer architectures are well-suited to process control. For large processes, they allow successive decomposition of the process control problem into manageable elements. One of the principal problems with single-computer control of large processes is the complexity of functions which must be co-resident in the central computer. The need to handle real-time process control activity in the same computer that performs analysis and computation represents a performance bottleneck. It also presents a significant software development problem, requiring great development control and co-ordination, and difficult testing of software.

There are economic advantages in DCCS hardware costs. During the period of their initial acceptance, there was an advantage that a relatively low-cost investment could result in a proof-of-principle prototype. Of even greater significance, though, is the point that it is generally less costly to distribute the intelligence in smaller computers than to centralise intelligence in high-performance computers.

Another major benefit of a distributed system is its almost inherent characteristic of fault isolation. A failure in one portion of a distributed control system can normally allow the remainder of the control system to continue to operate, albeit at some lower level of performance. In addition, the operational portion of the system can assist in reporting and diagnosing the fault. Other significant benefits include the availability of distributed control stations (operability) and the ability to add control functionality without major system redesign (extensibility).

4. DESIGN ISSUES OF DISTRIBUTED COMPUTER CONTROL SYSTEMS

DCCS design starts with traditional engineering practice: the major elements include specification, design, optimisation, implementation, testing and maintenance (including extension). These elements are strongly interdependent. For example, selection of the hardware and software technology to be used in a DCCS has a strong impact on the corresponding methods of implementation, testing, and maintenance.

DCCS design begins with a preliminary definition of the system criteria, including the technical requirements, costs and schedules. Technical requirements encompass safety, functionality, performance, reliability, operability and extensibility. Requirements definition is now normally an iterative procedure, in which the design of the process is itself influenced by the capabilities of the control system! For a large, long-term project, much of the control system functionality may not be precisely defined. Because extensions can be accommodated without fundamental changes to the remainder of the system, DCCS technology is well-suited to such applications.

The performance of the control system is usually a major criterion. In process control systems, the rate at which the system is sampled and controlled should be at least twice the rate at which significant changes can occur (Nyquist criterion). In this context, a control system which functions at least at that rate will be said to be operating in *real-time*. In order to ensure that a given performance level will be achieved, control systems are increasingly simulated or modelled during the design process.

Another key criterion is reliability. This includes a wide range of issues including the fault-tolerance of the system (Neumann, 1986) (e.g., degree of fault isolation, response to software errors), its long-term maintainability, method of fault recovery (e.g., partial versus total system restart), and the need for system diagnostics for on- or off- line operation.

In the absence of additional constraints, DCCS designers must select from an enormous array of currently-available technology. These options are categorised here into eight groups (Fig. 6).

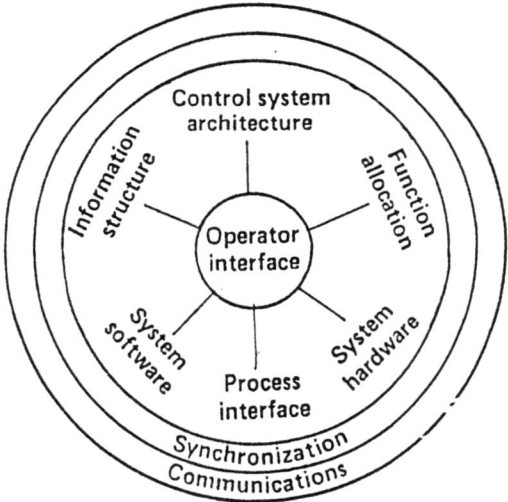

Figure 6: Considerations in the design of DCCSs

Communications. The collection of hardware and software which comprises the communications scheme linking the various electronics systems, forms the backbone of any distributed computer control system. It has a fundamental impact on all other design issues and on the ultimate ability of the system to meet its design criteria. Selection of the communications system is also critical since it is one of the most difficult DCCS elements to replace or extend functionally, once it is in use.

There has been increasing acceptance in recent years of the seven-level Open Systems Interconnect (OSI) reference model (Figure 7) for interprocessor communications, developed by the International Organisation for Standardisation (ISO) in Geneva. This model has provided a framework for the development of communications standards. These can be implemented in either hardware or software, depending on their complexity. An important contribution of this model is the establishment of a universally-understood terminology for describing communications functions. It must be mentioned, however, that there is much debate as to whether such open systems can ever meet the requirements of real-time distributed systems.

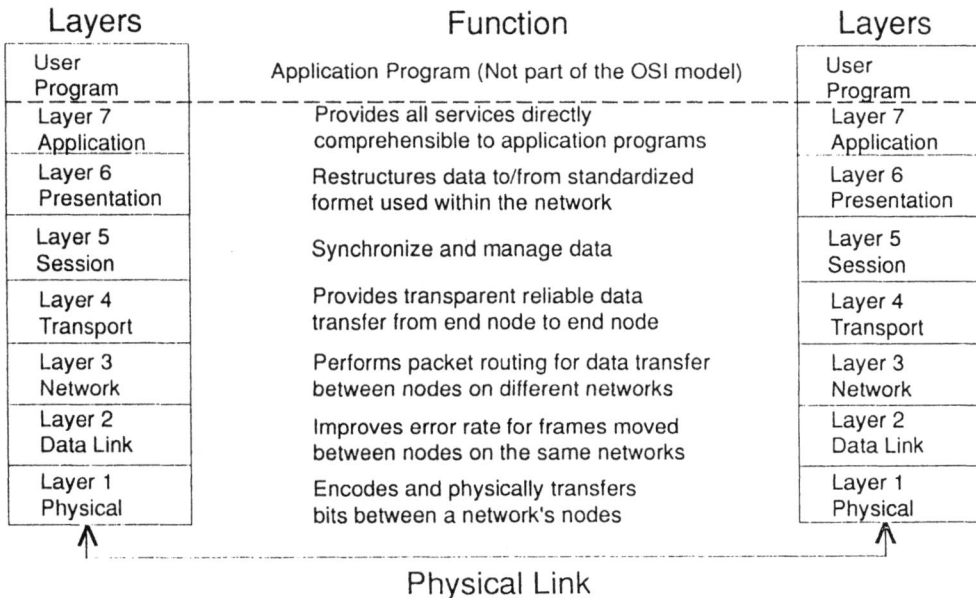

Layers	Function	Layers
User Program	Application Program (Not part of the OSI model)	User Program
Layer 7 Application	Provides all services directly comprehensible to application programs	Layer 7 Application
Layer 6 Presentation	Restructures data to/from standardized formet used within the network	Layer 6 Presentation
Layer 5 Session	Synchronize and manage data	Layer 5 Session
Layer 4 Transport	Provides transparent reliable data transfer from end node to end node	Layer 4 Transport
Layer 3 Network	Performs packet routing for data transfer between nodes on different networks	Layer 3 Network
Layer 2 Data Link	Improves error rate for frames moved between nodes on the same networks	Layer 2 Data Link
Layer 1 Physical	Encodes and physically transfers bits between a network's nodes	Layer 1 Physical

Physical Link

Figure 7: The 7-layer OSI communications reference model

The DCCS designer must therefore decide whether to opt for a communications system which conforms to available standards, proposed standards, or perhaps de-facto standards (such as commercial LANs from the office-automation industry). The choice will, clearly, affect the final cost and long-term extensibility of the system. The designer must also decide whether to use separate communications systems for communication at low levels (e.g., to device controllers) and at high levels (between computers). Chapter C1 will investigate these aspects in greater detail.

The designer must also choose the medium for data transfer (e.g., wire or fibre-optics); whether data is to be transferred serially or in parallel; and whether communications system access is to be by demand, controlled by a master, or given via token-passing. In an issue related to architecture, a fundamental decision as to bus versus point-to-point communications is required. The use of separate versus a common communications scheme for devices and processors is also important.

Architecture. The architecture of a DCCS is based upon the communications system characteristics and the required process-control functionality. Decisions as to whether a multi-level or a uni-level system are to be used depend more on the characteristics of the communications system than on the process. However, the topology of interconnection among the hierarchy of computers will greatly affect functionality. Network topologies can include node-to-node, ring, bus-oriented, star, cube, etc. The issues of which functions are to be loosely- versus tightly-coupled impact performance and fault isolation characteristics. Another architectural consideration is the aspect of how real-time state information is to be acquired. Are devices polled, are they accessed through interrupts, or are they represented by state information which can be accessed when required?

Process Interface. The nature of the process interface must be understood and specified. The adoption of a standard can substantially reduce the software and hardware development cost, but may restrict the availability to the process system designer of control elements and sensors. Device characteristics and interface requirements must be identified in advance, as far as possible. For example, a device which generates a need for attention every few milliseconds can seriously degrade the performance of certain interrupt-driven architectures. Again, the design of the communications system is highly-dependent on the process interface structure. A highly-intelligent interface can clearly reduce the amount of traffic on the transmission system.

Function Allocation. The allocation of functions within the control system directly impacts its performance and often its fault-tolerance. Within a multiprocessor system, handling of a function may be centralised or distributed. The allocation of functions can be done statically during implementation, pseudo-dynamically during system start-up, or dynamically during system operation, according to process activity. True dynamic allocation is much more difficult to implement, but has the advantage of allowing better use of system resources. Simulation and complex analysis of statically-allocated systems is often required to ensure that performance criteria are met.

Hardware. Major decisions are, naturally, made during the communications and architectural design of the system. However, within any given architecture, several hardware decisions must still be addressed. There is the issue of whether to standardise on a particular computer manufacturer or interface standard. This may involve balancing the cost of long-term system maintenance against flexibility in optimising performance and functionality. The decision to standardise is important, but the choice of level at which to standardise is critical. Open architectures, which specify performance levels but not the methods of achieving them are often preferred to closed architectures at the initial, specification phases of large projects.

218

Software. Software for DCCS applications can be complex, and accounts for an increasingly-greater portion of the cost of such systems. Initially, the DCCS designer must choose programming languages, operating systems, database managers, and communications packages. Where standards-based hardware subsystems have been adopted, this task is often eased by the corresponding availability of standard software. However, in general, software for specific applications must still be developed, whether through traditional programming techniques or by using more-advanced program-writing systems. A specific software project management methodology must be identified and set in motion. Quality assurance of software through the specification, design, implementation, testing and maintenance phases is a recognised necessity today in the development of any large system. The DCCS designer must generally also accommodate requests for high-level functionality which may be provided by pre-packed software such as decision-support systems and resource-scheduling systems.

Data. The organisation of data within a distributed computer control system is as important as the selection of the communications system. Designers must identify the types of data required (e.g., state information, command logs, device set points) in each node of the system, how and when this data is acquired, its structure, whether it is of local or global interest, and what type of access controls are required. As the size of distributed systems increases and performance criteria are raised, decisions must be made as to what data should be in central, as opposed to distributed, databases. When data must appear concurrently in more than one distributed location, issues of data consistency (i.e., all data instances having same value) and validity (data must reflect the current system state if it is to be useful) must be addressed. The need for retaining "old" data must be very carefully examined, as it is common experience that large percentages of databases are filled with "write-only" information! As with most other areas of DCCS, there is no currently identified "preferred" scheme for dealing with this issue, although schemes using data tagged with time and age information appear viable. Journalling of significant events, transactions, and commands may also be required for maintenance, safety, or legal reasons. Chapter C1 will take these issues further.

Human Computer Interfaces. The most important contact between operations personnel and many processes is through the process control system. Safety and ergonomic factors predominate in the design of the primary operator interface, often taking the form of a control "console". The capacity of this device to communicate quickly and clearly with an operator is essential to good system operability, both real and perceived. Where a console or control panel is used infrequently (e.g., for maintenance), it is important that it provide the minimum required functionality and that its design does not lead to unnecessary operator errors. Where an operator interface is used more continuously, it is important that it be ergonomically designed.

Information displays, often based on colour graphics systems, must be clear and properly-oriented to minimise fatigue. Important data must be immediately accessible and the method of operator input should discourage improper system commands. Command entry must be simple and cross-checked by the system for validity. Often the control system designer involves operations personnel in the design of the operator console system. Much thought must be given to the question of alarm-handling. Traditional concepts of bombarding the operator with a mass of alarm indicators are being superseded by intelligent alarm systems, with built-in priority schemes, etc.

5. NEW DIRECTIONS AND EMERGING TRENDS IN IMPLEMENTATION

For fairly obvious reasons, DCCS designs have undergone rapid evolution since such systems first appeared. First, the sheer availability of new technology is a driving factor, but not necessarily just because of its interest value. In designing systems for long-term operation and extensibility, newer technology, particularly that based on standards, is more likely to be available throughout the life of the system, to aid in maintaining and extending it. In addition, such technology generally becomes available at comparatively lower cost, whilst older technology becomes obsolete and scarce. It is easier to recruit high-quality engineering talent if system designs are based on non-obsolete technology. The reduction in DCCS costs motivates the use of new technology with improved cost/performance ratios.

The second major factor in DCCS evolution relates to process-related requirements for new functionality. Pressures are increasing for improved productivity, reduced waste, lower operational manpower requirements and greater reliability. To meet these criteria, DCCS designer are required to put more functionality into the control system. Readily-available standard hardware and software for common functions may be adopted in order to provide the time, budget, and resources needed to develop new functionality. In many cases, the need for more rapid response to changing process conditions leads to higher performance requirements in DCCS implementations. This motivates the use of higher-powered computers, particularly now in the form of workstations and RISC-based architectures, and improved communications facilities.

The third major factor affecting new DCCS designs results from issues in the non-technical arena, related to economics, education, sociology, ethics, and political influences. Environmental issues are becoming ever-pressing and safety of paramount importance.

Returning to the first point, emerging developments in communications in particular will have major impact on the systems of the future. The most significant of these developments is the emergence of bus-oriented interconnection techniques in which, potentially, all the computers involved and, in many cases, the process interfaces themselves, may use common physical communications media. Whilst the use of a single media would be questionable from aspects of reliability, it is very evident that significant sections of processes and plants will be able to (and in many cases, already do) share common communication systems - typically arranged in an hierarchical fashion (see figure 5). This implies that any computer in a section can access any other computer or device in that section directly, without going through an intermediate processor or series of processors. The system performance is improved by eliminating much routing delays and allowing greater distribution of functions among processors. Extensibility is also improved, since adding more nodes or extra computational power is simply accomplished through additional connections to the common bus. A key design factor, of course, will be the interconnection of the various sectors.

The emergence of communication standards will ease the task of interconnecting between different manufacturers' products. MAP, for example, (Rodd, 1989) suggests a system architecture in which a large number of nodes use common communication buses. While bus-bandwidth considerations must still be addressed, such architectures point to a clear trend in today's systems. They are characteristic of third-generation computer control systems (Fig. 8). Another major characteristic of third-generation systems is their

220

interconnection into the management information systems (MIS) which are used by operations and higher-level management.

In summary, many emerging technologies are having major influences on DCCS designs. Advances in communications and VLSI-technology are leading to DCCS architectures characterised by highly-distributed intelligence, down to the process sensor level. Hierarchically-connected systems are being replaced by designs in which a hierarchy of process computers communicate over one or a limited number of bus-oriented communications system(s). Sophisticated techniques, such as image-processing, are becoming pervasive, particularly in CIM applications, whilst emerging low-cost PC and LAN technologies are filling the continued need for economical implementations of moderate-performance distributed control systems (Suski, 1988).

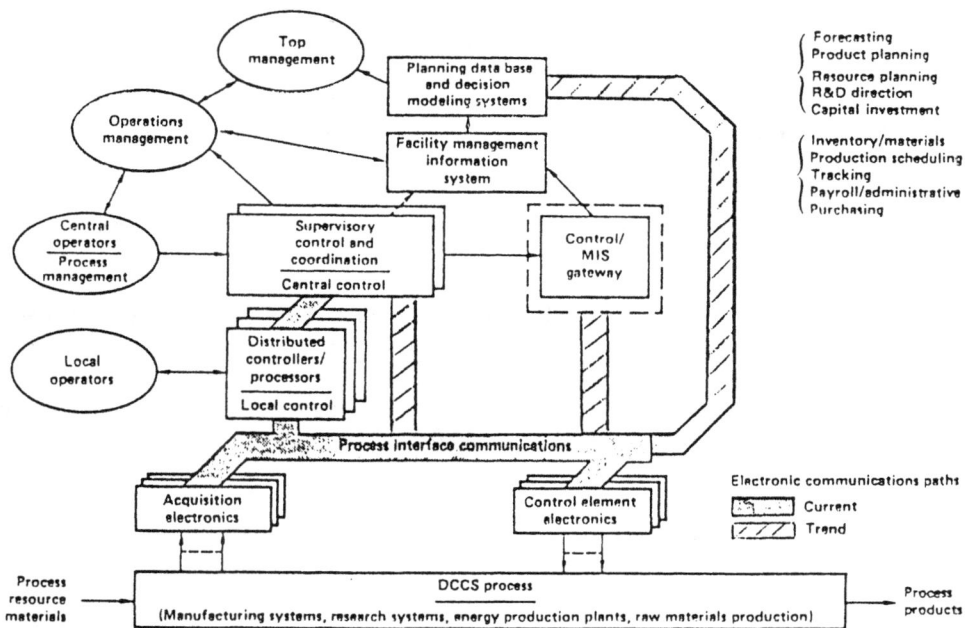

Figure 8: Towards 3rd generation distributed computer control systems

Fundamentally, an array of low-cost, rapidly becoming standardised hardware is available to the DCCS designer. The situation with software, however, is not so optimistic, and we are still in the early stages of seeing well-defined, easy-to-adapt software platforms and readily-available re-usable software. Real-time operating systems still lag at least 2 decades behind the hardware, and, truly distributed, real-time databases remain a twinkle in the eye of many researchers.

221

REFERENCES

Astrom, K.J., "Process Control - Past, Present and Future", IEEE Control Systems Magazine, August 1985, pp.3-9.

Lane, J.M., "Digital Computer Control of a Catalytic Reforming Unit", ISA Trans., Vol. 1, 1962, pp.291-296.

Stout, T.M., "Computer Control of Butane Isomerization", ISA Journal, Vol. 6, 1959, pp.98-103.

Suski, G. and Rodd, M.G., "Current and future issues in the design, analysis and implementation of DCCS", Proc. IFAC Workshop on DCCS, 1988, Pergamon Press, pp.1-14.

Mayr, O., "The origins of Feedback Control", Scientific American, October 1970, Vol. 223, No. 4, pp.110-118.

Neumann, P.G., "On Hierarchical Design of Computer Systems for Critical Applications", IEEE Transactions on Software Engineering, September 1986, Vol. SE-12, pp.61-70.

Rodd, M.G. and Verbruggen, H.B., "Expert systems in Advanced Control", Proc. American Control Conference, Purdue Univ., 1991.

Rodd, M.G. and Deravi, F., "Communication systems for Industrial Control", Prentice-Hall, 1989.

Williams, T.J. and Kompass, E.J., "Distributed Digital Industrial Control Systems: Accomplishments and dreams", Vol. 0., Proc. IFAC World Congress, Pergamon Press, 1990.

Larson, R.E. and Hall, W.E., "Control Technology Development during the 1st 25 Years of IFAC", Proc. IFAC 25 Years, Pergamon Press, 1980.

Towards Intelligent control: Integration of AI in Control

H.B. Verbruggen   A.J. Krijgsman   P.M. Bruijn

Delft University of Technology,
Department of Electrical Engineering, Control Laboratory,
P.O. Box 5031, 2600 GA Delft, the Netherlands.

1   Introduction to Intelligent control

There is a vast, growing interest in the application of Artificial Intelligence methods in all fields of engineering. Possible applications of AI methods in control engineering can be found in nearly every stage of the analysis of processes and the design and implementation of control algorithms ([Verbruggen 1991]). They range from pure advisory systems for the designer or operator (off-line applications) via diagnosis and supervisory systems (on-line applications), to in-line direct control applications. However, the real-time demands imposed by in-line control applications add an extra dimension to the problems arising when using AI methods in this field. This contribution concentrates on in-line and on-line (supervisory) control applications and highlights the differences between the methods described, called "intelligent methods", and the methods previously applied in solving control problems.

A number of names have been introduced to describe control systems in which control strategies based on AI techniques have been introduced: intelligent control, autonomous control, heuristic control, etc. Why are these systems denoted by the controversial name: "intelligent"? What is the difference between these methods and conventional or more advanced and sophisticated controllers which have been applied in control engineering applications in the past? It seems rather presumptuous to call these systems intelligent after many decades have been spent on developing sophisticated control algorithms based on solid theoretical frameworks, such as linear system theory, optimal and stochastic control systems and their extensions, such as adaptive and robust control systems.

There are many ways to solve problems depending on the amount of knowledge available, the kind of knowledge available, the theoretical background, experience, personal preference, etc. After a pioneering phase in which simple process models, operator's experience and controllers with predetermined structure dominated the scene, control engineering was dominated for decades by the developments of linear (optimal) control theory. The need for mathematical models of the processes to be controlled stimulated research in

223

fields such as system identification, parameter estimation and test signal generation. By formulating the requirements imposed on the controlled processes in a mathematical form (in terms of performance criteria or cost functions) it was possible to derive many control algorithms whose parameters are analytical functions of the process model and the parameters of the cost function. Thus, a closely coupled design procedure was developed which automatically generated the desired control algorithm (structure, parameters). Knowledge about the process and intelligence of the designer woulde be completely captured in the off-line design procedure.

In a case where the process cannot be described by linear models and/or the requirements are not translated to simple criteria and quadratic cost functions, no analytcal solution can be found, and the design is translated to a numerical optimisation problem. There is a free choice of the structure and the number of parameters of the control algorithm to be determined. This procedure can be called a "loosely coupled" design procedure. There are more degrees of freedom in the design phase, and experiments on simulated processes support the designer in choosing the right controller structure and parameters. In this case mathematical models are also required, but knowledge and experience of the designer are necessary in the final stage of the design procedure as well based on extensive simulations.

A very natural approach, gaining more impetus in control engineering applications is the so-called "model-based predictive control method", in which the design procedure is based on:

- the prediction of the process output over a certain prediction horizon, based on a model of the process or on previous in- and output signals

- a desired process output behaviour, often called the "reference trajectory", over the same prediction horizon

- the minimisation of a cost function, yielding a control signal (process input), satisfying constraints on magnitude as well as on rate of change of the control signal and other signals related to the process. The control signal is calculated over a so-called "control horizon" which is smaller then the prediction horizon.

- only the control signal for the next sampling instant is applied, after which the whole procedure is reiterated.

The complete uncoupling of the different steps allows the application of all kinds of process models, reference trajectories and cost functions, and allows the inclusion of constraints. Due to these properties the application of some AI techniques such as artificial neural nets fits this approach quite naturally.

In the previous approaches, reasoning about the process behaviour and the desired overall behaviour is based on the formalism of mathematical descriptions. The level of the power of abstraction is rather high. In case where the mathematical model of the process is not available or can only be obtained with great effort and cost, the control strategy should be based on a completely different approach. Instead of a mathematical, a behavorial description of the process is needed, based on qualitative expressions and experience of

224

people working with the process. Actions can be performed either as the result of evaluating rules (reasoning) or as unconscious actions based on presented process behaviour after a learning phase. Intelligence comes in as the capability to reason about facts and rules and to learn about presented behaviour. It opens up the possibility of applying the experience gathered by operators and process and control engineers. Uncertainty about the knowledge can be handled, as well as ignorance about the structure of the system.

Most interesting is the case in which during operation new facts and rules are discovered and implemented, and knowledge is adapted to changing circumstances.
We are talking about a completely different approach in which the attainments of control theory (stability analysis, observability, controlability, etc) do not apply. That is the reason that many people in the control theory community are not convinced of the possible applications of these methods. Most control engineers have a strong mathematical background and are biased to an over- emphasis on theory. They feel that a process should be modelled always mathematically before a controller can be designed. Often they feel that unless a linearised model can be derived from which a quadratic performance index can be minimized, a reasonale level of control cannot be achieved. Other people, confronted with difficult to describe or partly unknown processes, are willing to apply these methods, because experience based on real process behaviour can be implemented, regardless of whether the system is linear, non-linear, time-variant, etc. In those cases control theory can hardly provide them with adequate solutions.

In this contribution a motivation is given to show why the methods based on control theory developed thusfar are not able to handle all real-life problems, even when they are enhanced by adaptation of the controller, or when the design is based on robust control methods.
In the next section it will be demonstrated that in an increasing number of applications we are forced to introduce alternative methods to control these systems. The alternative methods, however, are quite common in the daily lives of human beings.

The following methods are introduced, with emphasis on real- time applications:

- Knowledge-based Systems (KBS), based on expert systems (ES). This is a symbolic reasoning approach, commonly used for advisory systems in which the knowledge of experts in a small field of expertise is made available to the user.

- Artificial Neural Networks (ANN), which are learning systems based on subsymbolic reasoning.

- Cerebellar Model Articulation Control (CMAC) methods which are quite simular to ANNs, and also learn the system behaviour from aplied inputs and measured outputs

- Fuzzy Control (FC), which is very well suited handling heuristic knowledge to control a system

In many cases a combination of methods is most convenient, leading for instance to Fuzzy Expert Systems and Neural-Net-controlled systems supervised by a knowledge-based system. Intelligent control components can be used in various ways. The following configurations are distinguished:

225

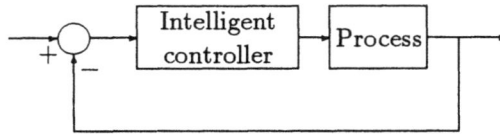

Figure 1: Direct intelligent control

- Direct control configurations (see Fig. 1), in which one of the above-mentioned intelligent control methods is included in the control loop. As in many conventional control configurations, the controller can be built up of different parts, i.e. a control and a process identification part.

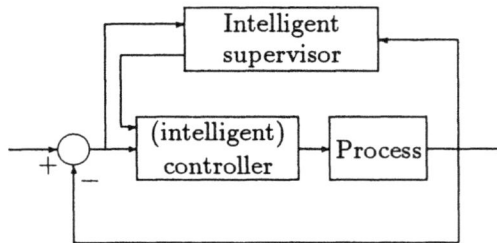

Figure 2: Indirect intelligent control

- Supervisory control configurations (see Fig. 2) in which the controller is either a conventional or an intelligent controller which is supervised by an intelligent component, for example a KBS.

Various examples of systems controlled by different intelligent components will be given. It will be shown that a combination of intelligent components is most usefull, because the methods support each other in many ways and are complementary in various aspects. However, some problems are still unsolved, for instance the stability problem in using the proposed intelligent methods.

Generally, control algorithms use a very fine quantization in order to approach the continuous case. This is most usefull when the system is well known and the requirements are very strict. However, why should such a fine quantization be used when there is much uncertainty and lack of information about the process to be controlled? Why not using a knowledge-based model with a rough quantization? The rough quantization can easily be translated into symbolic variables and offers an opportunity to incorporate other symbolic knowledge in the controller(for instance, from operators). Besides, accurate control is not always necessary, but is only required in the neighbourhood of the desired behaviour. We are no longer bound to fixed controller strategies for all circumstances in which the process occurs.

What is interesting is that a more interesting "landscape of control" will come into being. How do we define the landscape of control? This landscape can be depicted only for very simple algoritms, but it illustrates very well the many additional possibilities introduced by intelligent controllers.

226

Assume a control strategy which produces a control output u as a function of e (the error signal between desired and measured process output). In a classical proportional controller every u is calculated with a certainty of 100 %, given a measured or calculated value e. With a very fine quantization a thin wall illustrating the relationship 1 (elswhere

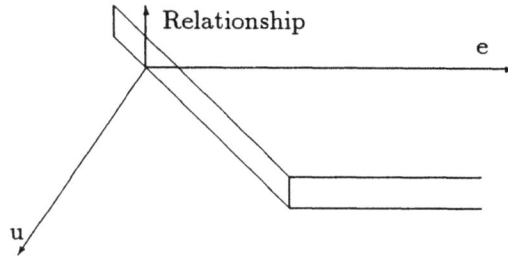

Figure 3: Landscape of control: classical proportional

the relationship is zero) between e and u is given (see Fig. 3). Using a nonlinear relation between e and u, a curved thin wall will be generated. By the introduction of a KBS, the influence of quantization and precision becomes visible. The relationship between e and u

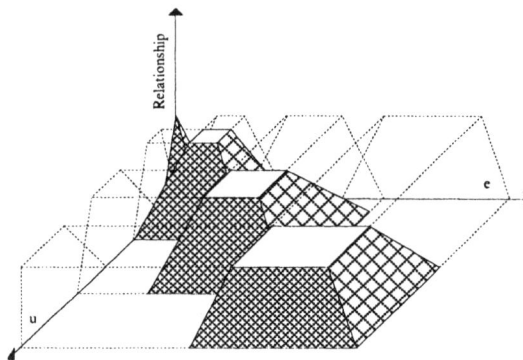

Figure 4: Landscape of control: fuzzy rule-based

is illustrated by a number of linked blocks with different bases but with the same height. Introducing uncertainty by fuzzy sets yields a more interesting landscape: pyramids or topped pyramids which are partly overlapping dominate the scene (see figure 4. Because a number of rules are fired at the same time with different influence on the output signal a nearly continuously changing u-signal is obtained despite a roughly quantized e-signal. The values of the u-signal depends on the defuzzification method used, the shape of the pyramids and the logic being used. More interesting landscapes are created when for different areas of the e, u-plane different KBS are introduced, or when the relationship is assumed to be uncertain and depending on gained knowledge. In that case the landscape will change dynamically: some structures in some areas will be emphasized and change their shape, while others will fade away, and new structures will even come into existence!

227

2   The need for Intelligent Control

In the past control theory forced the application of control engineering to adopt a Procrustean approach, where the application is made to conform to the requirements of the control system. In many cases real life is different, and so are the real life applications in the process and manufacturing industries. It can be stated that modern production and manufacturing methods introduce a number of interesting problems which cannot easily be solved by conventional control methods. Examples are:

- frequent changes in product throughput

- frequent changes in product mix and individualisation of products

- introduction of more advanced and highly complicated production methods

- increases in production and manufacturing speed, using fewer product buffers and flexible, lightweight mechanical constructions

- the introduction of plant-wide control systems which integrate in one system tactical, managerial, scheduling, operational, monitoring, supervision and control tasks.

As a result control problems will be more dificult to solve, because of:

- changing conditions and operating points

- highly nonlinear and time-varying behaviour

- inherent instability of the total process or some subprocess

- increase of required speeds in relation to the dynamics of the process

- increase of the complexity of the process models (interactions, influence of higher-frequency dynamics)

- incomplete or excessive data. Some data is unreliable, noisy or incomplete

- the mixture of qualitative and quantitative knowledge

- mixed continuous/batch operations

Therefore new methods based on an entirely different approach were required to handle these problems. Artificial Intelligence methods emerged at the same time, and seemed to be able to solve some of the problems mentioned above. A number of trends tendencies caused the growing interest of control engineering people in intelligent control systems based upon AI techniques.

- The restrictions of conventional control methods based on mathematical models and the high emphasis on linearized systems

228

- The success of operators in controlling a number of practical control problems which are difficult to model, by applying a heuristic approach guided by a long-term experience. This approach yields satisfactory results in most cases results, but is very sensitive to changing conditions, and the know-how of a small number of operators, and has to be build up gradually when new problems arise or new installations are being set up.

- A multi-layered hierarchical information processing system has been defined for plant-wide control, based on various levels of decision making. On each level people are responsible for their part of the automation level. However, their decision should be based, not only on the informatiom available on the level concerned, but also on information provided by the adjacent levels. There is a growing need for concentrated and qualitative information, which is adapted to the people responsible for the decisions on each subsequent level. Intelligent systems, such as expert systems, should provide means to overcome the communication barriers which always seem to exist between different disciplines.

- The field of system analysis, design and simulation has developed considerably over the past decades, and has been concentrated in software packages and their accompanying toolboxes. Two types of knowledge are required from the user: specific knowledge related to a particular software package, and specific knowledge related to a certain analysis or design method. Expert systems can advise people to use these packages in a more-convenient and proper way.

- In practical control algorithms the code for the actual control algorithm plays only a minor role. There is much logic built around the algorithm, taking care of such activities as: switching between manual and automatic operation, bumpless transfer, supervision of automatic tuning and adaptation, etc. Continuous control and sequencing control are merging. A knowledge-based system can be used as an alternative to logic, sequencing and supervision code.

- The selection of the right control configuration and the right controller parameters can be implemented by a knowledge-based system. In addition, it can be advantageous to reconsider the resulting control configuration when the assumptions concerning the process operation no longer hold. Automatic strategy switching can be performed by an intelligent supervisory system.

- The successful proliferation of AI techniques in many engineering applications by the introduction of expert systems and artificial neural nets. Many hard- and software tools are available to facilitate the application of these methods. There are, however, still many problems to be solved when these tools are incorporated in real-time applications ([Rodd 1991]). Fuzzy sets have been successfully introduced to manage uncertainty.

However, as in many other developments taking place over a short time period and introduced as a panacea for solving all types of problems, the control community should also be warned against a too-rash application of these methods. The use and abuse of these methods should be carefully considered, keeping in mind that many of the implicit attainments of control theory are not available in intelligent control. Finally, a well-considered combination of conventional and intelligent methods, should be selected.

3   Intelligent Control Components

In this section a very concise introduction is given of the various components from the broad field of AI which can be used in real-time on- and in-line applications.

3.1   Expert Systems

Essentially, expert systems are collections of 'rules of thumb' - certain aspects of human knowledge captured and stored in a computer. In an expert system the knowledge is more clearly represented and separated from the programs (called the inference machine) which operate upon it. However, expert systems cannot do anything that cannot be done in a conventional computer. They can be programmed in many languages, although special AI-languages like PROLOG and LISP exist and can run on normal PC's or workstations. There is a parallel between ES and human thought processes and the analogy between ES and deductive human reasoning is obvious. The programming style of ES is not based on a functional or procedural based traditional approach, neither on an object-oriented approach, but on rule-based programming. The rules define actions which should occur, depending on specific conditions which have taken place. The rules are triggered when certain patterns appear in memory, instead of actually being called by other functions. The processing of the rules is undertaken by an inference engine, which allow the computer to infer information that is not explicitly stored in the database. Other contributions will highlight specific properties of Expert Systems, however, in this contribution those issues will be emphasized which are concerned with the specific application of ES in a real-time environment. In intelligent control applications we are not talking about rule-based expert systems which have access to limited, or shallow knowledges. The rules are purely emperical and specified by the designer of the system. Model-based and knowledge-based expert system are, however, most useful in intelligent control applications. In model-based expert systems the rule based system is supported with knowledge about the functioning of the system the expert system is developed for. In the field of fault detection and fault diagnosis a model of the system is used to infer the cause of the problem and to confirm an initial hypothesis. When the expert system has facilities to determine where to look for more information, we are talking about knowledge-based expert systems. Both model- and knowledge-based expert systems are often called deep knowledge systems. Although many expert system shells are commercially available for different hardware configurations, few of them are useful for real-time control applications. In the first place there is the important point to obtain knowledge about the process being controlled. Some of the knowledge provided by the user is based on known process behaviour and some knowledge may be inferred by the in-line measurement performed during process operation. In the second place the ES need to be integrated in the process control environment. This implies that the ES should be able to handle numerical algorithms, sensor-based acquisition, real-time performance, etc. In the third place the ES should be able to handle qualitative data and control laws. Few ES provide adequate provision for handling uncertainty. The introduction of fuzzy systems is most useful, but provisions should be provided to protect the system from unpredictable and unmaintainable behaviour. The question of real-time expert systems will be discussed in section 4, and the question of uncertainty handling in the next paragraph.

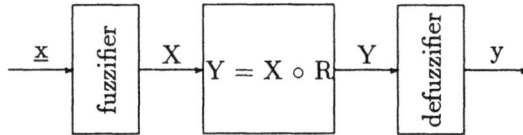

Figure 5: *Fuzzy inference.*

3.2 Fuzzy Control

Application of fuzzy set theory in control is gaining more and more attention from industry. An application of fuzzy control in a cement kiln was probably the first application of fuzzy control.

The theoretical approach of fuzzy inferencing is based on the idea of relations rather than on the idea of implications. For example, a fuzzy controller is a function of more than one variable, which is a combination (union) of the functions representing the individual fuzzy rules. The "direction" of inference is defined by the form of the fuzzy rules. A typical rule $'i'$ can be stated as:

$$\text{IF } x_1 \text{ is } X_1^i \text{ AND } \ldots \text{ AND } x_N \text{ is } X_N^i \text{ THEN } y \text{ is } Y^i$$

in which x_i is the input variable and X_i is the fuzzy set the variable is mapped on. The result, however, describes a relation represented by a multi-dimensional function (a function with range $[0,1]$ of $(N+1)$ variables):

$$R(x_1,\ldots,x_N,y) = \bigvee_i R_i(x_1,\ldots,x_N,y) \tag{1}$$

where

$$R_i(x_1,\ldots,x_N,y) = \bigwedge(\bigwedge(\mu_{X_1^i}(x_1),\ldots,\mu_{X_N^i}(x_N)),\mu_{Y^i}(y)) \tag{2}$$

where $\mu_{X^i}(x_j)$ is the membership funtion of the fuzzy set X_j, and as can be seen from (1) and (2) the resulting fuzzy system no longer has a 'direction': it is a relation. This is why, for example, fuzzy models can be used for predictions as well as "inverse" control [Brown 1991].

By applying the "compositional rule of inference" using a fuzzification of the inputs, which is in fact also a relation presented by a function of $(N-1)$ variables, a fuzzy output is obtained. In figure 5 this fuzzy inference is shown schematically. In practical applications however, a more efficient method is usually used [Harris 1989]. This method is based on the assumption that the fuzzification of an input is represented by a singleton. The application of the compositional rule of inference, using the fuzzified input, results in a fuzzy output. The fuzzified input is represented by the union of all the singletons describing the individual (numerical) inputs. The same fuzzy output can be obtained by determining for every fuzzy rule $'i'$ the individual fuzzy output

231

$$\mu_Y^i(y) = \bigwedge(\bigwedge(\mu_{X_1^i}(x_1(t)), \dots, \mu_{X_N^i}(x_N(t))), \mu Y^i(y)). \tag{3}$$

Combination of those individual fuzzy outputs by applying a union results in the complete fuzzy output of the system

$$\mu_Y(y) = \bigvee_i \mu_Y^i(y). \tag{4}$$

Note that the inputs in (3) are represented in a numerical way ($\mu_{X_j^i}(x_j(t))$) and the output is represented in a fuzzy way ($\mu_{Y^i}(y)$). Several defuzzification methods are available to translate the fuzzy output into a crisp, numerical output value. Based on this practical method of applying fuzzy rules and inference, a generalization of the inference mechanism can be made allowing fuzzy rules and inference to be applied in the way that rules and inference mechanisms are used in 'conventional' expert systems.

A fuzzy system is able to manipulate qualitative data by a kind of inference engine, presented by a membership function, and to infer qualitative data. However, in a control engineering application quantitative data are measured which should be fuzzified (e.g. 0.3 moderate and 0.7 small) before further manipulation. Otherwise qualitative data should be defuzzified to quantitative data before being applied to a process. The fuzzy reasoning for the functions AND and OR can be implemented in various ways. In table 1 several of these implementations are summarized. It can be shown that dependent on

	AND	OR
Lukasiewicz	$\max(p + q - 1, 0)$	$\min(p + q, 1)$
Probabilistic	pq	$p + q - pq$
Zadeh	$\min(p, q)$	$\max(p, q)$

Table 1: *Possible implementations of AND- and OR-operation.*

these implementations and the number of fuzzy sets defined on e, Δe and u, the fuzzy controller canbe compared with a PID controller. However, by changing the number of fuzzy sets or the AND and OR implementations a desired nonlinear behaviour of the controller can be obtained.

3.3 Artificial Neural Networks

The ability to learn should be a part of any system that exhibits intelligence. Feigenbaum (1983) has called the "knowledge engineering bottleneck" the major obstacle to the widespread use of expert systems, or knowledge-based systems in general. It refers to the cost and difficulty of building expert systems by knowledge engineers and domain experts. The solution to this problem would be to develop programs which start with a minimal amount of knowledge and learn from their own experience, from human advice, or from planned experiments. [Simon 1983] defines learning as: *any change in a system that allows it to perform better the second time on repetition of the same task or on another task drawn from the same population.*

232

Simon's definition of learning is very general. It covers various approaches to learning such as adaptive systems and artificial neural networks. The process of (structured) knowledge acquisition is also a part of learning. This kind of learning is done by human beings, in contrast to *machine learning*.

Memory plays an important role in this area of learning. The learning of a human being is strongly correlated to memorizing history. Experiences from previous events and actions can (and must) be used to improve control actions in the future. Learning from experience is therefore one of the main reasons for research into learning components like neural networks and associative memories.

The idea of subsymbolic reasoning using neural networks, which is defined as the counterpart of symbolic (e.g. rule-based) reasoning is not new. The idea was originally meant as an attempt to model the biophysiology of the brain, i.e. to understand the working and functioning of the human brain.

A biological neuron forms the basis upon which artificial neural networks are based. Each of those neurons is composed of a *body*, an *axon* and a large number of *dendrites*. These dendrites form a very fine "filamentary brush" surrounding the body of the neuron. The axon can be seen as a very fine long thin tube which splits into branches terminating in little *endbulbs* almost touching the dendrites of other cells. The small gap between such an endbulb and a dendrite of another cell is called a *synapse*. The axon of a single neuron forms synaptic connections with many other neurons. Impulses propagate down the axon of a neuron and impinge upon the synapses, sending signals of various strengths down the dendrites of other neurons. The strength of the signal is determined by the *efficiency* of the synaptic transmission. A neuron will send an impulse down its axon if sufficient signals from other neurons impinge upon its dendrites in a short period of time. A signal acting upon a dendrite may be either *inhibitory* or *excitatory*. A biological neuron fires, i.e. sends an impulse down its axon, if the excitation exceeds its inhibition by a critical amount, the *threshold* of the neuron.

A human brain consists of many neurons, called a neural network, consisting of many interconnected neurons. In artificial neural networks (ANN) a layered structure of artificial neurons is used. The artificial neuron is a model of a biological neuron and is used to mimic its basic behavior. It consists of inputs, x_i, (axons), which are weighted, w_{ij}, (synapses) and fed to (dendrites) a summer, z_j, (body). An output, y, (axon) is produced after passing a threshold, represented by an arbitrary linear or nonlinear function, called transfer function, $f(z_j)$.

The neuron just described is called the McCulloch-Pitts Neuron. When a binary threshold is used as transfer function this neuron is called a perceptron. In an ANN the artificial neurons are called nodes. They are usually organized into layers. Data is presented to the network via the input layer, hidden layers are used for storing information and an output layer is used to present the output of the network. In most cases ANNs have adaptable weights. Learning rules are used to adjust the interconnection strengths, wij. These rules determine how the network adjusts its weights using some error function of any kind of criterion. In a strict forward network information flows from the input layer, through a number of hidden layers to the output layer. In case of feedback connections, information reverberates around the network under control of an algorithm taking into account a convergence algorithm. In that case the network can be described by a nonlinear dynamical system. The application of a neural network can be devided into two main phases:

- the *training phase*, in which the weights of the network are adjusted by a number of input (and output) signals in order that the network performs as desired.

- the *recall phase*, in which the network computes an output for a given input in the range of signals which has been trained.

In learning there are two essentially different methods: supervised and unsupervised:

- *Supervised learning*, which occurs when the network is supplied with both the input values and the correct output values, and the network adjusts its weights based upon the error of the computed output.

- *Unsupervised learning*, which occurs when the network is only provided with the input values, and the network adjusts the interconnection strengths based only on the input values and the current network output.

The number of algorithms developed and parameters to be chosen is very large resulting in a large number of names connected to the various networks introduced. Among the parameters to be chosen are:

- the number of inputs to the network

- the number of layers (minimal one representing a linear system, called flat network)

- the number of nodes (the total number and the number by layer)

- the type of the transfer function $f(z_j)$: linear, nonlinear, binary

- fixed or adaptable weights

Among the algorithms to be chosen, points to be considered are:

- forward or feedback connections

- supervised or unsupervised learning

- choise of the learning law

With the introduction of multilayered neural networks more complex learning could be performed using the *back propagation learning rule*. This rule was introduced by [Rumelhart 1986]. In literature many applications of these networks for pattern recognition and classification have been shown. An extensive overview of all possible neural network configurations is given in [Khanna 1990].

Artificial neural networks are very appropriate tools for modeling and control purposes. The ability to represent an arbitrary nonlinear mapping is an especially interesting feature for complex nonlinear control problems. ([Cybenko 1989], [Cotter 1990], [Billings 1992]). The most relevant features for control concerning neural networks are:

(i) arbitrary nonlinear relations can be represented

(ii) learning and adaptation of unknown and uncertain systems can be provided by weight adaptation

(iii) information is transformed into an internal (weight) representation allowing for data fusion of both quantitative and qualitative signals

(iv) because of the parallel architecture for data processing these methods can be used in large-scale dynamic systems.

(v) the system is fault-tolerant so it provides a degree of robustness.

In [Narendra 1990] many control schemes are proposed. Possible applications are: modelling of (non)linear processes, prediction of process outputs and inverse control of processes.

3.4 CMAC

The CMAC (Cerebellar Model Articulation Controller) [Albus 1975a] algorithm is based on the functioning of the small brain (cerebellum), which is responsible for the coordination of our limbs. Albus originally designed the algorithm for controlling robot systems. It is essentially based on lookup table techniques and has the ability to learn vector functions by storing data from the past in a smart way. The basic form of CMAC is directly useful to model static functions. However in control engineering one is mainly interested in controlling and modelling dynamic functions. To give CMAC dynamic characteristics the basic form has to be extended, either by adding additional input information related to previous in- and outputs, or by inherently feeding back the output.

CMAC is a perceptron-like lookup table technique. Normal lookup table techniques transform an input vector into a pointer, which is used to index a table. With this pointer the input vector is uniquely coupled to a weight (table location). A function of any shape can be learned because the function is stored numerically, so linearity is out of the question.

The construction of the CMAC algorithm

The CMAC algorithm can be characterized by three important features, which are extensions to the normal table lookup methods. These important extensions are: distributed storage, the generalization algorithm and a random mapping.

In control engineering many functions are continuous and rather smooth. Therefore, correlated input vectors will produce similar outputs. Distributed storage combined with the generalization algorithm will give CMAC the property called generalization, which implies that output values of correlated input vectors are correlated. Distributed storage means that the output value of CMAC is stored as the sum of a number of weights. The generalization algorithm uses the distributed storage feature in a smart way: correlated input vectors address many weights in common.

Because of an unknown range of a component of the input vector, a complex shape of the input space or a sparse quantized input space, it is very difficult to construct an appropriate indexing mechanism. These problems often result in very sparsely filled tables. The CMAC algorithm deals with this problem in a very elegant way. The generalization algorithm generates for every quantized input vector a unique set of pointers. These pointers are n-dimensional vectors pointing in a virtual table space. CMAC uses a random mapping to map these vectors into the real linear table space. The essence of random mapping is similar to a hash function in efficient symbol table addressing techniques.

Using a random mapping implies the occurrence of mapping collisions, which gives rise to undesired generalization. However, because distributed storage is used, a collision results only in a minor error in the output value. The major advantage of using random mapping is its flexibility, because irrespective of signal ranges and the shape and sparseness of the input space, the size of the table can be chosen related to the number of reachable input vectors.

An output value of the CMAC algorithm is calculated by simply summing the weights selected by the input. The weights can be adapted by comparing the output to a desired output (control application) or a measured output (modelling application). Adaptation is done in a cautious way using the delta learning rule which is comparable with the back propagation procedure.

4 Control Engineering Applications of AI Techniques

Although expert systems, artificial neural nets and fuzzy control are being increasingly applied in a variety of applications and many successful stories are being recorded (most of them on simulated processes or pilot plants), in general the most promising engineering applications appear to lie in:

- Real-time control. The application of knowledge-base systems, fuzzy set logic and neural nets as part of a real-time control system, substituting a conventional or model-based control algorithm. In some cases the operator actions are modelled

- Supervisory control. A knowledge-base system monitors a single control algorithm or a number of control algorithms, or switches the control strategy depending on some major changes in the system (throughput, product-mix, operating point). In most cases a knowledge-based system on a supervisory level will supervise knowledge-based controllers on a lower level.

- Alarm monitoring and fault diagnosis. Knowledge-based systems are used to monitor process variables to detect malfunctioning and to provide a diagnosis to avoid alarm situations or to advice remedial actions either to the operator or to an automatic system to guide the system to a safe operating point under the given situation.

- Design environments for control engineering. Expert Systems have been used for process identification and for designing lead-lag networks. Another useful application is the use of an expert system as an advisory system for the user of complicated analysis, design and simulation packages. The fourth area will not be treated in this contribution because emphasis will be given on in- and on-line control applications.

236

The fourth area will not be treated in this contribution because emphasis will be given on in- and on-line control applications.

4.1 Real-Time Control

In this paragraph a number of in-line applications of intelligent control is described. Traditionally, controller design has been based on fundamental information (described in terms of mathematical relationships) as well as on experimental data. The control structure and the control algorithm was based on this information and derived by analytical design or extensive simulations. Important issues are: the way the control algorithm is implemented (robustness, sensitivity) and the calculation time needed to evaluate the output of the controller especially in time- critical circumstances. It was natural that in complicated situations (no quantitative information or mathematical model of the process available) in which (additional) qualitative information is available the incorporation of an expert system embedded in the control loop as an alternative control method has been explored. This situation (see figure 1) is called direct expert control or direct knowledge-based control. The controller design is based on the experience of the operator and design engineer, as well as on direct observations of the process variables. The relationships between the variables may be known, or assessed in terms of qualitative variables. This approach is particularly useful for highly nonlinear and (partly) unknown systems in the sense of a mathematical model. Because of the lack of fundamental knowledge and possible blind spots in the available knowledge provided by the experts, the direct expert control system should be supervised by a supervisory system (see figure 2) which is able to reason over a longer time-period and is able to detect malfunctioning or biased operation. Due to strict time constraints in a real-time environment and due to various levels of reasoning (information over a small or large time period) it has been shown advantageous to split the direct and supervisory knowledge-based controller into several subcontrollers, each acting on different information and time scale. In [Jager 1991] a system is described consisting of 4 subcontrollers:

- a reflex controller based on a few rules and instantaneous measurements of the process

- a model-based controller which forces the process to the desired state based on a prescribed trajectory

- an in-line adaptation mechanism, adapting some of the rules of the previous subsystems based on instantaneous measurements

- an on-line adaptation mechanism which adapts the rules of the subsystems, based on trend information and desired overall behaviour of the process

It has been shown that the knowledge-based system operates smoother when the rules are based on fuzzy sets and logic. The system operates reasonably well for various processes even when very few information is available concerning the system to be controlled. The behaviour can be ameliorated by adding learning facilities. In that case the system may start from a more adequate initial condition of the rules of the expert systems and will remember previous successful operations. Research is going on the generate automatically new rules and to fade away rules which are not adequate for a given situation.

In a direct knowledge-based (fuzzy) control system there is no attempt to use or derive a rather precise model of the process to be controlled. Most of the rules are based on rough or fuzzy information about the signals influencing the systems, i.e. a phase plane consisting of e and Δe is tesselated quite roughly in an irregular pattern which is finer near to the desired state and coarser far from this state. In order to avoid oscillations or undesired bias, special attention should be paid to the situation where the process output approaches the desired output (zooming). At first sight it seems perhaps incredible that with a very rough quantization still good results can be obtained. However, adaptation of rules and smoothing by defuzzification methods is mainly responsible for these results.

Artificial Neural Nets and CMAC algorithms are inherently very suitable for in-line control, once they have learned a process model or a control behaviour. They are very fast and robust, but need a level of quantization of the signals concerned which is quite small compared to direct (fuzzy) expert system control.

Modelling using neural networks
Modelling dynamic systems using neural networks needs a special configuration in which information about the history of the input and output of the plant is used. In figure 6

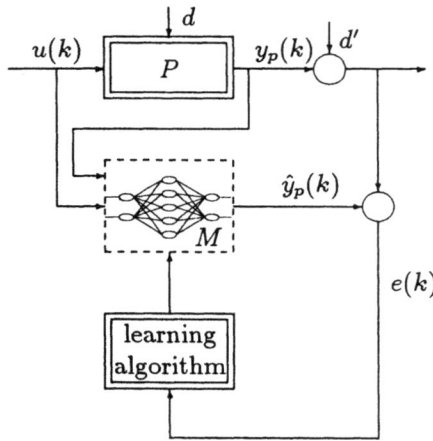

Figure 6: Plant identification

the general identification structure using neural networks is depicted. Of course many types of neural networks are suitable for use this configuration. The general objective of this configuration is to describe a general analytic system Σ with a nonlinear difference equation:

$$\Sigma : y(k+1) = f(y(k), \cdots, y(k-n); u(k), \cdots, u(k-m)) \tag{5}$$

in which $n + 1$ is the order of the system, with $m \leq n$. If insufficient time-lagged values of $u(t)$ and $y(t)$ are assigned as inputs to the network, it cannot generate the necessary

238

dynamic terms and this results in a poor model fit, especially in those cases where the system to be modelled is completely unknown. Therefore we have to use model validation techniques to be sure that the model developed is acceptable and accurate enough to be used in a control situation in which the control action depend on the model. Generally it is 'easy' (although time consuming) to train a neural network which predicts efficiently over the estimation set, but this does not mean that the network provides us with a correct and adequate description of the system. It is difficult to get analytical results but in [Billings 1992] some model validity tests are introduced which were proved to be useful experimentally.

Neural control

Using a neural network as a general 'black box' control component is extremely interesting. There are several possibilities for neural control, which can all be thought of as the implementation of internal model control (figure 7), in which a model M is used to control plant P using a controller C. The structure is completed with a filter F.

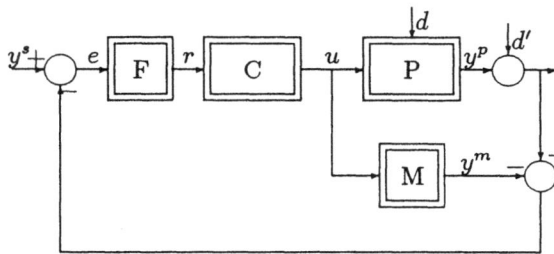

Figure 7: Internal Model Control

- **inverse control:** the model of the plant to be controlled is used to train the controller in a so-called "specialized learning architecture". The model, obtained by observation, is used to generate error corrections for the controller. Under special conditions this kind of control can be guaranteed to be stable ([Billings 1992]): the system has to be invertible and monotonic. In the next section on CMAC an example of this kind of control using the inverse of a system is shown.

- **model-based control:** the model of the plant to be controlled is used in a conventional (e.g. predictive) control scheme to calculate the proper control actions. One of the possibilities is to implement an iterative method to calculate the inverse. Other possibilities are the use of a neural model as the basis of an on-line optimization routine to calculate the optimal control action.
  A special situation exists when the controller is also a neural network, while the neural model (emulator) is used to generate the Jacobian of the process. This gradient information is then used to adapt the controller. The method is known as the "error backpropagation" method [Narendra 1990], [Narendra 1991].

- **reinforcement control:** no neural model is derived and used but the controller is adapted according to a criterion function which evaluates the desired behaviour.

The effect of this kind of control is that limitations concerning the linearity of the system to be controlled are released. The choice of the best approach, and determining when the training procedure is completed are the main issues of concern.

In the section about CMAC an example is given of neural-based control using an inverse model of the system to be controlled. Examples of control based on error backpropagation can be found in [Hertog 1991].

The CMAC algorithm can be used in the same way as artificial neural networks for modelling and control. The results of the in-line application of these algorithms in real-time control have been described in the contribution on CMAC.

4.2 Supervisory Knowledge-based control

In most industrial control applications, supervisory control forms an important bridge between local control and optimizing control and thus is an important ingredient in plant-wide control. It is here that the merger of discrete and continuous control is seen. Expert systems can be used as an alternative or extension to logic and sequencing and can be intermingled with the logic surrounding a DDC algorithm or a more advanced algorithm. One can distinguish a wide variety of work taking place in this area including:

- the automatic tuning of controllers

- the supervision of adaptive and other sophisticated controllers such as intelligent controllers

- the supervision of the controller structure in a multiloop control environment, i.e. strategy switching and changing the control configuration by interchanging the control actions between control and manipulated variables.

The automatic tuning of PID controllers has been successfully solved in many applications by the use of expert system techniques. The approach advocated by [Kraus 1984] utilizes pattern recognition and mimics the procedure used by an experienced process engineer. The system is composed of a transient analyzer which determines some key parameters as damping and oscillation frequency of the closed loop system, followed by an adjusting mechanism which uses a collection of empirical rules to tune the PID controller. Retuning of the parameters can be done automatically or at the initiative of the operator. Another approach is based on relay feedback, see [Åström 1988]. This auto-tuner is based on the idea that crucial information for tuning a PID-controller according the rules of Ziegler and Nichols is presented by the ultimate gain and frequency of a closed-loop system on the verge of instability. The ultimate frequency can be obtained by relay feedback. A number of measures is taken to keep the oscillation within specified limits. The gained information on frequency and gain is used to adjust the parameters of the controller. Next, control is automatically switched to PID-control.

Sophisticated control algorithms such as adaptive control and intelligent control do not work correctly without extensive supervisory logic. It is difficult to know beforehand all the effects of parameter changes of the process and the controller, of the level and frequency content of the noise and of the coupling of interactive loops. Whilst many parameters used in the controllers are chosen during the design stage, a significant number of them will have to be adapted during the actual operation of the system. A knowledge-based supervisory controller can be set up to manage this task, based on the performance of the separate parts the controller consists of. In addition, rules can be relatively easily added when more experience has been built up. An interesting application of a supervisory knowledge-based system monitoring and (re)tuning a Unified Predictive Control (UPC) algorithm, is given in [Soeterboek 1992, Krijgsman 1991a]. As a result of the unified approach the UPC algorithm has many design parameters, among which polynomials. In [Soeterboek 1992] rules of thumb are derived that can be used to select initial settings of the controller based on the process model and the desired closed-loop behaviour (the servo and regulator behaviour can be tuned independently). These rules of thumb have been implemented in an expert system. When information about the process model or the desired behaviour is missing, the expert system will ask the user additional information. Finally, the system will provide initial settings, some of them based on provided information, other ones will be preset on default values. After this off-line parameter initialization, these values are implemented in the controller and the real-time control situation is initiated, supervised by an expert system. First, this expert system supervises the stability of the control loop, taking measures to ensure a stable response. Secondly, the performance of the system is monitored and fine tuning is provided to tune some of the parameters in order to fulfil the requirements. Finally, the expert system is looking for any further improvement by fine-tuning the parameters once again to fulfil also additional requirements of lower importance.

4.3 Alarm Monitoring and Fault Diagnosis

Both process industry and vendors of process control systems are seeking solutions for this very urgent area. Since we are creating more and more complex plants, which still require operator intervention in case of faults or alarms, there is a definite need for a tool to lighten the task of operators, to provide them with timely information and to generate even automatic procedures to bring the system back to a desired or safe state.

Alarm reduction is a very important issue, because an avalanche of alarm massages and conflicting indications can confuse operators and delay corrective actions. As is only too well known from some of the well-publicised accidents which have occurred in the last decade, the major problem which results lies in expecting a human operator to take the appropriate action when faced with such a situation. An expert system can investigate patterns of alarm messages, including their time sequences, and, knowing the dynamics of the plant under control examine the cause-and-effect relations and the trends over the recent time period. On this basis, the operators can be advised to take the most appropriate action. Having achieved much success in fault detection and reducing the number of alarms presented to the operator, the next step is fault diagnosis. Fault diagnosis is based on process knowledge, and on experience of experienced operators and process engineers. An expert system can be applied which forms an opinion as to the actual cause of an error and provides messages describing the possible faults and recommending remedial action. It is clear that the problem becomes very complicated when multiple faults should be taken into account. In case of uncertainty alternatives will have to be presented and at that point, the experience of an operator is called upon to make a value decision as to what action should be taken.

It is often possible to detect in an early stage possible malfunctioning of (a part of) the system. Alarm prediction is a technique by which it is possible to prevent alarming conditions by using predictive models. By continuously monitoring the various operating conditions of the plant and the application of knowledge and experience of previous situations, measures can be taken to prevent alarm conditions from arising. Also minimal outage of the plant can be obtained by using expert systems to develop appropriate maintenance schemes.

Process Fault Diagnosis is gaining much attention in process control during the last years [Patton 1989]. The early detection and localisation of faults is necessary to obtain a high level of performance for complex systems. A possible approach is to use very accurate (mathematical) models, in which not only the parameters have to be accurately estimated but also a transformation is necessary to the real physical quantities involved, i.e. when a time constant is estimated, this time constant consists of at least two quantities of which one or more can cause a fault which has to be detected.

Also global models can be used. Techniques are being developed to build such global models of the process by means of qualitative reasoning methods [Kuipers 1989]. It is being suggested that even with relatively incomplete knowledge, there is normally sufficient information in a qualitative description to predict the possible behaviour of an incompletely described system. Most promising are methods which combine the analytical problem solving techniques, the qualitative reasoning approach and heuristic knowledge in the form of fault trees and fault statistics.

A major problem is how to integrate information processing tasks and knowledge representation in an object oriented way. Looking at the information processing flow four entities can be recognized:

- the reality or outside world, represented by the real plant

- a plant representation; a portrayal of the reality

- the analysis task: updating the plant representation by observing the reality

- the synthesis: planning and control actions to influence the reality so that it behaves as desired

Artificial neural networks and heuristics can be used to represent (part of) the plant. Virtual sensors can be introduced to describe numerical routines like filtering and estimation, yielding nonmeasurable variables, etc. Next, this information is used to update the plant representation. In the synthesis part the behavior of the plant is observed and it is decided to choose a target state. Control actions are planned to reach this target state in a well-defined way [Fusillo 1988]. In [Stephanopoulos 1990] and [Terpstra 1991] a plant representation scheme is proposed in which regions, fault-modes and views or scopes are recognized. A region comprises of a set of constraints defining the operating state in which the model is valid. The plant can be represented in different regions. Each fault-mode represents one fault, defined by: a deviation of the model relative to a normal model. The same system can be used in various contexts. Each context highlights different aspects of the same component. Views offer a "filter" which selects the apsects which are relevant to that view.

The plant is divided into sections, units, subunits, components, etc. Each composes of objects which have the responsibility to fulfil a function. The Multi-Level Flow Modeling (MFM) method deals with goals and functions and can be used to describe the process for fault detection and diagnosis see [Lind 1990].

5 Real-time issues

In a time-critical situation, extra demands are necessary when using intelligent systems in the lower levels of automation: the supervisory and control levels. In these layers the expertise of process operators and control system designers should be integrated with the time-varying information obtained directly from the process by measurements.

Real-time behaviour is often easier to recognize than to define. As discussed by [O'Reilly 1986] many definitions of real time exist, mostly related to "fast"; meaning that a system processes data quickly. In [O'Reilly 1986] a formal definition of real time is offered:

- *a hard real-time system is defined as a system in which correctness of the system not only depends on the logical results of a computation, but also on the time at which the results are produced.*

The most important item is the response time, When events are not handled in a timely fashion, the process can get out of control. Thus, the feature that defines a real-time system is the system's ability to guarantee a response before a certain time has elapsed, where that time is related to the dynamic behaviour of the system. The system is said to be real time if, given an arbitrary event or state of the system, the system always produces a response by the time it is needed.

In a real-time intelligent control problem several critical issues arise because the process moves on in time and therefore any process data which is acquired looses its significance and value after a certain period of time. We are bounded by the physics of the system and are only able to influence the behaviour by control actions over a certain time-span. Often the time history of a variable, the trend, is very important and is translated into a few vital parameters, which will be used to influence control decisions.

5.1 Real-time issues in expert systems

Using expert systems in a real-time environment, we have to face several critical issues.

- **Non-monotonicity**: Incoming data do not remain constant during the complete run of a system. The data can loose its validity as time passes and therefore any reasoning based on this information has to be time-related. All dependent conclusions should be removed in that case.

243

- **Temporal reasoning**: In real-time applications time is all-pervasive. As a result, any direct or supervisory control system must be able to reason about temporal events. The point at which we return information to the process must be at a correct time - as required by the dynamics of the process. In the real-time knowledge-based system reasoning takes place over a certain time-slot (one sampling period for direct knowledge-based control and several time-slots for supervisory control). Actual data and to a certain extent data from the near past and predicted data over a small time horizon are taken into account.

- **Interfacing to the external world**: A real-time knowledge-based controller must be integrated with a variety of conventional software (signal processing, application specific I/O, etc.).

- **Asynchronous events:** In most systems there is a combination of synchronous events (sampling of process variables) and asynchronous events (caused by disturbances or event-driven operations). Asynchronous events are unscheduled but should be served in most cases instantaneously. Whilst ultimately one will more towards architectures which are far more synchronous we have still to accept that our real-time control systems will have to cope with unexpected events.

- **Focus of attention**: Once we accept that we are putting our expert systems in control of real processes, we have to accept the consequences of managing those processes under the most critical conditions. This requires our expert system to shift its attention from its normal mode of operation into a highly critical narrowly focussed mode. It is therefore highly recommended to divide the expert system, which is controlling and supervising a plant, into several subsystems each with its own area of attention.

- **Safety**: As we are dealing with real, live processes including human operators, we are responsible for the safety of the plant. Hence we have to strive towards fully deterministic and verifiable systems or we have at least to provide a secondary, reliable, safety - control system.

In general expert-system shells available for diagnosis and advisory purposes, no features are available for temporal reasoning or the other issues mentioned above.

In real-time control an important parameter is the sampling period which has to be chosen. The minimum sampling period which can be used is defined by the amount of time needed by the system to make the necessary calculations. *Progressive reasoning* is an appropriate tool in a real-time expert system [Lattimer 1986]. The idea behind progressive reasoning is that the reasoning is divided into several subsequent sections go into more detail about the problem.

The progressive reasoning principle is very well suited to process-control purposes. With the application of a direct digital controller using AI techniques, a problem arises when the sampling period is chosen smaller than the time needed to evaluate the rule base. This will cause a situation where the inference engine cannot provide a hypotheses. Since each hypothesis is indirectly coupled to a control signal the expert system will not be able to control the process. To solve this problem, the rule base is divided into a set of different knowledge layers, each with its own set of hypotheses and therefore its own set of control signals. The minimum length of the sampling period which can be used is now defined by the time needed to evaluate the first knowledge layer. This will generate a fairly rough conclusion about the control signal. The next layers are used to produce better control signals using more information about the process to be controlled. In a typical example using this progressive reasoning principle the lowest (fastest) layers of the system are used to implement a direct expert controller, while the next layers are used for supervisory and adaptive control.

Several attempts have been made to develop expert-system shells which are able to deal with the real-time issues. G2[1] is probably the best known commercially available real-time expert system shell, to be used at the level at which response times of seconds are required. For higher speeds more dedicated solutions have to be used like DICE (**Delft Intelligent Control Environment**) [Krijgsman 1991b].

5.2   Real-time issues in neural computing

Neural networks are very simple processing units. As a result of the highly parallel structure, several hardware implementations of neural network architectures have been developed and are currently subject of research. The big advantage is the relatively simple calculations which have to be performed. The calculation effort in a feedforward and recall situation is rather low and short computation times are possible. Training of the ANN takes a lot of time in evaluating the examples to be processed, but in many cases this can be done in an off-line situation.

Nowadays, a number of commercially available tools for neural computing are available. These tools are mainly used in an off-line situation, providing the user with high-level graphical interfaces (NWorks for example). Some hardware boards are available, on which neural-net configurations can be run at very high speeds.

The CMAC algorithm is a not particulary computationally intensive algorithm, but memory consumption is rather high. With the current move towards very cheap memories this methods becomes more and more appropriate. Hardware implementation of the algorithm could considerably speed up the calculation times.

5.3   Fuzzy control hardware

The use of fuzzy control modules in real time gives rise to the same problems as in an expert-system environment. For fuzzy control purposes many calculations using fuzzy membership functions have to be carried out. Some dedicated hardware and software (e.g. Togai) is currently available to speed up calculation times.

---

[1]G2 is trademark of the Gensym Corporation

6 Comparison

In the previous sections, intelligent control configurations and intelligent control components are introduced and described. Each component has its own strengths and weaknesses. According to the available knowledge (theoretical, empirical, experimental...) a proper choice for the use of a specific intelligent control module must be made.

In general there are a number of criteria which influence the choice of an intelligent control component:

- knowledge description

- ability to adapt

- explanation facilities

- quantization level

- ability to learn

- ability to process qualitative data

According to these items a comparison can be made which is summarized in table 2. Expert systems and fuzzy-control components are summarized as symbolic AI, while the

Features	Symbolic AI	Neural networks
variables	logic	analog
processing	sequential	parallel
	discrete	continuous
knowledge	programmation	adaptation
level	high	low
knowledge	explicit formulation	no formulation
learning	learning deals	numerical learning by
	with rules	synaptic weights
explanation	explanation facilities	no explanation at all

Table 2: Comparison of AI methods

other intelligent control components are indicated as neural networks. Of course we must be aware that this is a very strict separation wheras in practice the situation may be more confused.

In cases where we are dealing with problems for which experts are available, their knowledge (e.g. heuristics) can be implemented using symbolic tools like expert systems. But in a case where no real expert is available and knowledge about the system to be controlled has to be collected by experiments on the system itself, other tools are more appropriate: neural nets.

7 Concluding remarks

In the preceding it is shown that there is a need for intelligent control where traditional control methods, even the ones introduced recently and based on mathematical models, fail to control (partly) unknown or highly nonlinear systems. Control methods based on AI techniques can fill in the gap between practice (diffcult to model or to control processes) and theory (no generic methods available based on mathematical models). In many cases a combined approach is desired because knowledge about the process model is only partly available and should be learned from experiments on the process itself. One possibility is the integration of small subsymbolic models controlled by symbolic systems, e.g. supervisory knowledge-based control of a neural-net controlled process.

It can be stated that Artificial Intelligence techniques can be used at any level of control and for any task: as off-line, as well as on-line, or in-line. At any level, from design to actual control, these techniques can be used to store knowledge and overcome limitations of existing methods and algorithms. Both Artificial Intelligence methods: symbolic reasoning and subsymbolic learning, have their own benefits and disadvantages. The choice which method must be used depends on the type of problem you are dealing with.

In this contribution emphasis has been given to real-time applications of AI techniques, which are quite challenging because of the time-critical situation: we have restricted ourselves to problems of the lower levels of process automation. Also on higher levels of a plant-wide control system many applications of intelligent systems can be expected and are already implemented. Although many experiments show that intelligent control is quite promising we want to warn about too much dispersion of efforts, because new developments should be embedded in an integrated control system and should be extensively verified.

References

[Albus 1975a]   Albus J.S. (1975). *A New Approach to Manipulator Control: The Cerebellar Model Articulation Controller (CMAC)*. Transactions of the ASME, pp. 220-227, September.

[Åström 1988]   Åström, K.J. and T. Hägglund , *A New Auto-Tuning Design,* Proceedings IFAC International Symposium on Adaptive Control of Chemical Processes, Copenhagen, Denmark, 1988.

[Billings 1992]   Billings S.A., H.B. Jamaluddin and S. Chen (1992). *Properties of neural networks with applications to modelling nonlinear dynamical systems*. International Journal of Control, Vol. 55, No. 1, pp. 193-224.

[Brown 1991]   M. Brown, R. Fraser, C.J. Harris, C.G. Moore. *Intelligent self-organising controllers for autonomous guided vehicles*. Proceedings of IEE Control 91, pp. 134-139, Edinburgh, U.K., March 1991.

[Cotter 1990]   Cotter N.E. (1990). *The Stone-Weierstrass Theorem and Its Application to Neural Networks*. IEEE Transactions on Neural Networks, Vol. 1, No. 4, pp.290-295, December.

[Cybenko 1989]   Cybenko G. (1989). *Approximations by superpositions of a sigmoidal function*. Mathematical Control Signal Systems, Vol. 2, pp. 303-314.

[Fusillo 1988]   Fusillo, R.H. and G.J. Powers, *Operating procedure synthesis using local models and distributed goals*. Computers and Chemical Engineering, Vol. 12, No.9/10, pp. 1023-1034, 1988.

[Harris 1989]   C.J. Harris and C.G. Moore. *Intelligent identification and control for autonomous guided vehicles using adaptive fuzzy-based algorithms*. Engineering Applications of Artificial Intelligence, vol. 2, pp. 267-285, December 1989.

[Hertog 1991]   Hertog, P.A. den (1991). *Neural Networks in Control*. Report A91.003 (552), Control Laboratory, Fac. of El. Eng., Delft University of Technology.

[Jager 1991]   Jager, R., H.B. Verbruggen, P.M. Bruijn and A.J. Krijgsman (1991). Real-time fuzzy expert control. *Proceedings IEE Conference Control 91*, Edinburgh U.K.

[Khanna 1990]   Khanna, T., *Foundations of neural networks*. Addison Wesley Publishing Company, Reading, USA, 1990.

[Kraus 1984]   Kraus, T.W. and T.J. Myron, *Self-tuning PID controller uses pattern recognition approach*. Control Engineering Magazine, June 1984.

[Krijgsman 1991a]   Krijgsman, A.J., H.B. Verbruggen, P.M. Bruijn and M. Wijnhorst (1991). *Knowledge-Based Tuning and Control*. Proceedings of the IFAC Symposium on Intelligent Tuning and Adaptive Control (ITAC 91), Singapore.

[Krijgsman 1991b]   Krijgsman A.J., P.M. Bruijn and H.B. Verbruggen (1991). *DICE: A framework for real-time intelligent control*. 3rd IFAC Workshop on Artificial Intelligence in Real-Time Control, Napa/Sonoma, USA.

[Kuipers 1989]   Kuipers, B, *Qualitative Reasoning: Modelling and Simulation with Incomplete Knowledge*. Automatica, Vol. 25, No. 4, pp. 571-585, 1989.

[Lattimer 1986]   Lattimer Wright M. and co-workers (1986). An expert system for real-time control. *IEEE Software*, pp. 16-24, March.

[Lind 1990]   Lind, M., *Representing goals and functions of complex systems, - an introduction to multilevel flow modeling*. Internal report, Technical University of Denmark, Institute of Automatic Control Systems, 1990.

[Narendra 1990]   Narendra K.S. and K. Parthasarathy (1990). *Identification and Control of Dynamical Systems Using Neural Networks,* IEEE Transactions on Neural Networks, Vol.1, No. 1, March, pp.4-27.

[Narendra 1991]   Narendra K.S. and K. Parthasaraty (1991). *Gradient Methods for the Optimization of Dynmical Systems Containing Neural Networks.* IEEE Transactions on Neural Networks, Vol. 2, No. 2, pp. 252-262, March.

[Patton 1989]   Patton, R., P. Frank and R. Clark, *Fault Diagnosis in Dynamic Systems: Theory and Applications.* Prentice Hall, New York, 1989.

[O'Reilly 1986]   O'Reilly C.A. and A.S. Cromarty (1986). Fast is not "Real Time". in *Designing Effective Real-Time AI Systems.* Applications of Artificial Intelligence II 548, pp. 249-257. Bellingham.

[Rodd 1991]   Rodd M.G. and H.B. Verbruggen (1991). Expert systems in advanced control - Myths, legends and realities. *Proceedings 17th Advanced Control Conference*, Purdue, USA.

[Rumelhart 1986]   Rumelhart, D.E., G.E. Hinton and R.J. Williams, *Learning Internal Representations by Error Propagation.* Parallel Distributed Processing, vol. 1, D.E. Rumelhart, J.L McClelland and the PDP Research Group, eds., Cambridge, MA:MIT Press, pp. 318-362, 1986.

[Soeterboek 1992]   Soeterboek, A.R.M., *Predictive Control - A Unified Approach.* Prentice Hall, 1992.

[Stephanopoulos 1990] Stephanopoulos, G., *Artificial Intelligence ... What will its contributions be to process control?* Chapter 4.7 in: The second Shell process control workshop. Solutions to the shell standard control problem. Editors: D.M. Prett et al., Butterworths, pp. 591-646, 1990.

[Simon 1983]   Simon, H.A. (1983) *Why should machines learn?* In: Michalski et al., Machine learning: An Artificial Intelligence Approach, Palo Alto, CA: Tioga.

[Terpstra 1991]   Terpstra, V.J., H.B. Verbruggen and P.M. Bruijn, *Integrating information processing and knowledge representation in an object-oriented way.* IFAC Workshop on Computer Software Structures Integrating AI/KBS Systems in Process Control, May 29-30, Bergen, Norway, 1991.

[Verbruggen 1991]   Verbruggen, H.B., A.J. Krijgsman and P.M. Bruijn , *Towards Intelligent Control.* Journal A vol. 32 no. 1, 1991

PART III

REAL-TIME ISSUES

Road Map.

The third part of the course deals with *Real-time Issues*. The text shows in what kind of real-time issues are important for control applications. Thereby special attention will be given to real-time networks (what are the requirements, etc.) and to real-time aspects in expert systems. The target field primarily is process control, .

There are two chapters:

Real-Time and Communication Issues. In this chapter the important issue of communication is discussed. For real-time automation, a good and solid communication between the components of the control system is vital for its correct functioning. Here an introduction is given to the elements of communication systems and the requirements for these communication systems in process control applications are given.

Real-Time Expert Systems. From a comparison between real-time systems and existing expert systems, the requirements for real-time expert systems are derived. An insight in architecture for real- time expert systems is given, showing the reader how the specified requirements can be met. The chapter concludes with some examples on real-time expert systems architectures.

This road map should enable the reader to make up his own choice in this basic course, depending on his skills and future use.

REAL-TIME AND COMMUNICATION ISSUES

M.G. RODD

University of Swansea, Wales, Singleton Park, Swansea, SA2 8PP, U.K.

1. INTRODUCTION

In most automation applications we are required to control real-world, physical processes - processes which have their own intrinsic time-related properties. A chemical reaction, once started, proceeds according to its own dynamics, and the chemistry takes little notice of the controlling computer. Likewise, once a piece of metal is removed by a milling-machine, it cannot be replaced! These processes operate according to their real-time dynamics, and as a result, any controlling system must recognise these inherent real-time characteristics. A real-time computer control system is thus one in which the correctness of the system depends not only on the logical results of computations, but also on the time at which the results are produced. "Real-time" does not necessarily imply "fast", but does mean "fast enough for the chosen application".

It is important, therefore, to realise that in many stages of automation, two aspects of the control require real-time to be taken into consideration. In the first place, there are many situations in which the system must respond according to the requirements of the plant. One could consider, for example, a robot feeding a tool into a numerically-controlled milling machine. If, after the robot begins to move the tool towards the milling machine, the computer controlling the latter suddenly finds that it is not in a position to receive that tool, then the operation must be aborted. This means that a message must be sent back to the robot to halt its operation. In a situation like this, there is no way that a multitude of layers of communication protocols could be negotiated in the time available before that message needs to get through. Likewise, in an alarm situation, where something goes wrong, responses must take place at a very high rate, in order to cope with the situation.

In the second place, the question of real-time synchronisation of parts of the factory must also be borne in mind. Thus, for example, a robot picking up a component which is being conveyed by an automatic guided vehicle or by a conveyer belt, must know exactly when it is to undertake its picking task. Whilst current solutions require extensive interlocking between, say, the end effectors of the robot and an indexing system running on the conveyer belt, this does rather defeat the objective of having a communication channel between the intelligent controllers of these devices.

Finally, the question of real-time data consistency is important. The point here is that if operators are to get a true picture of the total operation of their plants, then what they see on their screens must correspond to what is actually happening on the plant. However, if the communication system requires a long time for messages to be passed up the various communication channels, then it is clear that operators simply cannot see a picture on their screens of what is actually happening at that moment in time.

Basically, this all boils down to considering the question which is often referred to as "Reality in real-time control". Essentially, it must be emphasised that feedback control loops can be designed to operate successfully only if the time dynamics of the process to be controlled are sufficiently understood. This statement is just as true for continuous control as it is for discrete control aspects, where the dynamics of a cell, consisting of a variety of machine tools, are complex and probably non-linear. Fortunately, it all results in what has been referred to as "the first law of applied control". This simply says that the controller must respond to the speed of the plant! In other words, it is the plant which determines the pace of operation and not the computer. Thus, it must follow that control systems must be designed to match the requirements of the plant, and not necessarily those of an international communications standards organisation!

2. REQUIREMENTS FOR REAL-TIME AUTOMATION

In any situation where time is crucial, the following points must be considered when the control system is designed. Of course, in the context of large systems, we are assuming a distributed system, consisting of a multiplicity of computers communicating over communication channels (see Chapter C1). Essentially, the features which should be available include:

- A distributed real-time control system
- Real-time process monitoring
- Access to consistent, global databases
- Availability of multi-level supervisory controllers
- Availability of programming support tools and
- An effective and efficient information management system.

Arising directly out of these points is a fundamental philosophy in real-time control systems, which is that data has a meaning *only* when it is associated with time. This means that any communication system which cannot support time information is largely useless in many applications. It also indirectly implies that all nodes or computers in the distributed system must have a globally-agreed real-time clock. To provide a consistent picture of the plant to the operator or control system, all data must be linked to its time of creation, etc. It is essential that the real-time database is consistent in real time, and this can be achieved only by having all data stored together with the time at which it was created. It has also been found in many cases, that the length of time for which that data is valid - sometimes called the *validity time* of the data - also needs to be stated.

These points give rise to a consideration as to what structures are necessary to support communication systems which can respond in real-time. It is very clear, for example, that a maximum delay time through the communication system must be guaranteed. It is common experience that in most manufacturing and automation systems, a *guaranteed maximum delay time* across a network of 5 milliseconds appears to be reasonable, although in many new plants, 1 millisecond or better might well be required. In the process-control industry, it has often been estimated that a delay of up to 20 milliseconds is tolerable, but more recent information has shown that the lower this delay can be held, the higher the degree of control which can be applied. (Incidentally, in many cases this also results in a reduction in the mathematical algorithms required to determine plant parameters.)

Two critical factors emerge from these discussions. First of all, delay times must be *deterministic*. In other words, they must be guaranteed under all conditions, particularly under maximum loading. It is clear also, that the *delay times must be matched to the individual plant's response.* Delay times are related, not merely to the electronic characteristics of the transmission media, but also to the protocol requirements. It is clear that a heavy, 7-layered approach to protocols can be very questionable, because of the sheer amount of time that it takes for information to move up and down the various stages. The protocol thus adopted should offer embedded real-time, in that it should inherently handle the timing information. In fact, one can be so bold as to say that many of the protocols selected by MAP/TOP from those based on the OSI model (ISO, 1984), are of little value in real-time systems, since they require a positive acknowledgement and retransmission (PAR) structure. These protocols essentially consist of sending and receiving systems which operate on the basis that after a message is sent, the sender will wait until an acknowledgement is received. If a "time-out" occurs, it will keep trying to transmit until it is successful. In a poor environment, if such PAR protocol is introduced at every level, a substantial degradation (and variable delays) can occur, hence affecting performance. The worst case, of course, occurs under alarm conditions, when one node might issue an instruction to the rest of the network, saying that an erroneous condition has arisen. If all other nodes on the network have to receive this message to act, then they will all generate acknowledgement signals and the communication network will be flooded out by the replies. If, in fact, a CSMA approach (such as used in Ethernet) is adopted, of course, the system will become completely bogged down.

A more reasonable approach will be based on strict real-time, which assumes that all data is *time-stamped* and also has an *explicitly defined validity time*, during which the data is valid. Once this time has run out, the data is discarded. In such a system, the sender must specify whether an acknowledgement is required and the receiver has to be told if such a task must be undertaken. If the sender does not specify a need for an acknowledgement, then no action is taken. The sender will retransmit the message only if it has requested an acknowledgement and none is forthcoming.

Thus, whilst within the OSI-based protocol suites, non-PAR systems can be selected, great care must be exercised in choosing the right tools for the job. One must always remember that the OSI-model was *not* developed for use in *real* real-time environments!

3. REAL-TIME COMMUNICATIONS

3.1. Background

A critical component of any modern control system is clearly the communication system -

used to move information around the plant. There is little doubt that as we move towards more integrated control, the communication system becomes extremely important. However, with industrial communication systems being a relatively small component of the total plant cost, the market available to producers of such communication systems is extremely small. It is not surprising, therefore, that industrial communication systems have to build upon those technologies which are utilised elsewhere in the communication field. As special as we might feel our applications are, we have to accept that we are addressing a small sector of the market and, in general, communications between our computers will be dominated by the needs of the larger data-processing and data-communications industry.

In the past, however, this reality did not seem to be a problem. After all, process engineers were used to using communication systems running at 110 bits per second, and now were being offered systems with bandwidths of 10 megabits per second! However, the mismatch between the *actual* application requirements and the available products is now increasingly becoming painfully obvious. Suddenly, we are being faced with situations in which we are utilising to the full the communication systems, and discovering that the available networking technologies, as elaborate as they might seem at first sight, simply cannot support our real-world time-critical control requirements.

The following section addresses some of the fundamental issues which must be considered in the designing communication systems for use in so-called real-time applications.

However, before looking at the communication systems appropriate for supporting industrial control systems, it is important that we look back at the nature of the process which is being controlled - not from the point-of-view of control algorithms, but from the point-of-view of implementing these very controlling algorithms! Industrial communication engineers must have a good understanding of the requirements of their applications before they can design an appropriate communication system.

As we have moved towards distributed computer control structures, it has become evident that we need some form of hierarchy in order to structure these systems. We have to harness geographically related processes and, wherever possible, reflect the natural integration which results. We also find that as we develop an appropriate form of hierarchy in our distributed computer control systems, the various levels of automation define the information requirements. In practice, the so-called five-layer model, as illustrated in figure 1, seems to become relevant, reflecting as it does most practical situations. This simple model, reflects, in essence, the concept of the 2nd generation of DCCS introduced in the chapter on Computer Control Systems, of this text. It also reflect the logical model of a 3rd generation System.

A further requirement of the distribution is to cope with the problem of system failure. During the early days of computerisation we learned, to our cost, the disasters which result from the "eggs in one basket" syndrome. The loss of an essential computer could be devastating. We have learned that the only way of tackling this is to distribute the system, so that the failure of any particular piece of software or hardware, or, communications equipment, will lead to a situation which at least can be contained.

Clearly, it is important to emphasise that fundamental to such integrated systems and their hierarchical implementation is the data which is the "lifeblood" of any distributed system.

258

Equally important is the fact that the "veins" which carry this lifeblood are the *communication channels* which we install between our computing nodes. As we will see later, the nature of these "veins" has a tremendous influence on how we move information around the plant or, in a grander way, between plants using various public services!

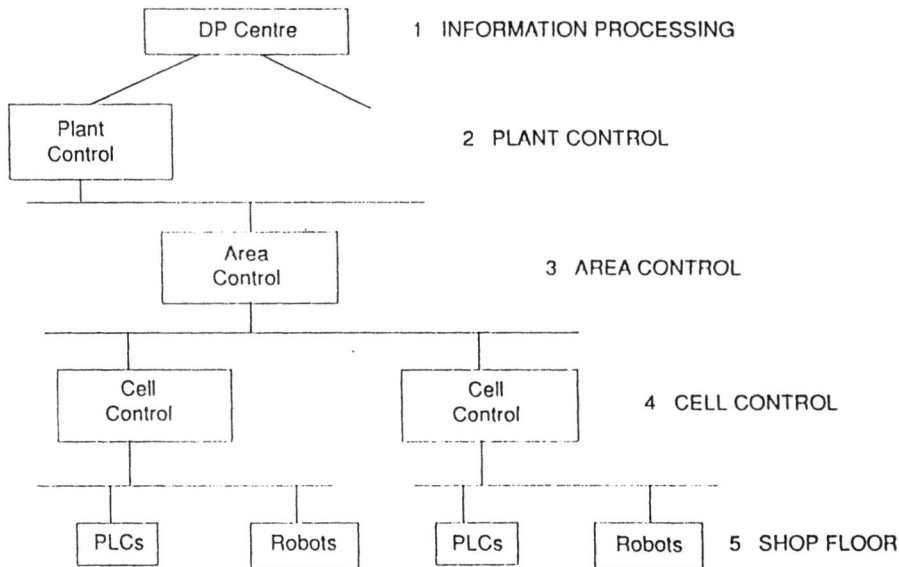

Figure 1: The simplified model of a distributed computer control system

3.2. The real-time problem

To design a control system one must appreciate the essential temporal nature of the process that one is controlling! In essence the problem is one of distinguishing between an "on-line" computing system and a "real-time" computing system. This distinction has become extremely blurred over many years, and it is vital that we return to some fundamental understanding.

We need to repeat here the opening remark in this chapter - in a real-time computing system the correctness of the system depends not only on the logical results of the computation but also on the time at which the results are produced. "Real-time" does not necessarily imply "fast", but does mean *fast enough for the chosen application*. We must draw a vital distinction between real-time and on-line: this is particularly critical as so often the terms are confused.

In essence, in time-dependent, real-time processes (which most manufacturing and continuous process are), data is judged not only on its actual value but also on the time at which the value is obtained. For example, a sensor measuring an attribute of a dynamic system should specify not only the actual value of that attribute but also when that parameter was sampled, since, of course, the system is time-varying. Data in any distributed network is thus valid only for a given period of time and, indeed, has meaning only if the time of creation is specified. This forces designers increasingly towards the concept of atomic units of data in which the value and the time of creation are inseparable. Fundamental to all this is that we are "completing control loops". The designer of a simple loop controller assumes that the variables which are being measured are obtained when they "think" they are. However, if we put a transmission system between the point of measurement and the point at which the measurement is utilised, that transmission system has its own characteristics. These characteristics must be considered in solving the control algorithm problem. There is no point in comparing the value of a temperature sensor with a required set point unless we know at what point in time we are actually measuring that temperature.

On the other hand, in the *on-line* situation, we are merely required to be able to have discussions directly with our computer. If that computer takes a little bit longer to handle, say, a request, then this is not serious, except for maybe causing some irritation. In the real-time system, any delay could cause the plant to be in serious error - essentially because the transmission system forms part of the control loop!

It is clear, therefore, that the data which flows around our distributed computer control system will, in many cases, be time-dependent and not merely value-dependent. It is important, therefore, that we look at the nature of the data in our real-time control system when we select methods of transmitting and storing that data as the links in integrated computing systems.

CHARACTERISTIC	REAL-TIME	ON-LINE
Response time	Hard	Soft
Pacing	By Plant	By Computer
Peak Performance	Must be Predictable	Degradable
Time Granularity	Less than 1 mSec	About 1 Sec
Data Files	Small-to-Medium	Large
Data Integrity	Short Term	Long Term
Safety	Critical	Non-Critical
Error Detection	Bounded by System	User Responsible
Redundancy	Active	Standby

Figure 2: Real-time versus on-line

Professor Hermann Kopetz of the Technical University, Vienna, proposed some years ago in a public lecture that the characteristics shown in figure 2 give us a good clue as to the difference between "real-time" and "on-line". He and many other workers in the area have also suggested that there are various degrees between these two extremes. However, it is felt here that this distinction is not really necessary and that a simpler and more reasonable approach would be to define a real-time process as one in which the temporal characteristics are absolutely defined, whereas an on-line situation is one in which these can be degraded without any serious loss of performance.

As a result, it is suggested that in real-time control, data normally has meaning only when it is associated with time. This implies, in the first place, that all data must be *time-stamped* at the time of creation in order to provide that time information. Of course, the direct consequence of this is that all nodes in a system must have access to a globally agreed real-time clock. Again, work by many authors has pointed to ways in which such clocks can be created. In essence, they require a good-quality local clock at each computing node, with

260

some form of synchronisation - by means of frequently transmitted time messages, a separate timing channel, or access to some internationally transmitted standard. (Kopetz, 1987).

However, merely time-stamping the data does not solve all the problems. A particularly important issue which must be considered, and which will be referred to later, is that of consistency. Not only must data be consistent in value, but this consistency must be referred back to time. Two aspects of this are illustrated by the following examples.

In the first place, consider the implementation of redundant sensors. Say, for example, one is measuring a single temperature variable but using three redundant temperature probes, with each probe producing its "version" of the temperature. A further controller must make use of this temperature information, and the question arises as to which temperature value to use. Whilst in the past this would probably have been handled by a voting mechanism, clearly if each probe transmits its information separately over the communication network, then there is no way for the controlling node to know how to implement this voting, since it will not know when each value has been created. The only way to cope with the problem is by time-stamping those temperature variables.

The second example is a very complex one, relating to the problem of an operator getting a consistent picture of an overall plant. Consider that the operator might be at least two or three "networks" away from the actual plant; information which is then presented to him on the screen will have to travel across the communication networks before it can be displayed. Information which is obtained locally could well get to him before information which is obtained from a remote site. If, however, he (or an *accompanying artificial intelligence system*) is required to base a decision upon the data presented: this is an impossibility unless the actual time at which all the data was created is available to him, thus ensuring that consistency can be checked. (Rodd, 1989(a))

In summary we can conclude that our communications systems form an integral part of our control systems. These are, in fact, real-time systems which require real-time data! Not only must data be related to time, but, like the control algorithms themselves, time-domain determinism is vital at all stages!

3.3 Levels and functions

Referring back to Figure 1 it is important to investigate the data requirements at various levels, and from this one begins to see important patterns emerging. (Rodd, 1989(b))

At the higher levels, we see that the functions which are typically required include

- File transfer
- Electronic mail
- Remote file editing
- Status reporting to operators and higher levels of management
- Data acquisition from lower levels
- Supervisory control of lower levels, and
- Program transfer

In essence, at these levels, we are moving *bulk data* around, typically between relatively large computers. This could be between Computer Aided Design (CAD) systems, or sending plant status information through to management computers running production scheduling or manufacturing materials requirement planning. At this level we can see the need for efficient, but safe and reliable, transfer of bulk data. However, if we are, additionally, utilising this information for control decision-making, then it is necessary also to be able to send time-tagged information across this system, otherwise no useful control decisions can be made! However, we normally accept at this level of networking that delays will occur, particularly when large files are being transferred. It is clear also that we could almost consider this an *on-line environment*, provided that there is some form of access to time-stamped data where necessary.

Moving to the lower levels, we see a different collection of functions. Here we are really getting into the real-time control world, and the following characteristics emerge

- Minimal, defined message delay times
- Deterministic behaviour - particularly under crisis situations
- Real-time synchronisation
- Assured data consistency
- Inherent message redundancy
- Message broadcasting (single station sending information to many users), and
- Regular and frequent data updating.

Essentially, we are talking here about highly efficient data transmission. Speed is important, but only to the extent that it must match the response of the plant. Thus a very slow plant will not require very high transmission rates. However, what it will require are deterministic rates. Information which flows at this level is typically relatively small in quantity, but critical in terms of its determinism. For instance, if a program does not complete on time, this could lead to serious error conditions. It is important also to realise at this stage that the network system must work most efficiently under the worst possible conditions. Our communication systems are now truly part of the control loops!

Looking at our networking (or, in fact, internetworking) strategies in this way, gives us a clue as to how they may best be implemented, and what technologies are appropriate at which level. That there are different functions cannot be doubted, but a fundamental criterion has to be that where data is associated with time then the communication systems must inherently support this.

4.   COMMUNICATION SYSTEMS FOR DISTRIBUTED
      COMPUTER CONTROL

Supporting distributed computer control systems will be communication systems, and their very nature will have a great influence on what is moved around the plant, and how. We have too often been guilty of assuming that because we appear to have very large bandwidths available to us on our networking systems, we are free to move as much data as we wish around a plant. In reality, however, once we try to move this data reliably and efficiently we suddenly discover that our communication systems start to become bottlenecks in themselves! Also, we suddenly discover that the nature of the communication systems can greatly influence the efficiency of moving the data. Particularly relevant, too, is that once we realise the temporal nature of data in process control then we have to look at the time characteristics of our communication systems.

Stepping back from the performance aspects of our systems for a while, a fundamental issue in the development of any interconnected computer structure is that of incompatibility. Essentially, in a distributed computer system what we are trying to do is to permit various computers from various manufacturers to share various pieces of data. Whilst from the point-of-view of the supplier it would be desirable for all our computers to be from their own company, and therefore be totally compatible, we soon realise that in the distributed computer control world we are at the mercy of a variety of suppliers, each of whom has particular strong points. Thus, certain suppliers can provide very fine data processing hardware and software, whereas others specialise in hard-nosed, factory-floor compatible controllers. The difficulty in producing a distributed system is to bring all these bits and pieces together. The compatibility issue revolves around hardware, software and, of course, data. Naturally, we would like to be free to design our own communication systems and indeed our own programs. However, the reality is somewhat different!

In practice, solving the compatibility problem is extremely difficult and we are only beginning to make some impact on it. The key has to be standardisation - whether or not we or the suppliers acknowledge it. Naturally, at the upper data communication level much progress has been achieved of late in the move towards OSI-based protocols, and in other fields, too, the success of standardisation is evident - Ethernet and TCP-IP (Rodd, 1989(b)) are good examples of standards which have met their desired marks.

Of course in the hard world of process control standardisation is critical, but unfortunately, as was said previously, we are relatively small as an industry, when compared to the data communications field. Whilst we would like to set our own standards there is no way that we can economically go down this route and therefore we tend to be the "step-children" of the data communications industry.

Realising the problems of standardisation, General Motors initiated its MAP exercise some years ago (MAP, 1987) and at the same time Boeing Corporation developed its TOP initiative. In both cases the idea was to produce a profile of protocols which would be appropriate to their particular areas of application - in the case of MAP, manufacturing, and in the case of TOP, technical offices. It was realised right from the start that there was no point in going it alone, and that the way ahead would be to select an agreed, and internationally-acceptable protocol profile and within the various layers, to select certain protocols. It was also recognised that, where necessary, they would have to inject extra emphasis to direct the move towards the development of actual products.

Although MAP has attracted much criticism, it has undoubtedly been significant in that it has brought to the fore the need for standardisation. The exercise has also hastened the development of protocols which are appropriate (at least to some degree) to manufacturing and the process industry. It is clear, however, that the products which are resulting from this exercise are in many cases too expensive and too complicated to be of present value in the hard world of automation. Also, as will be highlighted later. The acceptance of the OSI-principles has resulted in protocols which are in direct conflict with the real-time application needs. However, at the same time, it must be acknowledged that certain aspects of the MAP exercise have undoubtedly proved to be extremely important.

Of particular significance, for example, is the development, through MAP, of MMS - Manufacturing Messaging Services (ISO, 1988). MMS provides standardised mechanisms for sending messages around a process or manufacturing plant. The standard itself defines a core of services, and then it is left to specialised groups to provide so-called "companion standards", which expand on these services for particular application areas. It is probably fair to say that the MAP standardisation exercise should have started at the MMS level, leaving many of the lower-layer, and somewhat technologically-related issues for later, in that many of the lower layers, such as the physical layer, are highly influenced by technical achievements. Current issues, for example, whether MAP should use Ethernet or Token-passing might well eventually be rendered redundant.

The point is, though, that there can be little doubt that the standardisation exercises are extremely critical to the design of future distributed computer control systems. They will also have serious impacts on how we handle data within DCCSs. We cannot ignore them: they are economically critical and simply to neglect the MAP exercise, for example, would be extremely naive. To look closer at MAP, to take what is good out of it and then to assist in the development, or redevelopment, of some of the layers which are proving to be troublesome, makes eminent sense.

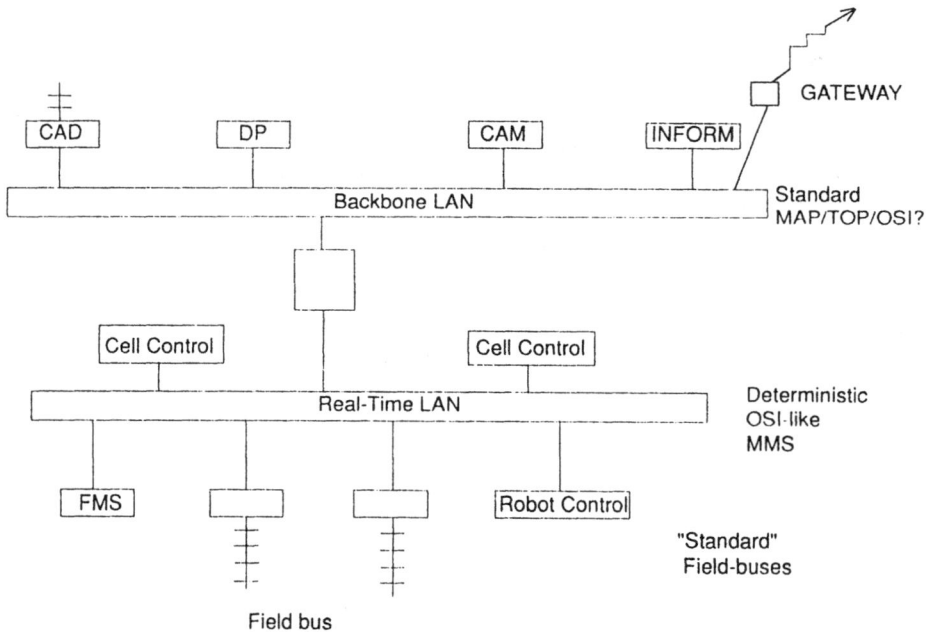

Figure 3: Network hierarchy

Another aspect of the trend towards standardisation fits in extremely closely with the idea of development hierarchies within integrated control systems. Exercises, such as MAP, have acknowledged that we will ultimately see a high-capacity data backbone running through the plant, fed by bridges or gateways from less-capable, cheaper networks, which in turn could well be fed by very low-cost, field-bus type networks. After much debate and academic discussion as to the ideal structure, there does appear to be an increasing consensus that our systems will ultimately move to a structure like that illustrated in figure 3.

In this figure one sees the idea of having a powerful backbone LAN which can in turn talk, via gateways, to wide area networks. The backbone will also be able to communicate with more deterministic, high-speed networks - which we have chosen to call "Real-time LANs". Whilst it is clear that current OSI proposals do not cope with the requirements of such real-time systems, there is no doubt that in the future they will. It is interesting to note also that many important companies operating in this area are envisaging models very similar to that proposed in figure 3. It is critical also to note that below the real-time LAN level, there is a move towards so-called "Standard" field-buses and here the current international attempts to standardise these low-level buses are particularly relevant.

Important throughout this discussion is that whilst various components are not yet defined through International Standards, there is no doubt that in time they will be. More important, though, is the acceptance of the OSI model and the acceptance of some of the higher-level services such as MMS, FTAM, and the Directory Services. (Rodd, 1989(b))

We cannot ignore international standardisation: as much as we would like to reject many of the proposed standards, for economic reasons we simply cannot. What we have to do is work with the and alongside standards - striving towards systems which do meet fully our real-time requirements.

5.   IMPLEMENTATION ISSUES

5.1. In the beginning there was OSI

As we mentioned previously, the drive for open computer communication systems provided the stimulus for much work in the standardisation of OSI protocols directed at a variety of applications. Some of the most significant advances have been accomplished in the Manufacturing Automation Protocol and Technical Office Protocol (MAP/TOP) arena, and these profiles provided the initial exposure to OSI for most potential industrial users. The ensuing development phases were, however, accompanied by disillusionment on the part of many users - the common grievances being that resulting products were:

- Too expensive
- Too complex
- Too cumbersome, and
- Too slow for time-critical applications.

Certain problems may be attributed to the fact that the OSI-based technologies have yet to mature; however, further restrictions are imposed by the inherent nature of the OSI 7-layer model and by the protocols which have emerged (figure 4).

Despite these, there are very good economic reasons to adopt, albeit in modified forms, the OSI concepts (Izikowitz, 1989). The economic reasons, we repeat, are simple - the computer world is dominated by the broad data-processing and data-communications industry. The industrial sector is minute in comparison.

Just as we have turned to computing devices originating from the data-processing world, so we, in the industrial control sector, are forced to borrow from there for our communication products.

Also, just as MAP (MAP, 1987) and TOP looked towards OSI (ISO, 1984) for their selected profile standards, so must we, as we strive for systems which will cater for our real-time applications. We must now reassess the OSI-based systems, using the very valuable work developed through the MAP initiative, to find real solutions. The key, though, is to appreciate fully the system requirements in supporting real-time process control.

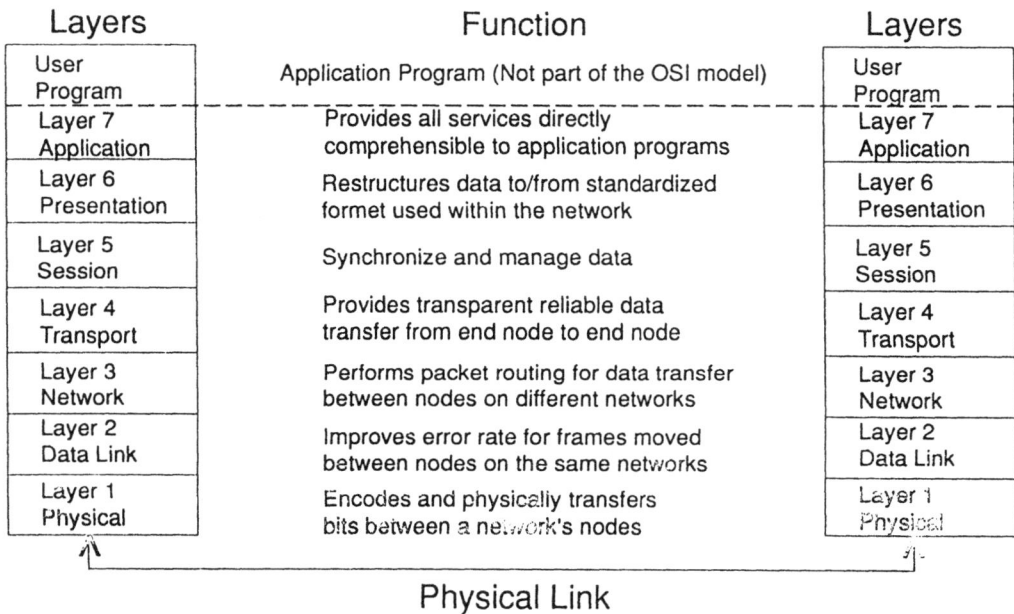

Layers	Function	Layers
User Program	Application Program (Not part of the OSI model)	User Program
Layer 7 Application	Provides all services directly comprehensible to application programs	Layer 7 Application
Layer 6 Presentation	Restructures data to/from standardized format used within the network	Layer 6 Presentation
Layer 5 Session	Synchronize and manage data	Layer 5 Session
Layer 4 Transport	Provides transparent reliable data transfer from end node to end node	Layer 4 Transport
Layer 3 Network	Performs packet routing for data transfer between nodes on different networks	Layer 3 Network
Layer 2 Data Link	Improves error rate for frames moved between nodes on the same networks	Layer 2 Data Link
Layer 1 Physical	Encodes and physically transfers bits between a network's nodes	Layer 1 Physical

Physical Link

Figure 4: The 7-layer OSI reference model

5.2 System requirements for real-time control

In real-time distributed control there are some very important factors which must be considered and which, in the past, have been largely neglected, although well-discussed by such authors as Kopetz (1982) and MacLeod (1984). Amongst the more-critical factors are time, reliability and redundancy. As has been pointed out earlier, the adequacy of a control

266

system to meet its requirements must be judged in terms of time concepts, and this is particularly the case when one is involved in high-speed digital control systems based on sampling techniques.

It is, therefore, recognised that communication systems (and their supporting structures, such as the messaging systems) which are to be used in real-time distributed systems, must be designed with time constraints always in mind. The rest of this section discusses some of the points which must be considered when designing such a system. (Rodd, 1989(c))

(i) Autonomous operation. In order to implement fault tolerance and redundancy in a truly distributed system, the system controlling functions should be distributed. Consequently, every node on the network should be capable of working autonomously! It is, therefore, necessary to incorporate functions such as self-checking and emergency-handling at every node. This contradicts the well-accepted "master-slave" philosophy which is currently most common in distributed systems. It must be conceded that in the typical "master-slave" situation, the failure of a "master" may be catastrophic since it may bring the whole system to its knees. Of course it is recognised that redundant "masters" may alleviate this short-fall to some extent. It is suggested, however, that the inherent benefits which distributed control systems provide, cannot be achieved with a "master-slave" approach. Indeed, it can be argued that it is impossible to obtain any real fault-tolerance in a system without inherently autonomous nodes!

(ii) Data. As was emphasised earlier, in time-dependent processes data is judged not only on its actual value, but also on the time at which the value was obtained. Real-time data in any distributed network is thus valid only for a given period of time, and indeed has meaning only if the time of creation is specified! This forces designers increasingly towards the concept of atomic units of data, in which both the value and the time of creation are inseparable. This implies that all data values in a real-time situation should be time-stamped at the source at which that data is created. Indeed, this concept goes further, and as has been pointed out (Kopetz 1982), the accurate time-stamping of data provides many other advantages. It provides, for example, a way of handling distributed database consistency, as well as a means of removing data from a network in the case of late or invalid messages - thereby assisting in network management functions.

(iii) Distributed global time information. In order to support autonomy, fault-tolerance and provision of time-stamping, it is evident that the provision of a global, physically distributed, time reference is essential (Lamport 1985). It is our belief that every node on a network should have access to a defined time value. We suggest that every station on the network should contain a local clock which is periodically synchronised both to other local clocks as well as to a globally-agreed clock reference.

With the assistance of this distributed time reference, each node may be furnished with a system-wide view which provides the capability to execute functions across the network in real time! The availability of globally agreed time facilitates the synchronisation of operations as well as the execution of various error-detection/correction procedures (Wensley et al. 1978). For example, in, say, machine monitoring, when a series of events occurs, it is possible to get the system back to a stable condition only if the time at which various error signals occurred is known.

267

Although it is naturally acknowledged that one cannot achieve absolute global time synchronisation, work by Kopetz and Ochsenreiter (1987) has shown that we can, indeed, go a long way towards such a goal. Various methods have been proposed and shown to work in practice. The simplest of these methods is to provide each node with its own receiver of an international radio timing signal, or to provide a secondary, star-configured network running around the distributed system carrying real-time information. However, Kopetz and others have shown that it is possible to include, within the communication protocols, sufficient timing information to support the time-based global synchronisation.

(iv) **Guaranteed transmission time** . It is suggested that a fundamental requirement of a real-time distributed system is that all information should be transmitted within well-defined time parameters. It is unacceptable to design a control system based on, say, sampled-data theory, and then to support this by a communication system in which the period taken to transmit certain variables from one part of that control system to another is virtually undefined! Thus, it becomes essential to provide a communication system which can guarantee, at all times, an upper bound for the time taken for a message to go from one application process to another. It is important here to emphasise the "application process". Although we might say, for example, that a token-passing network is to some degree deterministic, if we support this network by a set of OSI protocols which are themselves naturally non-deterministic, then guaranteed performance cannot be ensured!

Also, the performance of the supporting communication system should be guaranteed under all conditions - particularly in emergencies. It is important to realise that many protocols perform at their worst at times of heavy load - for example, when errors are detected. Indeed, it is suggested that under critical or emergency conditions, the performance of the network must be absolutely deterministic in order to ensure orderly administration of the situation.

(v) **Primitives for real-time communication** . It is suggested that in the real-time environment many of the current, reliable, connection-oriented methods for communication are inappropriate. It can be seen, for example, that any PAR-based protocol (which involves Positive Acknowledgement with Re-transmission on time-out) is unacceptable since this can lead to messages actually arriving after being invalidated by the passage of time. Various aspects of the communication primitives which should be considered include:

- The passage of time, as this may invalidate the information in the message.
- The question of where error detection should take place. If, for example, a temperature sensor is responsible for communicating its information back to the central processor, there is absolutely no point in that temperature sensor trying to handle the detection of any error - since there is nothing it can do about it anyway! Consequently, in many cases, error detection by the sender might be totally inappropriate.
- Communication traffic. This should neither be dependent on transmission error rates, nor become unpredictable in the case of high error conditions.
- Broadcast messages. In many real-time distributed control systems, the broadcasting of messages is of more importance than point-to-point communications! Error signals, system resets etc. are essential, as is the reception of common variable values by many users.

(vi) Datagrams. For many of the reasons mentioned above, it is suggested that a *datagram service* (often called "send-and-pray" since there is no intrinsic acknowledgement of message receipt) is appropriate for real-time communication systems. The rationale for this proposal is the following:

- No inherent acknowledgement is required, resulting in a lower overhead than a connection-oriented approach.
- Communication traffic depends totally on the message quantity and is determined at system design time.
- Services can easily handle multi-cast and broadcast.

In summary, although one can force a connection-oriented service to look like a broadcast service (eg. by each node, in turn, replying to a global message), it seems that much benefit will be gained by moving away from connection-oriented services towards broadcast datagrams in order to meet many of the real-time communications requirements.

(vii) State-driven versus event-driven systems. This long battle continues, but there is evidence to show that a state-driven approach to real-time control is increasingly being accepted as more appropriate than an event-driven system - despite the relatively constant heavy traffic which results (MacLeod 1984). In storing and transferring data within a distributed system, one can either use all the sampled values necessary to describe the data (i.e. the state) or only represent the changes (i.e. events) which result from transitions from some previous states. (Events, of course, represent changes of states, and states represent information during the intervals between events!)

In real-time distributed control, a state-driven approach seems to offer many significant advantages:

- State processes are never required to wait for inputs since they continue to use the states which have already been received, as long as they are still valid.
- Redundancy is intrinsic.
- Relatively loose synchronisation may be supported, since any data value available always contains a time-stamp.
- It appears easier to handle error recovery and error checking in a state-based approach, since one has a complete picture of the system.
- Finally, as we strive towards consistency in the distributed databases within a system, if one has the states of the critical processes constantly available, then every database in the distributed system can check its consistency against time.

For these and many other reasons, it is suggested that a largely state-based approach is more appropriate than an event-based approach in a real-time distributed computer control system. Furthermore, the communication systems should support this approach.

5.3. OSI and real-time communications.

From a first glimpse of the OSI structure and its mode of operation, one immediately identifies a rift between the open systems approach, and the characteristics we believe are fundamental to real-time distributed computer control systems. Of prime importance are the

269

lack of facilities to support a "state"-based communication system and the costly and non-deterministic overheads incurred in implementing features which have limited uses in a real-time communication environment.

(i) Broadcast and Multicast of Data. A state-based data communication system involves the broadcasting of data which is frequently sampled at periodic intervals. Inherent in this communication mechanism are the features of a "fast as possible" datagram service which directly supports fault tolerance, and recovery of communication errors. Indeed, even the implementation of a simple PAR (Positive Acknowledge with Retransmission) protocol is seen as an unnecessary overhead in real-time environments. After all, as was mentioned earlier, a simple sensor trying to communicate with its controller really can do nothing about an unsuccessful attempt at sending a message.

The very nature of OSI communications, with the creation of associations between peer entities at corresponding layers and the maintenance of a virtual circuit, inhibits the multicasting and broadcasting of messages in the system. Furthermore, the peer-to-peer layered structure implies a confirmed connection-oriented mechanism which is impractical - if not impossible - to support using a multicast communication scheme. Owing to this feature alone, it seems improbable that the OSI model in its current form will support the completely unacknowledged, connectionless broadcasting of data.

(ii) The Express Transfer Protocol. The emergence of a new LAN protocol, XTP (eXpress Transfer Protocol), designed as a "real-time" replacement for the existing network and transport protocols, reveals the existence of a real need for functionality not currently met by OSI.

XTP is designed to enable data transmission at the high speeds supported by the emerging advanced local area networks - namely the 100 Mbits/s fibre distributed data interface (FDDI) and the new 16 Mbits/s token ring. The protocol supports confirmed multicast communications with a flexible addressing scheme which allows address formats to be specified by the implementor. Furthermore, protocol benchmark tests of XTP indicate that existing protocols such as OSI TP4 and TCP/IP (Transmission Control Protocol/Internet Protocol) require "five to ten times more processing power than XTP in order to attain equivalent transmission speeds"..

Chips implementing XTP were scheduled to be available in 1991 and, although XTP is by no means an ideal solution, it must be seen as a step in the right direction towards a standardized real-time communication system.

The replacement of the network and transport protocols, however, does not address the problem of the connection-oriented service provided by MMS.

(iii) Factors affecting OSI determinism. The penalty incurred as a consequence of using a layered communication protocol is that the protocol processing requirements at every layer result in the formation of queues of messages awaiting service. This is compounded by the presence of multiple application associations (in a single system) with one or more receiving stations - these application associations requiring the establishment and maintenance of virtual circuits between the communicating entities, as well as the allocation of computer resources. The ramification of such a dynamic system is that it is inherently non-deterministic, in that the response time of any one message is dependent on the state of the

270

communication system at the given instant in time. Furthermore, the number of associations maintained by any one station affects performance by reducing the number of buffers available per association for receiving data, thus making it necessary to use flow control on incoming data. (An association here refers to establishing a "virtual current" between two users.)

In order to increase the determinism - or place an "acceptable" upper bound on the non-determinism of the system - one must address the issues of reducing the message queues as well as the processing overheads associated with the virtual circuits.

(iv) ***Restricting the number of associations.*** One possible method of enhancing performance degraded by multiple associations could be to limit the number of active application associations at any one time. Consider, for example, the extreme case of limiting the number of application associations to one per station - the implications are less than favourable. In the event that an application is required to transmit real-time data and if the destination is already involved in a transaction, the sender may not set up the required association, and as a result, the data may not be transferred in time to be of use. The situation is further compounded by the fact that the destination may not be able to close its current association until it holds the token. The sender would thus be required to wait at least as long as the token cycle time before the data may be transmitted.

It is interesting to note that this scenario is not uncommon in current implementations, since the number of possible associations is implementation-dependent and directly related to the computing resources of the system.

Reducing the number of associations maintained at every node will, however, improve the performance of an individual association by making more buffers available per association. In the case of a DCCS, this may not be a pertinent factor since messages are typically short and are not subject to segmentation.

(v) ***Static Associations.*** The overhead incurred by the negotiation and establishment of associations may be reduced by setting up all the required associations (virtual circuits) at the time of system configuration. These associations would remain static for the lifetime of the particular configuration. Under these circumstances it would be necessary to know all possible communicating entities during operation of the system under all conditions - a rather complex task (if at all possible).

A static configuration makes no assumptions regarding the frequency of message transmission and is still prone to the non-determinism of message queues produced by "concurrent" associations.

5.4. EPA/Mini-MAP - is this the solution?.

Much interest has been expressed in EPA/Mini-MAP - and certainly several vendors are claiming products in this area. The Enhanced Performance Architecture (EPA) is a reduced-protocol profile providing limited services to the application process in an attempt to improve the speed of communication (Rodd, 1989 (b)). The architecture comprises a 3-layer stack consisting of the physical, data link and application layers, incorporating MMS as the

271

main application service element. Any additional functionality (usually found in layers 3 to 6) must be implemented by the user in the application processes of the system.

(As a point of clarification it should be noted that a MAP/EPA node provides both the full and reduced protocol profiles, whereas a node with only the EPA is called Mini-MAP.)

There are, however, several critical issues yet to be resolved in EPA. For example, the use of MMS (Manufacturing Messaging System) in EPA necessitates a Logical Link Control mechanism different from that used in MAP. Owing to the fact that MMS is a connection-oriented protocol, the reduced services offered by the EPA protocol profile require the logical link control to provide additional functionality before MMS may be used. MAP specifies LLC Type 1, whereas the EPA specifies LLC Type 3.

It is obvious, therefore, that LLC is a significant area of incompatibility between MAP and EPA. Furthermore, the reduced stack does not conform to the OSI 7-layer model which defines the interaction between adjacent layers. Since Mini-MAP normally uses 5 Mbits/s carrier-band transmission media, as opposed to 10 Mbits/s broadband in the MAP backbone, a bridge or router is required for their interconnection, resulting in additional latency and potential congestion. Consequently, an application process on a full MAP stack requiring information exchange with an application process on an Mini-MAP stack, requires the use of a MAP/EPA node as an interface.

6.   CONCLUSIONS

It is evident that limiting the operation of the OSI stack, as well as using static associations, leads to a "proprietary" system that comes close to negating the very aim of open systems: additionally, even under a restricted stack, the real-time requirements would not be met. From the discussions, it is clear that the MAP profile of standards is unsuitable in its current form for real-time applications.

The advent of the MAP Enhanced Performance Architecture (EPA), was deemed to be the solution to most of the user grievances but has not received wide-spread acceptance. It is incompatible with MAP at the lower layers and, for the real-time users, it is still inherently non-deterministic. Furthermore, the well-known underlying real-time architecture of a distributed system is not readily supported by either MAP or MAP-EPA.

However, we should return to the fundamental issue - we will, by necessity, have to use OSI-based communication systems. It is a question now of bending the standards to fit our real needs - and, of course, bending the ears of the standards-making representatives! (Rodd, 1990)

REFERENCES

ISO DIS 9506. 1988. "Manufacturing Message Specification".

ISO IS 7498. 1984. "Information Processing Systems - Open Systems Interconnection - Basic Reference Model", American National Standards Association Inc.

Izikowitz, I., Rodd, M. G. and Zhao, G. F. 1989(c). "A Real-time OSI-Based Network Is it possible?", Proc. IFAC Workshop on DCCS, Pergamon Press.

Kopetz, H. , Lohnert, F., Marker, W. and Panthner, G. 1982. "The Architecture of Mars", Report MA82/2, Technical University of Berlin.

Kopetz, H. and Ochsenreiter, W. 1987. "Clock Sychronization in Distributed Real-time Systems", IEEE Transactions on Computers, Vol.C-36, No.8.

Lamport, L. and Melliar-Smith, P. M. 1985. "Synchronizing Clocks in the Presence of Faults", Journal of the ACM, Vol. 32, No.1.

MacLeod, I. M. 1984. "Using Real-time to Achieve Coordination in Distributed Computer Control Systems", Proceedings of the IFAC World Congress, Budapest .

Map 3.0 Specification, May 1987.

Wensley, J., et al. 1978. "SIFT: Design and analysis of a fault-tolerant computer for aircraft control", Proceedings of the IEEE, Vol. 66, No. 10.

Rodd, M.G., Izikowitz, I and Guo Feng Zhao. 1990 "RTMMS - An OSI-based Real-time Messaging System", Int. Journal on Real Time Systems, Vol. 2, pp.213-234.

Rodd, M. G. 1989(a). "Real-time Issues in Distributed Databases for Real-time Control". Proceedings IFAC Workshop on Real-time Distributed Databases, Pergamon Press.

Kopetz, H. and Ochsenreiter, W. 1987. "Clock Sychronization in Distributed Real-time Systems", IEEE Transactions on Computers, Vol.C-36, No.8.

Rodd, M.G., and Deravi, F., (1989(b), "Communication Systems for Industrial Control", Prentice Hall.

273

REAL TIME EXPERT SYSTEMS

Alfons Crespo Lorente
Depto. Ingeniería de Sistemas, Computadores y Automática
Universidad Politécnica de Valencia

1 REAL TIME SYSTEMS

1.1 Definitions

A real-time computer system can be defined as a system that performs its functions and responds to external, asynchronous events whithin a predictable amount of time.

There are some complementary definitions of real time systems:

- The correctness of the system depends not only on the logical results of the computation but also on the time at which the results are produced [Stankovic88,90]

- A RTS is a systems that has to respond to externally-generated input stimuli whitin a finite and specified period. [Young82]

The above definitions determine the predictable character of the system and the fast response time that is implicit in a real time system.

There are several classification schemas for real-time systems. From the response point of view, tasks can be classified as [Stankovic88]:

- Critical tasks are those that must meet their deadline, otherwise a catastrophic result might occur.

- Essential tasks are those tasks that are necessary to the operation of the system , have specific timing constraints, and will degrade the system's performance if their timing constraints are not meet. However essential tasks will not cause a catastrophe if they do not finish on time. There are a large number of them. They must be treated dynamically (modes, conditions, contingencies, etc.). [dynamic scheduling policy]

274

- Unessential tasks may or not may deadlines, and they execute when do not affect critical or essential tasks. Many background tasks, long-range planning tasks, and maintenance functions fall into this category.

Each task can have a set of time constraints:

- Deadline: Maximum delivery time of the actions

- Periodicity: time between two consecutive executions a task

- Preemptive or nonpreemptive execution of a task.

- Type: critical, essential or unessential (hard or soft)

- Incremental: given a fast answer with the possibility to refine the answer for the rest of its computation time

- temporal resources used.

The new generation of RTS will be large, complex, distributed and operate in uncertain environments. Many of them will include one or more expert system components or other AI programs. Applications as complex process control, robotics, monitoring of large systems, etc., are some of these fields.

RTS can be built on top of:

- Conventional O.S. with extensions for Real Time. In this case a set of additional features are provided in order to manage tasks with priorities, memory locking, fast input/output, etc..

- Run time support provided by the system: This is used in Ada environments when embedded system are developed.

- Explicit real time kernels: The software runs using the facilities provided by a kernel of RTOS. (VRTX, DARTS, etc.)

Most of the commercial environments for real time applications do not cover all the requirements of modern RTS. Current operating systems are inadequate for the following main reasons [Stankovic91]:

1. timing constraints are not considered explicitly

2. predictable task executions are difficult to ensure

3. tasks with complex characteristics (precedence constraints, resource requirements) are not handled explicitly.

Moreover, in future complex real time systems requirements should be based in the following considerations:

275

- Tasks are part of single application with a system-wide objective. The tasks are known beforehand and thus can be analyzed to determine their characteristics.

- Predictability should be ensured so the timing properties of both individual tasks and the system can be assessed.

- Flexibility should be ensured so system modifications and on-line dynamics are more easily accommodated.

2 EXPERT SYSTEMS

Knowledge based systems is directed specifically toward complex applications for which no polynomial algorithms are know. So, computation times are usually highly variable and unpredictable.

According to the computation process in artificial intelligence applications, some of the main characteristics are:

- Symbolic operation

- Approximate processing

- Response time not predictable

- Need of specific environments

Foundations of **Real time systems** and **Expert systems** have a large number of deep conflicts being not easy to collaborate to produce the desired systems. Some of these incompatibilities are shown in the following table.

RTS	ES
No dynamic data	All objects are dynamics
No memory growing	The memory is managed in the environment
No garbage collection	Main mechanism in the memory manager
Predictable computation	Not guaranteed response
Known resources	Resources not known
Static environment	Dynamic environment

However, there is an important need of this kind of techniques in real time systems due to they decrease the cognitive load on users, increase the productivity without the cognitive load, can manage large number of information (sensors, actuators, evolution, etc.), and can be used when conventional techniques have failed.

An example of this kind of application could be a rotary cement kiln because of:

276

- Several chemical reactions are produced with different dynamic.

- Instantaneous and delayed measurement

- Large number of incidence.

- Variable evolution have an important influence in the system

- Several stable states can be reached

- Different products composition are obtained.

- Several expert operators: (Operation, experience, formation)

- It is integrated in a distributed system where information is shared by different activities

- New Sensors can be added to the system (surface kiln temperature scanner)

- New cement composition

- New operators formation

2.1 RTES requirements

Last years a considerable effort has been done in order to define the kid of computation and implementation of artificial intelligence techniques for real time applications[Laffey88] [Albertos92] [Iserman87]. The following points resume the main requirements of real time expert system with respect to conventional ones.

- Continuous Operation: The system completes a cycle based on data acquisition, system evaluation, and output of results. Inputs are periodic or aperiodic, so the system is ready to be executed when new information is added.

- Guaranteed response time: an important property of this systems is that task resource requirements be predictable. When complex, cognitive computational tasks are incorporated into traditional RTS, it is difficult to guarantee a priori about the computational requirements of the tasks.

- Temporal Reasoning: as result of apply these systems to complex processes, it is needed to consider the values evolution and manage predictions about values. In order to do it, a model for temporal data management and methods for reasoning about time should be defined.

- Nonmonotonicity: due to new information (incoming from sensors), as well as deduced facts, do not remain static during the execution. Each period of acquisition data values are updated and old information or decay in validity with time or is obsolete.

277

- Event asynchronous management: although in real time systems the activity is performed in an orderly predictable manner, it has to be able to react to external asynchronous events and schedule the corresponding activities.

- Interface to external environments: in order to integrate it in real time software and environments.

- Efficient computation: the performance limitations of AI techniques reduces the application of this systems. Inference algorithms suitable to be applied in real time environments are some of the aspects to increase this efficiency.

As conclusion of this set of requirements, some important topics must be taken into account preferably:

- Guaranteed response time and its relation with the system architecture, reasoning techniques, and scheduling policies.

- Temporal representation and reasoning related with the model and its efficiency.

- Integration in real time environments, or the use of suitable real time operating systems

3 RTES Architectures

Blackboard systems are being used to experiment knowledge based systems in real time frameworks and it has been seen as one of the best alternatives to match complexity and variety of actual applications [Hayes90] [Jagannathan90] [Erickson91].

A blackboard is a medium for asynchronous communication between otherwise in-dependent knowledge sources. Besides, the blackboards has also been used as a data structure for representing the current world model, the solving state, and, perhaps, past and future states. The world model is represented by means of classes or levels. A class or level defines the common behaviour af all instances (objects or nodes) of it. This behaviour is defined by means of a set of characteristics or attributes, that represents the data, and, in some cases, a set of operations or methods that is specify the view or set of messages that this object is able to react.

The blackboard as a medium for communication is mostly used as a broadcasting manner: all knowledge sources can write or read. However, in some proposals, KS do not read information, it is sent from the blackboard to all KS interested in a concrete data. Previously, when KS's inform to the blackboard its interest (which data require) when are created.

However, expert system need to be integrated with other software components to complete the overall system including critical and non critical tasks. One of the main problem emerged is the predictable behavior of the tasks with AI techniques. In order to improve predictability it is needed to tackle two levels as shown in figure 5.

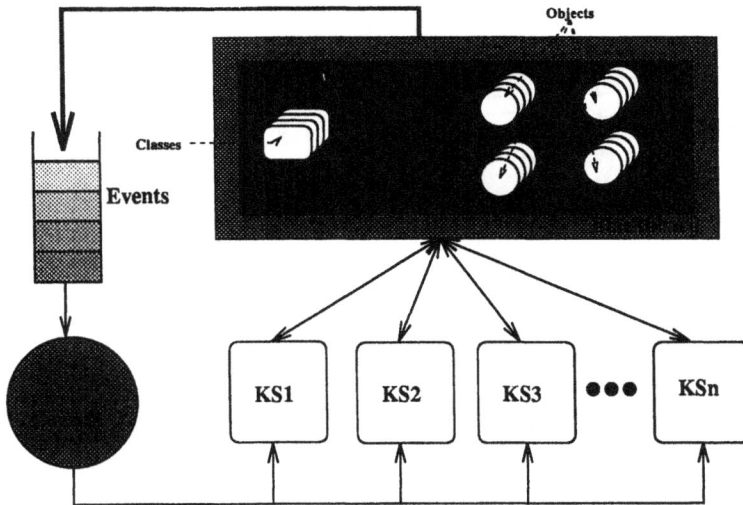

Figure 1: Blackboard Architecture

1. Global architecture: In a distributed environment some nodes perform real time activities implying AI components. Tasks in a node can be split into **critical tasks, other tasks (no critical)**, and **expert server tasks** where AI techniques relevant to a concrete problem are embodied.

   All these tasks run on top of a real time operating system that provides priority based scheduling. The priority of each task is determined based on periods, computation time and resources used.

2. Expert task architecture: which provides architectural foundations to be applied in complex environments with temporal constraints. The expert task can be based in a blackboard architecture with extensions for real time problem solving.

The first level is solved by means of traditional techniques in the field of real time systems using appropriated scheduling algorithms in real time operating systems. On the other hand, the second level has to be considered in a separate way in order to manage its predictability. In the rest of this lecture we will concentrate our study in the second level.

3.1   Black-Board Architectures

The blackboard model consists of three main components:

- **Blackboard Database**: It stores the application objects organized in multiple levels (classes), providing mechanism to define classes, create instances and manage them. The main problems addressed are related to the concurrent access of several parallel inference tasks, temporal object representation and reasoning, and efficient methods to use objects in memory.

279

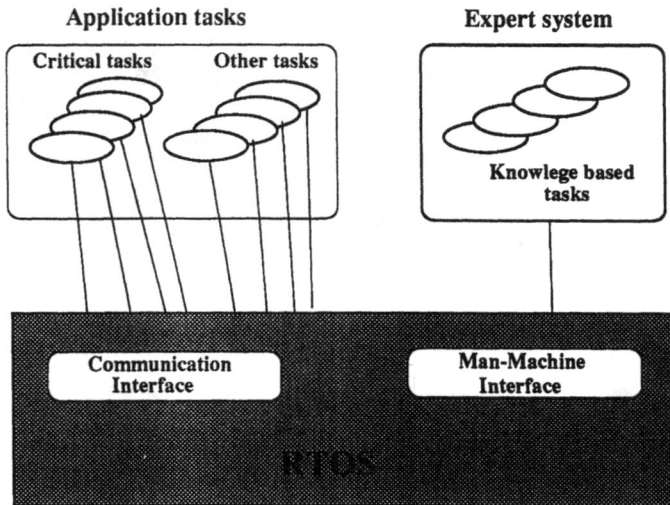

Figure 2: Global Architecture

- **Knowledge Source Formalism**: which includes a domain specific knowledge. Global problem is partitioned into several knowledge sources (KS) each one contributing to solve a subproblem. KS are represented as procedures, sets of rules or logic assertions. The main aspects to be applied in real time systems are related to efficient inference methods, interruptability and parallel execution.

- **The Control Component**: The main objectives of control are to manage goals in response to dynamics of the environment, improve system robustness, and maximize use of resources. Both BD and KS emit events to the control (stored in an agenda) in order to notify changes in the system. Control permits select the most appropriate knowledge sources to solve a problem allocating the used resources at time to complete the reasoning process taking into account the deadlines. Interesting points are: selection of most appropriated events, to guarantee the response time, parallel execution of KS.

Blackboard systems are event event-driven. Event are signaled as result of blackboard activity or inference processes. Events are managed in an agenda where Control selects the most appropriated to be handled.

3.2 Blackboard Database

Blackboard database (BBDB) provides mechanisms to object definition, instantiation and management. BBDB in real time systems offers the following features:

- *User class (levels) with temporal attributes definition*: classes define the common behavior of applications objects. A class has attributes defining its characteristics some of them can be temporal. For instance:

280

```
class TANK is defined by
    atemporal attributes:
        DIAMETER,
        HEIGHT,
        SENSORS
    temporal attributes:
        LEVEL,
        ALARM
    end TANK
```

In this case some attributes are constant or we do not want to remember old values, and others are temporal where we can store old values (know the evolution of a level, in which instants an alarm was produced) or perform predictions about it (in 30 minutes the level will be high, or an alarm will be signaled before one hour).

- *A model of temporal representation*: In order to maintain the temporal information a model must be defined. Next section will describe models to represent temporal information and functions to manage it. The temporal model has to be assume the real time constraints considering: i) time where old facts were produced can be known, so it is possible to reference all past fact with respect to an absolute clock; ii) future facts can depend on other facts and it is possible that a concrete time not be known; iii) efficient methods to manage temporal facts.

- *Object instantiation and management*: Objects are instances of defined classes. BBDB has to allocate objects into memory and manage it efficiently. The main problems arise due to dynamic creation and deletion of objects, management of obsolete information when temporal information is used, and predictable methods to remove old objects no longer referenced. When high level environments (Lisp, Smalltalk) are used as support to develop this features, it is tackle by the environment but user can customisize some parameters.

- *Updating and retrieving* : User can define methods to manage attribute (slots) values in classes, or BBDB provides a set of primitives to access or modify attribute values. In the last case, methods to get or set a value a uniform to all objects and efficient algorithms can be defined in the BBDB structure.

- *Uncertainty management*: Real time systems applied to control of complex systems need some kind of uncertainty management related to the acquired or deduced information. A large number of proposal to represent and manage it can be seen in the literature. This topic is the central point of other lectures and will be develop on it.

- *Temporal primitives to reason about time*: Past and future facts are maintained in object attributes. BBDB should provide methods to retrieve past information (at a concrete time, before another value) calculate trends, put future values and inform to the KS interested when this prediction is fulfilled.

281

- *Concurrent access control to objects*: Since several KS can be executing at time, it is necessary to protect information from concurrent access. So, internal mechanisms for lock and unlock resources should be provided as well as some kind of real time protocol to resource management. In this way, the BBDB can be considered as a multiple server where priority ceiling protocol can be applied in order to meet real time constraints.

3.3 Knowledge Sources

In general, KS tend to combine a collection of activities. These activities are grouped together according to different criteria. A KS can be instantiated as consequence of the determination of a goal and run until the goal is reached. In order to apply this approach to real time systems, each KS is structured in an independent task (integrated by the KS, inference engine of the appropriated formalism, and the local memory) called **agent**.

Each agent is able cooperate in a concrete reasoning process to a global solution. Sequential execution of agents provides the global answer to a goal.

The system can be seen as a collection of **agents** (reasoning modules) which share a common blackboard data base and communicate with each other by signaling events.

One important point is related to the efficiency of the reasoning algorithms, it is proved that forward chaining in production systems is exponential time depending on of the depth of the inference tree. Several algorithms has been provided to real time pattern matching RETE [Forgy82] and TREAT . Although in both cases cost evaluation is unpredictable (in terms of number of joins operations), it is possible to limit the cost establishing maximum number of number of matches, number of instances of a pattern, and number of instances of a relation.

RETE and TREAT algorithms build a network compiled from **if** parts of rules and inputs are changes in the working memory as result of modifications in the blackboard data base.

Progressive deepening is an alternative method to keep computing within time bounds. This approach evaluates first level of depth, then second, then third, and so on, until maximum time is reached or a satisfactory solution obtained.

Real time A^* algorithm is a modification of the known A^* taking into account the time as a constraint in the tree search.

3.4 Control Component

The control module plays one of the most important roles when blackboard are applied to real time system. Control reads from event queue and creates agents to solve a goal. The problem can be considered very close to the scheduling in a real time operating system: there are several goals to be evaluated with a deadline associated to each one, so tasks, as consequence of each event management has to be created and executed before a deadline giving the appropriated resources (computation time and access to objects).

Several methods can be used in order to manage a deadline:

- Priority Scheduling: In this case, events or goals have defined a priority based in the importance of the situation, so control select the most priority tasks to be executed. Based in the priority, queries to BBDB are internally handled in the proper way.

- Deadline Scheduling: based in the deadline of a response to an event. It can be managed like a dynamic priority that changes over the time.

- Progressive reasoning: In this case, several KS model a subproblem with different level of depth and, consequently, with different time-consuming. Taking into account the average time of each one, the deadline, and the activities to be executed, control can decided which of them is instantiated. This kind of evaluation can be complementary to progressive reasoning described before.

- Plan definition: taking into account the behavior of the expert task as defined before, it is possible that each or several executions a plan with regards to the global answers could be created. The planning selects and sequences tasks achieving one or more goals and satisfying a set of constraints like: shared resources, deadlines, etc.

  Scheduling selects alternative plans, and assigns resources and times for each activity of the plan so that the temporal constraints and shared resources are obeyed.

  When the plan is created, each task may specify: duration, start and end time, periodicity, resource requirements, relation with respect to other tasks, and other special constraints.

  Moreover, the plan should consider the global deadline of all tasks, so, taking into account the different versions of each subproblem (deepening levels), several strategies can be developed. If, as it is convenient, the system has to provide an answer with a determined quality (depending on the level of deepening evaluated), the plan could start from a minimum quality result, increasing, if there is more time, the level of depth in order to reach an equilibrate quality.

  The problem of build plans imply the creation of a complex graph and algorithms to analyze it. This problem requires large time consuming and should be limited to background activities.

4 Temporal Representation and Reasoning

The temporal representation and reasoning problem arises in a wide range of Knowledge-Based System (KBS) application areas, where time plays a crucial role such as in process control and monitoring, fault detection, diagnosis and causal explanation, resource management and planning, etc. In these cases, temporal data representation is needed in order to obtain conclusions about the problem. Moreover, temporal dependence of knowledge relations (temporal constraints and parameterization of their application) is possible. However, neither the several Expert System (ES) prototypes developed to handle these applications [ISERMANN87, LAFFEY88, PERKINS90, TSANG88, etc.] nor the several expert system development tools which have a certain temporal capacity offer a general solution to the problem and the temporal representation and reasoning capacity of the ES has an **ad hoc** treatment in each case, which is too restrictive for the wide range of temporal domains. Temporal constraint and parameterization about rule application is not usually considered and the dynamic of the problem has not a natural representation.

for temporal representation of the information and the properties for this representation. One group of these solutions was based on previous 'state models', where the successive states represent the temporal ordering. Another earlier proposal in this matter was the Kahn's 'time specialist' module. The temporal logic models, based on modal logic, constitute a more formal vision, where there is an explicit time representation, and the value of an expression depends on the instant it was calculated. The structure of the underlying time model depends on the properties of the temporal precedence relation, i.e.: lineal or branching, discrete or continuous, complete or not, convergent or unbounded. In this way, first order temporal logics are decidable in only a few very restricted cases, when time is assimilated to a finite subset of integers. All these approaches are too complex for their application in a KBS methodology.

Representing a temporal expression as an atemporal form in a many-sorted predicate calculus is not adequate for AI requirements. Time and its properties must be represented explicitly, and it is necessary to separate the atemporal expressions (expression-types) from their temporal components in which the expressions are true. Thus, these temporal models in AI have an explicit time representation, based on time-points or intervals. 'Expression-types', as well-formed formula in first order logic, represent problem data and are conceptualized by their properties in a specified time interval. 'Expression-tokens', as ¡expression-type, temporal-constraint-of-validity¿, represent the data facts. The logical value [T,F] of expression-tokens depends on the instant at which the interpretation is calculated

All these models constitute a useful frame of reference. However, they seem somewhat inappropriate for the specific problem of temporal representation and reasoning in a KBS methodology. The main drawbacks are related to expressiveness and structuring of temporal concepts, quantitative or qualitative specification of temporal dependence of the problem facts and knowledge relations, the representation of facts duration, the difficulty of reasoning about overlapped tokens in time and the limited causal relations.

In order to handle this problem in Artificial Intelligence (AI) areas (action-planning, qualitative reasoning about processes, explanation of the world at some earlier time, and prediction or temporal projection of the future) in a more specific way, several temporal logic-based models, from initial **situation calculus** [McCARTHY69], were defined [ALLEN84, McDERMOTT82, KOWALSKI86, SHOHAM88]. Thus, these temporal models in AI have an explicit time representation, based on time-points or intervals.

With a particular representation, the **Temporal Imagery** concept is related to representation and inference methods with large amounts of temporal information, and events and facts are arranged in a **time map**, based on temporal intervals [ALLEN83]

or time points [DEAN87]. Other models are included in the Qualitative Reasoning concept [DeKLEER84, KUIPERS84, FORBUS84], with a qualitative consideration of the problem dynamic, about which the laws of change are defined.

However, some important points are not completly solved when temporal reasoning is used under time constraints. The need of manage past and future facts and the number of possible variable relations can produce large operations for update and retrieve temporal information. Thus, an efficient management must be addressed.

4.1   Temporal information

Temporal information in control systems is relevant to know the evolution of the process and deduced variables and reason about it. The need of temporal information is related to past and future values. From a general point of view, a temporal value can be:

1. **exactly known**: when the time, where values were taken, is known (the alarm was 'on' from 10:20:03 till 10:28:45)

2. **partially known**: when a value is result a of sampling and this value is maintained till the next (the temperature at 10:23:22 was 120 degrees)

3. **known with an uncertainty**: when a variable will take a value in some instant of an interval (the kiln status will be HIGH before 20 minutes)

4. **dependent**: when the exact instant of a temporal value depend on other temporal facts (the alarm will be ON 20 minutes after the temperature be HIGH)

In terms of process control, 1 and 2 can be assumed as equivalent due to incoming values are produced at known instants (sampling period), and, if no more information is added, the same value can be maintained. This behaviour can be modeled by means of a *data persistence* as the time which the variable maintain its value. If this time is exceed a lack of information is produced and no value is assumed. When this persistence is ∞, the last value is always maintained.

Past values are always related to the instant they were produced, so, an absolute date can be used to store them. In this case there is not relation between temporal information, all of them are related to the clock and the relations must be deduced using clock values.

Future or prediction values introduce more difficulties due to the lack of knowlege about the exact instant they will be produced. In the third case, we predict that a status will be HIGH in some instant before 20 minutes, but we do not know the concret instant it will start, finish or its duration. So, a model to represent and manage prediction has to represent this uncertainty and be able to reason about it.

Besides, fourth case introduces more complexity because of the dependency with another fact, when a fact occurs the dependent fact is equivalent to the third case, if it does not occurs neither the second one.

285

Moreover, the system behaviour defines a set of temporal dependencies: a material in a cement kiln, from the input to the output takes about 10 minutes, so if we measure the temperature at the input, some prediction can be done about the future temperature in the middle and the output of the kiln, taking into account the process model, delays, etc. On the other hand, the system can take actions based in predictions in order to avoid problems in temperature.

In rule based systems, it means that rules should be fired with present and future facts and actions, in the right hand side of a rule, should be executed in the instant the values are deduced or when they are present.

The main relevant characteristics to real time systems can be stated as:

- The representation should allow past and future facts

- Past values should be attached to an absolute clock

- Future facts represent predictions that can be related with other temporal facts or to an absolute date.

- Future facts can be converted to current depending on its dependencies and the real time.

- Different granularity of time

- The model have to assume some persistence

4.2 Temporal models

In order to represent temporal information a model based on instantaneous dates (time points) or intervals. Both can be defined as:

- Point-based: basic information of a temporal fact is related to a precise point in time.
  The Temperature was 27 degrees at 10/9/91/12:12:00
  The Flow was 0.5 30 minutes after (Temperature,27,10/9/91/12:12:00).

- Interval-based: Basic information of a temporal fact is related to an interval of time in which the finding event occurred.
  Yesterday the temperature was 27 degrees
  The valve was open while the temperature was 27.

More adequate seems the Point-based model because of an interval can be decomposed in two points (begin, end). However, some events appear to be instantaneous (*yesterday, I parked my car in the street*), but this event can be more closely examined and we can find that **to park** is a set of sequential actions of a duration. Recursively, each action can be decompose in .....

286

286

4.2.1 Point-based models

Several models has been proposed in the liturature [Dean87] [Barber90] [Barber92]. The primitive is the **time-points** (tp). A tp represents a time instant in a discrete temporal line.

The main realtion between time point are:

- (BEFORE tp_i, tp_j): expresses that tp_i is temporally before tp_j

- (AT tp_i, td_k, tp_j): expresses that tp_i is temporally at tp_j

- (AFTER tp_i, td_k, tp_j): expresses that tp_i is temporally after tp_j

An additional concept is the **temporal duration**(td) as the temporal distance between two tp's.

The temporal relations relating two tp's and the duration td_k are:

- (BEFORE tp_i, td_k, tp_j): expresses that tp_i is temporally before tp_j plus a certain temporal duration td_k.

- (AT tp_i, td_k, tp_j): expresses that tp_i is temporally at tp_j plus a certain temporal duration td_k.

- (AFTER tp_i, td_k, tp_j): expresses that tp_i is temporally after tp_j plus a certain temporal duration td_k.

4.2.2 Interval-based models

In these models the primitve is the **interval**, related to two time instants where the event starts and finish its ocurrence. An interval is always considered as closed [Allen83].

The following relations between intervals can be expressed :

Relation	Symbol	Inverse	Example
X before Y	<	>	XXX YYY
X equal Y	=	=	XXX YYY
X meets Y	m	mi	XXXYYY
X overlaps Y	o	oi	XXX YYY
X during Y	d	di	XXX YYYYYYY
X starts Y	s	si	XXX YYYYYY
X finishes Y	f	fi	XXX YYYYY

287

Interval Relation	Equivalent time points
x < y	$x{\downarrow} < y{\uparrow}$
x = y	$(x{\uparrow} = y{\uparrow})\ \&\ (x{\downarrow} = y{\downarrow})$
x overlaps y	$(x{\uparrow} < y{\uparrow})\ \&\ (x{\downarrow} > y{\uparrow})\ \&\ (x{\downarrow} < y{\downarrow})$
x meets y	$x{\downarrow} = y{\uparrow}$
x during y	$((x{\uparrow} > y{\uparrow})\ \&\ (x{\downarrow} =< y{\downarrow}))$ or $((x{\uparrow} >= y{\uparrow})\ \&\ (x{\downarrow} < y{\downarrow}))$

4.2.3 Temporal reasoning

Temporal relations are represented in a **Time Map** (a graph) where nodes are the time points and the labeled directed edges show the temporal constraints between them:

Nodes: time points (tp_i)

Edges: Temporal constraints between time points: (**Before, At, or After.**

The set of management routines of a Time Map is called **Time Map Manager** (TMM). TMM functions manage the basic operations of creation, deletion, and temporal relations between temporal values. In order to manage future values in a proper way, a suitable solution is to provide a reason maintenance system (RMS) cooperating with the TMM. Both components allow to manage logical dependencies and coherence of temporal values.

The main operations of a TMM are the updating and retrieving temporal information.

The function for updating a temporal constraint between two time points. The updating process of a new constraint performs a consistency and non-redundancy checks. If the new constraint is consistent and non-redundant with the existing ones, the new constraint is asserted in the Time Map. Thus, the result of this operation may be:

- Contradictory, whether the temporal constraint to be updated is inconsistent with the information already represented in the TBB. In this case the TBB keeps with no modifications.

- Redundant, whether the temporal constraint to be updated can be retrieved from the ones existing in TBB, and so it will not be introduced.

- Updated, whether the temporal constraint is updated.

The retrieval of an specific temporal constraint between two nodes is stated as a path searching process between the two nodes throughout the rest of nodes in the Time Map. Due to the existence of many possible paths between the two nodes, the goal is to find a restrictive path according to the inquired constraint by combining the temporal constraints of the edges in the path. The above recovery processes obtains a response about a concrete temporal constraint between two time points.

5 Guaranteed Response Time

One of the main aspects that has to be considered is the guaranteed response time in the system. It has to be considered in each component of the system:

1. **Control:** to stablish a plan and execute it assuring that resources and temporal constraints are appropriated.

2. **Inference engine:** Given the best answer in time. Efficient organization and inference process.

   RETE and TREAT algorithms are the most efficent pattern matching algorithms.

   Progressive reasoning: several levels of deepening, using simple functions in the first levels and more complex functions in the upper levels.

3. **Knowledge Base:** Organized in the proper way to be used efficiently by the IE.

   Subproblem structuration: Regulators, subprocess, etc.

   Object Language definition: languages with multiple inheritance use graphs with exponential costs. Simple inheritance has a linear cost.

5.1 Knowledge Base organization

From response time point of view, each knowledge base can be structured in several versions. Each version gives an answer with a determined quality depending on its level (version i gives a worse quality response than version j). Of course, best answers require higher computation time. So, Each KS model a subproblem defining different or more or less complex ways of obtain an answer.

5.2 Control and scheduling

The control module plays one of the most important roles when blackboard are applied to real time systems. Control reads event queue and creates processes to solve a concrete goal. The problem can be considered very close to that of the scheduling in a real time operating system: there are several goals to be evaluated with a deadline associated to each on. So processes, as consequence of each event management, have to be created and executed before a deadline, giving the appropriated resources (computation time and access to objects).

Due to the global behavior of the server task [Paul91] [Crespo92] [Garcia92], when its period is reached, most periodic processes are suitable to be executed and some external events as consequence of external data can be queued. So, each time **Control** is going to be executed must build or revise an execution plan. This plan determines the set of tasks to be executed, the instant they have to be finish, the computation resources required, and, consequently, the version of regulator to be used [Sha87] [Sha89] [Sprunt89]. Between the set of possible plans to be generated, the one obtaining the best quality response in the most important tasks is selected. This plan is revised if new events are added when the plan is in execution. Associated to each operation mode a set of previous plans are created assuming the normal execution in order to reduce the time plan creation. The plan is divided into two steps: creation and execution.

Let be $\mathbf{T} = T_1, T_2, ..., T_n$ a set of tasks, where each T_i is characterized by a set of versions V_i each one requiring more computation time and better response quality. Task in $\mathbf{T}$ may depend on other task. A directed graph $\mathbf{G(T,R)}$ determines the existing task relation, where $R_{i,j}$ expresses the dependency between T_i and T_j, so T_j has to be executed after T_i. A maximum deadline $\mathbf{D}$ associated to $\mathbf{T}$ is defined. Within this time, a version of each task in the plan has to be executed. $\mathbf{D}$ is related to the service time of the expert task. From this deadline, it can be deduced:

$$\sum_{i=1}^{n} C_{min}(T_i) \leq D \qquad (1)$$

where $C_{min}(T_i)$ represents the computation time of the version V_1 of the T_i.

The user can define an importance to each task T_i which can be used to impose an execution order.

The goal is to obtain a plan with the best versions of each task constrained with the precedence relations, importance and deadline.

The precedence order is obtained from a topologic sort of R in the graph G. For this path some known algorithms ($O[\log(t, r)]$) can be used including the task importance in the node selection at each step. The main important point is the version V_i of each T_i.

6   Examples

In this section two architectures of real time expert system are described. First example shows the architecture developed in the project ESPRIT REAKT 5146, and presents a general purpose architecture. The second example describes a process control oriented architecture (RIGAS) that is being used for the control of a rotary kiln in a cement factory.

6.1   REAKT architecture

The main goal of this ESPRIT project REAKT [REAKT] is to develop a real time expert system environment and methodology.

The main features at the end of the second year are:

- Organization as an **Expert Task**
- Multi-process architecture
- Predictability

- High performance inference engine

- Temporal reasoning: past, present and future plus real time clock management

- Continous operation

- Focus of attention mechanism

- Integration with conventional software

The proposed architecture is shown in figure.

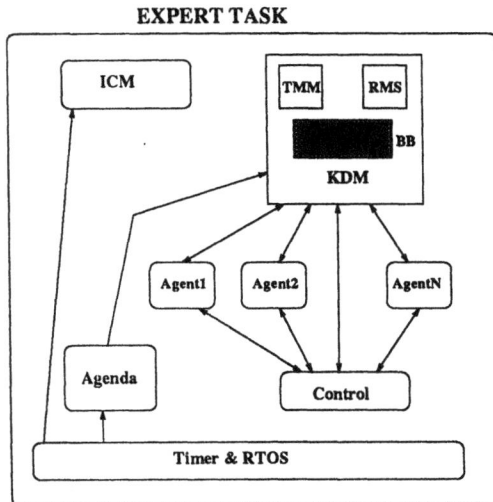

Figure 3: REAKT Architecture

The main components are:

- **KDM**: (Knowledge Base/Manager) is a temporal and concurrent blackboard where classes, objects, and knowledge bases are managed [Barber92]. Internally, a Time Map Manager (TMM) and a Reason Maintenace System (RMS) has been defined in order to manage temporal information. With this model as background, application objects are instances of classes with static and temporal attributes or slots. Each object in the blackboard has a pointer to the parent class description and a set of static and temporal slots. The structure of a temporal slot is formed by the following pointers:

    - object: which this slot belongs

    - current: pointer to the temporal value that current now

    - past: list of pointers to past temporal values

    - future: list of pointers to future temporal values

291

The TEMPORAL_VALUE class is defined by means of a **value** and the **begin** and **end** points as the instant limits where the value was, is, or will be taken. Both time points are pointers to *tp* (instances of TM_NODE) in the **Time Map**. Each *tp* has a pointer to the slot which it belongs, and three lists of *tp*'s with relation of BEFORE, AFTER, and AT associated with a temporal distance.

The following figure 4 shows the relations between these data structures.

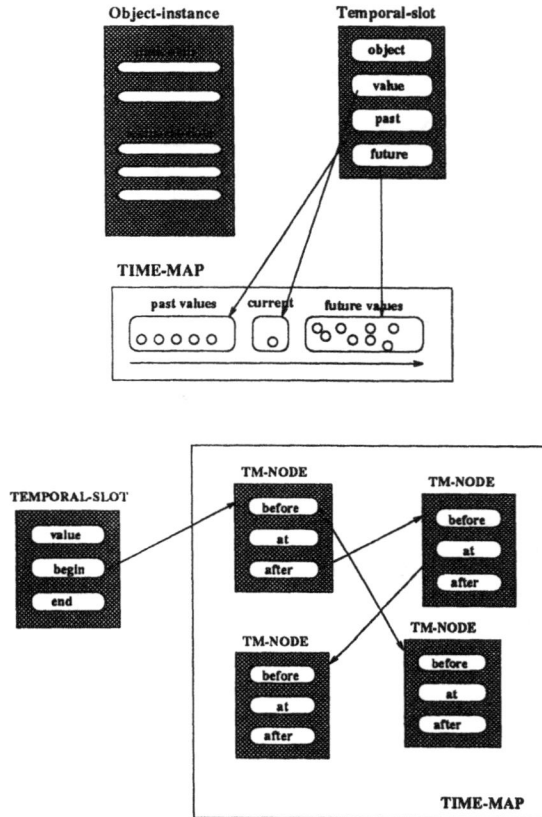

Figure 4: Temporal data structures

- **Agents:** are the knowledge based components. Each one is a process sharing data in the KDM. Agents are created and removed by the control component.

- **Control:** determines the internal execution. It receives events from otehr components, and as result, create intentions as a relation between a goal, the KS that have to be executed to solve it, and an execution plan to schedule all of them. The main characteristics to be considered are deadline of a plan and progressive reasoning.

- **ICM:** (Intelligent Communication Manager) allows to send and receive data from external inputs.

292

- **Timer:** control the system time, and allows the execution of delayed actions. It uses an agenda as base to organize the actions.

6.2 RIGAS architecture

This architecture starts from the real time point of view adding some rule-based features in order to include expert capabilities to the system [Crespo92] [Morant92] [Garcia90] [Rigas92].

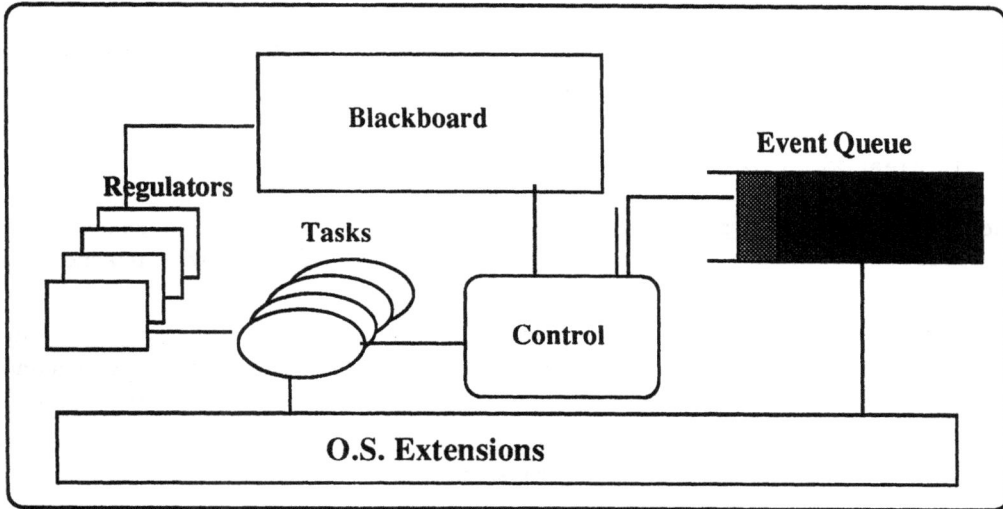

Figure 5: RIGAS Architecture

The system has the following components:

1. Scheduler: performs the selection of the tasks to be executed according to a scheduling algorithm taking in account deadlines, resources, time constraints, etc.

2. Event_Management: maintains the information related to the delayed and periodic events.

3. Regulators: are the knowledge components in the systems. Some regulators can be experts defining rule-based algorithms. Each expert regulator provides an inference engine in order to be independent.

4. Temporal_Object_Data_Base: contains the application objects. These objects are instances of predefined classes in the system with temporal features.

5. Man_machine_Interface: represents the the interface between the expert system and the user. The **MMI** shows graphical values, buttons to select options, panels explaining the inference, set points modification, etc. Moreover object, rule and requirement editors are incorporated in this module.

293

6. Communication_Interface: performs the external communication between the expert system and the data acquisition systems implementing protocols to guarantee a reliable and fast communication.

The global model is based on events. Events are emitted from several sources: hardware, tasks, rules, and objects. Events are collected in the agenda in order to manage periodic, instantaneous or delayed behaviors. When an event have to be known in the system the agenda inform to the scheduler this fact. The scheduler selects the tasks associated to the event, verify some time constraints and schedule it. The scheduler selects the most appropriated task or tasks to be executed.

Each task in the system is a thread of execution sending messages to the system objects (regulators, application objects, **MMI** or communication module).

6.2.1 Real Time Requirements

Requirements define the global reaction of the system. There are two basic requirements: **periodic** and **sporadic**, and two derived: **temporal periodic**, and with **precedence** relations.

A requirement specifies the response to the system to a single event, represented by some simple or composite action. The response can be dependent on some condition and a set of temporal constraints for the required actions specified as part of the requirement.

In an object oriented approach, *actions* specify the set of messages that are sent to the application objects in order to perform some activity.

Actions are sequential or parallel operations. Parallel actions are considered when the partial temporal ordering of the execution of the actions is not relevant to the final result.

6.2.2 Scheduler

Real time requirements are tasks or processes in the execution environment. These tasks have a set of constraints that should be guaranteed.

The scheduler has a predefined plan built taking into account the temporal constraints, resources used, and precedence relations between tasks.

The scheduler performs the following actions:

1. As result of an event occurrence (due to the clock or an event) the task or tasks related to the event are selected

2. Selected tasks and goals are selected in the planned graph and scheduled properly.

Parallel actions are considered as several subtasks (internally represented by a graph) with the same event or period and the same time constraints.

6.2.3 Regulators

The knowledge-based components are called **Intelligent Regulators (IR)** with reference to the traditional regulators in process control.

Each **IR**, as shown in figure 2, defines a set of problem solvers (rule-based or algorithm) in order to provide solution to a concrete subsystem.

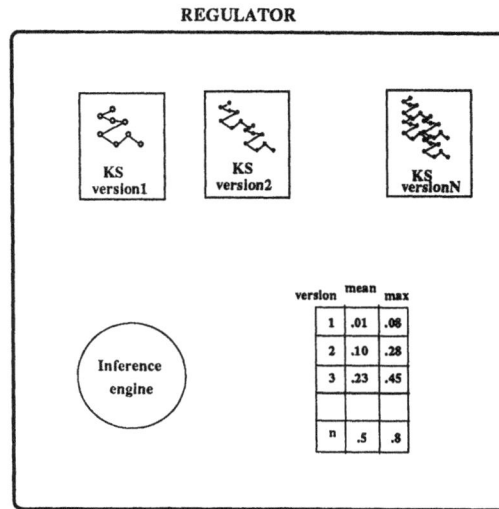

Figure 6: Regulator

The normal configuration of an **IR** is:

1. **Fast rule-based solver:**
   Using reduced set of variables, and simple funtions in the rules. A predicted time of execution is related to the solver.

2. **Complete rule-based solver:**
   Using all the variables and relations between them, and using complex functions in rules (temporal reasoning). As in the above solver, a predicted time is associated to the solver.

3. **Algorithm solver:**
   Classical regulator (PID, dead-beat, adaptative, etc.) could be defined in order to supply alternative results when no solution is reached by the above solvers. As in other cases, an execution time is associated to this solver.

As levels of rule-based solvers as the designer decide could be defined in the **IR**. Each level adds deep in the reasoning process with respect to the previous applied.

Moreover, the **IR** has defined an *inference engine* in order to evaluate the solution using the appropiate solver taking in account the response time.

295

References

[Albertos92] P. Albertos, A. Crespo, F. Morant, J.L. Navarro. *Intelligent Controllers Issues.* IFAC SICICA'92. Malaga. 1992.

[Allen83] J.F. Allen "Maintaining knowledge about temporal intervals". *Comm. of the ACM.* VOl 26, No. 11, 1983.

[Allen84] Allen J.F. *Towards a general Theory of Action and Time.* Artificial Intelligence, 23, 1984.

[Barber90] F. Barber, V. Botti, A. Crespo *Temporal Expert System: Application to a Planning Environment.* IASTED Symposium. USA. December 1990. 78-82.

[Barber92] F. Barber, E. Onaindía, M. Alonso. "Representation and management of temporal relations". *Novática.* 1992.

[Barber94] F. Barber, V. Botti, A. Crespo, DF. Gallardo, E. Onaindia. "A Temporal blackboard structure for Process Control". *1st IFAC Symposium on AIRTC.* Delft. 1992.

[Crespo91] A. Crespo, J.L. Navarro, R. Vivó. A. Espinosa, A. García. *A Real-Time Expert System for Process Control.* 3th IFAC Intl.Workshop on Artificial Inteligence in Real Time Control. September 1991. California. U.S.A.

[Crespo92] A. Crespo, J.L. Navarro, R. Vivó. A. Espinosa, A. García. *RIGAS: an expert server task in real-time environments.* 1st IFAC Symposium on Artificial Inteligence in Real Time Control. Delft 1992.

[Dean87] Dean T. and McDermott D. *Temporal Data Base Management* Articial Intelligence, 32, 1987

[Erickson91] W. Erickson, L. Braum. *Real-Time Erasmus.* Ninth National Conference on Artificial Intelligence. AAAI-91. Jul, California, 1991.

[Forbus84] K.D. Forbus. "Qualitative process theory" *Articial Intelligence*, 24 August (1984) pp.85-168.

[Forgy82] C. Forgy. "RETE: A fast algorithm for the many pattern/many object pattern match problem".*Artificial Intelligence.* 19, North-Holland. 1982

[Foxvog91] D. Foxvog, M. Kurki. "Survey of Real-Time and On-Line Diagnostic Expert Systems".*Proceedings Euromicro'91 Workshop on Real Time.* 1991

[García90] A. García, A. Espinosa, A. Crespo, J.A. de la Puente *QUISAP: an environment for Rapid Prototyping of Real-Time Systems.* EuroComp'90. Israel May 1990.

[García92] A. García, A. Crespo. *Towards an environment for complex real time control systems developing.* 17th Workshop on Real Time Programming. WRTP'92. 1992.

[Hayes90] B. Hayes-Roth *Architectural Foundations for Real-Time Performance in Intelligent Agents. Real Time Systems.* Vol 2, 1,2,

[Iserman87] R.Isermann (Ed.). *Procc. of the 10th world congress on automatic control.* IFAC-87, vol.6, (1987).

[Jagannathan90] V. Jagannathan, R. Dodhiawala and L. Braun. *Blackboard Architectures and Applications.* Academic Press. 1990.

[Klijgsman91] A.J. Klijgsman, R. Jager, H.B. Verbruggen, P.M. Bruijn. "DICE: A Framework for Real Time Intelligent Control", *3rd IFAC Intl.Workshop on Artificial Inteligence in Real Time Control.* September 1991. California. U.S.A.

[Kowalski85] R.A.Kowalski, M.Sergot. "A logic-based calculus of events." *New Generation Computing,*4 (1).(1985).pp.67-95.

[Kuipers84] B.Kuipers. "Commonsense reasoning about causality: Deriving behavior from structure" *Artificial Intelligence,* 24, pp.169-203.

[Laffey88] T.J. Laffey, P.A. Cox, J.L. Schmidt, S.M. Kao and J.Y. Read *Real-Time Knowledge-Based Systems. AI Magazine.* Spring. 1988.

[McCarthy69] J.M. McCarthy, P.J. Hayes. "Some philosophical problems from the standpoint of artificial intelligence." *Machine Intelligence,*4. Edinburgh Univ. Press. (1969).

[McDermott82] D.V. McDermott. "A temporal logic for reasoning about processes and plans." *Cognitive Science,*6. (1982). pp.101-155.

[Morant92] F. Morant, P. Albertos, M. Martinez, A. Crespo, J.L. Navarro "RIGAS: An intelligent controller for cement kiln control" *Symp. on AIRTC* Delft. 1992.

[Paul91] C.J. Paul, A. Acharya, B. Black, J.K. Strosnider. *Reducing Problem-Solving Variance to Improve Predictability.* Communications of the ACM. Aug. 1991, Vol. 34, No.8.

[Perkins90] W.Perkins, A.Austin. "Adding temporal reasoning to expert-systems building environments". *IEEE-Expert.* Feb.1990.

[Rauf89] P. Raufles *Toward a Blackboard Architecture for Real-Time Interactions with Dynamic Systems..* Proceedings of IJCAI 1989.

[REAKT90] Thomson, Syseca, Crin, GMV, UPV, Marconi, Etnoteam. *REAKT: Environment and Methodology for Real-time Knowledge Based Systems.* ESPRIT II 4651. 1990-93

[RIGAS92] RIGAS Reference manual. Internal Report DISCA-UPV.Enero 1992 (in Spanish).

[Sha87]   L. Sha R. Rajkumar and J.P. Lehoczky "Priority Inheritance Protocols: An approach to Real-Time Synchronization" *Technical Report CMU-CS-87-181*. Computer Science Department. Carnegie Mellon University, Pittsburgh, PA 15213. Nov. 1987

[Sha89]   L. Sha et al. "Mode change Protocols for Priority-Driven Preemptive Scheduling". *J. Real-Time Systems*, vol.1, 1989, pp.243-264

[SOHAM88] Y. Shoham. *Reasoning about change*. MIT Press.(1988)

[Sprunt89] B. Sprunt, L. Sha, J. Lehoczky. "Aperiodic task scheduling for hard real-time systems". *J. Real-Time Systems*, vol.1, 1989, pp.27-60

[Stankovic88] J.A. Stankovic and K. Ramamritham. "Misconceptions about Real-Time Computing: A Serious Problem for Next-Generation Systems". *IEEE Computer*. Vol. 21, No. 10, Oct 1988.

[Stankovic90] J.A. Stankovic and K. Ramamritham. "What is Predictability for real Time Systems?". *Journal of Real Time Systems*. Vol 2, 4, 247,254. 1990

[Stankovic91] J.A. Stankovic and K. Ramamritham. "The Spring Kernel: A new paradigm for Real-Time Systems". *IEEE Software*. May 1991. pp: 62-72.

[Tsang88] E.Tsang. "Elements in temporal reasoning in planning" *Procc. of ECAI-88 Munchen*. pp.571-573.

[Young82] S.J. Young, *Real-time languages: Design and development*. Ellis Horwood, 1982. Trad. cast. *Lenguajes en tiempo real*. Paraninfo, 1987.

PART IV

CAD Systems using Expert Systems

Road Map.

The fourth part of the course consists of one chapter, which deals with *CAD systems using expert systems*. Basic knowledge on control engineering is expected, and will not be discussed in this text. The target field primarily is process control, because of some applications chosen, but it has been kept sufficiently loose from this in order to be more widely useful.

User Interfaces and the Role of Artificial Intelligence in Computer Aided Control Engineering. This chapter shows the changing role of the engineer in designing control systems. Over the years this role changed from actually manually designing control systems to letting computers take care of the numerical aspects of designing in Computer Aided Control System Design (CACSD) and now also expert systems taking over (parts of) the design task. Starting from a model of computer aided design, the development in CACSD is shown. The developments of the computer hardware and the for control system needed software environments leads to the state-of-the-art and to artificial intelligence applications in CACSD.

USER INTERFACE ISSUES AND THE ROLE OF ARTIFICIAL INTELLIGENCE
in
COMPUTER-AIDED CONTROL ENGINEERING

Chris P. Jobling

Control and Computer-Aided Engineering Research Group
Department of Electrical and Electronic Engineering,
University of Wales, Swansea, UK.
Email: eechris@pyr.swan.ac.uk

ABSTRACT

Computer-aided control engineering is concerned with the application of computers to the control systems design process. In order to effectively support the design process, such computer aids must take account of the needs of human designers. In these lecture notes, computer-aided control engineering is examined from the human designer's perspective. A model of interactive computer-aided design is presented which highlights the importance of good human-computer interfaces. Next, the historical development of user interfaces for computer-aided control engineering, leading to a statement of the current state-of-the-art, is outlined. The role of artificial intelligence in user interfaces for control systems design is discussed and finally a look at the possible future development of human-computer interfaces for computer-aided control engineering is presented.

KEYWORDS

Computer-aided control systems design; computer-aided control engineering; human computer interfaces; artificial intelligence; expert systems.

INTRODUCTION

The emergence and rapid rise in computing power has important implications for all aspects of the computer-aided design of engineering artefacts. A successful interactive computing environment for engineering makes a specialised body of *knowledge* easily usable and easily accessible and therefore easily sharable. If it has an *appropriate* user interface it is easier to create, modify and experiment with systems (MacFarlane *et al.*, 1989).

In order to provide effective computer-based designer support, the user interface — the interface between the human designer and the computer — is particularly important. In these lecture notes we shall explore the user interface aspects of computer-aided control engineering (CACE).

So far, there have been three generations of CACE tools. The first generation was of stand-alone tools written to perform a single analysis, simulation or design task. For example the computation of a frequency response. The second generation was the integration of a group of analysis or simulation tools within a single package. This process has culminated in the development of sophisticated packages for each stage of the design process: the most notable examples being continuous systems simulation and computer-aided control system design (CACSD) for which packages have reached maturity.

In the third generation, for which examples currently exist only in research labs and universities, the aim has been to consolidate the work of the first two generations: to integrate existing packages so as to cover more of the design process within a single consistent framework; to introduce new techniques such as symbolic computation; and to integrate knowledge-based systems aspects of AI in order to make CACE tools easier to use and more useful for design.

The fourth generation will be a move towards the complete integration of CACE through the development of an open framework into which tools can be slotted and the use of advanced computer science techniques, for example interactive computer graphics, in the development of truly supportive environments for design.

During the development of CACE tools, user interface issues have taken second place to the development of new analysis and synthesis methods. This has created a false role for artificial intelligence (AI) which, rather than being used to encapsulate the knowledge needed for good design, has been used to overcome deficiencies in CACE. Instead of helping the designer to make best use of CACE, expert systems have been used to facilitate the integration of packages or to make tools accessible to novice or infrequent users. Some of these 'bad' applications of AI are discussed in these lecture notes but it will be argued that the solution to the problems addressed by such applications will be best overcome by a fundamental rethink in the way CACE packages and tools are developed. In particular, an enhanced role for interactive, direct manipulation graphical user interfaces will be emphasised. It will be concluded that the true role of AI in CACE is to encapsulate design knowledge and to provide tutorial or design assistance within a framework that itself provides adequate user interface and data management.

A MODEL OF COMPUTER-AIDED DESIGN[1]

Human beings and computers have different strengths and weaknesses. Among human beings' strengths are their ability to abstract, simplify and conceptualise; their ability to handle incomplete and ill-defined descriptions; their experience and common sense; their adaptability and flexibility; and their skills in pattern recognition and association. Their weaknesses include their limited short term memory; their slowness in executing complex procedures; their tendency to fatigue and distraction; the fact that similar stimuli (inputs) produce differing responses (outputs) which depend on mood; the fact they can only think about one thing at a time and their difficulties in retrieving information from their (essentially infinite) long-term

---

[1] This section is abstracted from a model of an interactive computer-aided design environment proposed by MacFarlane, Grübel and Ackermann (1989).

memories. Computers, on the other hand, are fast and reliable; have extensive and accurate short and long-term memories; are indifferent to fatigue; produce predictable responses to stimuli; can process enormous amounts of data; are capable of executing tasks in parallel and are capable of accurately performing extremely long and complex formally-specified procedures. A computer, however, cannot generalise, it has no means of conceptualising and no common-sense; it cannot deal with uncertainty; and it lacks flexibility and adaptability. In short, in purely *information processing* terms, a computer is superior to a human being but only human beings have the skills needed to *do design*.

Computer-aided design should be a partnership between the human and the computer which supports the human's strengths and exploits the strengths of the computer to the human's advantage. In this partnership the human will set and refine the design goals, argue from first principles in terms of abstract concepts, and handle ambiguity such as conflicts in the specifications and uncertainties in the description. The computer will handle complex analysis procedures and searches through complex data sets. It will also provide suitable *indicators* by which the human can make decisions and *drivers* by which he or she can manipulate the design objects.

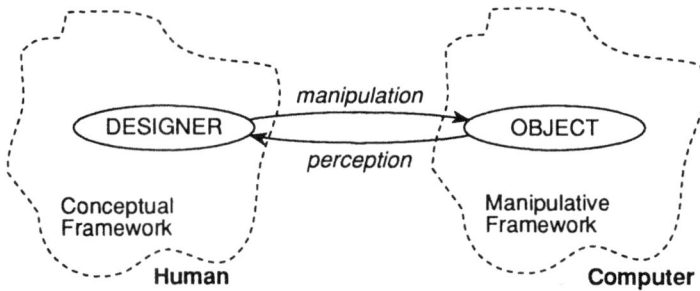

Fig. 1.   The human-computer partnership.

A simple model of this partnership is shown in Fig. 1. The machine passes data to the human by means of indicators and the human passes data to the machine by means of drivers. The human works in terms of a high level *conceptual framework* and the machine provides her with a powerful *manipulative framework*. In order to create a good symbiotic relationship between the human and the machine, the framework in which each operates must be brought together through a suitable set of indicators and drivers.

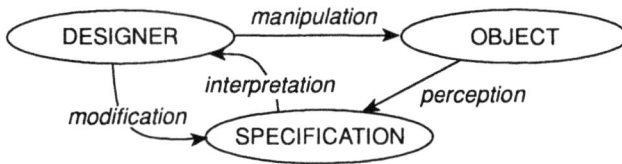

Fig. 2.   Feedback model of design.

Design, like a control system, is a feedback process by which the object being created and the specification against which it is manipulated are iteratively adjusted as the design proceeds. A model of the design process is shown in Fig. 2. The indicators in this model are used by the designer to help her to interpret the specification and to perceive how well the object satisfies it. The designer uses drivers to modify the specification and to manipulate the object being

305

designed so as to bring it closer to the specification. In the creation of an object, the designer must cope with both uncertainty and complexity. The best way of combining the talents of the human and the machine is for the human to handle the *uncertainty* in the design and for the machine to handle the *complexity*.

During the design process the designer deals with three things:
- objects;
- attributes of objects
- operations on objects.

Given these items, there are three ways that a design can proceed:
- a given attribute determines an appropriate object;
- a given object can be manipulated using the available operations on it to turn it into an object having a desired attribute;
- the set of objects can be searched to find an object with the desired attribute.

By considering the amount of analytical knowledge the designer has at her disposal at any point in the design procedure, we can broadly split the possible methods of working into those which are:
- *attribute centred*, or *analytical*, when an exactly computable result is possible;
- *operation-centred*, or *procedural*, when the designer is working from an extensive knowledge base but does not have an exact prescription for a solution;
- *object-centred*, or *experimental*, when the designer is working from a restricted knowledge base and is systematically searching for a solution.

These relationships are useful for relating the designer's conceptual framework to the computer's manipulative framework and are summarised in Fig. 3.

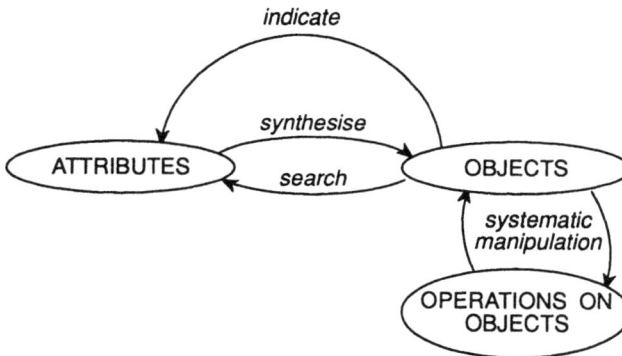

indicate

synthesise

ATTRIBUTES   search   OBJECTS

systematic manipulation

OPERATIONS ON OBJECTS

Fig. 3.   Design objects, attributes and operations.

Within an ideal computer-aided design environment the designer should be able to: build; analyse; browse; search; compare and evaluate; reason and hypothesise; synthesise; design; manipulate and modify; experiment; catalogue, store and retrieve. If these facilities are to be made available in an efficient and flexible way then the system must be: easily comprehensible to a single individual; wide in scope; modular with a manageable number of distinct parts; predictable in its behaviour; integrated and coherent in the ways that its different parts relate; helpful, with quick and efficient access to relevant information; tolerant of errors and supportive in enabling the effects of errors to be easily undone; extensible and adaptable; and self documenting. No current CACE environment provides all of these facilities, nor is there an environment that is completely 'user friendly'. However, user interface technology, within which we include AI techniques, has enabled some progress to be made towards the ideal.

THE DEVELOPMENT OF CACSD

The term *computer-aided control system design* (CACSD) has gained widespread use since 1982 and is defined (Rimvall, 1987) as:

"The use of digital computers as a primary tool during the [linear] modelling, identification, analysis and design phases of control engineering."

A 'waterfall model' of the control systems design 'life-cycle' is shown in Fig. 4. The rounded boxes represent the stages in the life-cycle. The 'disk-file' icons indicate the main 'deliverables' at each stage. Downward arrows show the development of the design, upward arrows indicate design iteration. Arrows are labelled with the processes that are performed during the movement from one stage to the next. Note that in real design iteration, it is possible, and often necessary, to jump back several stages. CACSD packages typically provide support for the analysis and the design of *linear* plant and controllers: that is only the central portion of the design cycle. Various other software packages are needed to:
• model the *nonlinear* plant, actuators and sensors,
• linearise the nonlinear model in order to provide a linear model for analysis,
• simulate the system through the various stages of design, and
• implement the control law in software.

The term *computer-aided control engineering* (CACE) is now gaining wide acceptance as meaning the *integrated* computer support for *all* the stages in the control systems design life-cycle. In other words, CACE provides 'whole product life-cycle' support, similar to that provided in *computer-assisted software engineering* (CASE)[2], and it is in these terms that we shall analyse its development.

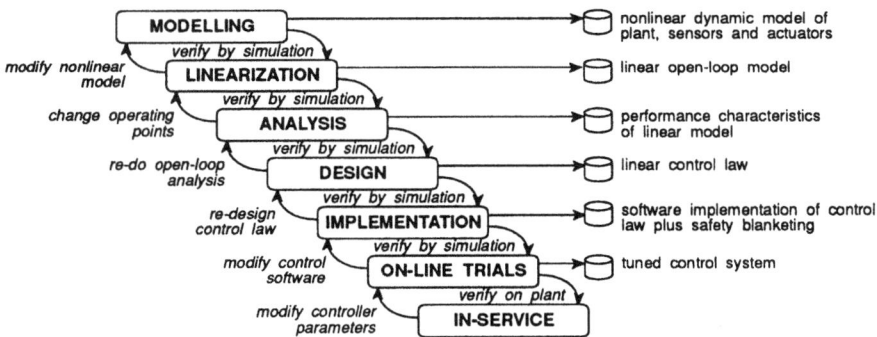

Fig. 4.   The control systems design life-cycle.

Figure 5 illustrates the development of CACSD and CACE packages over the last four decades. (This figure has been adapted and updated from the figure given by Rimvall (1987,1988)

---

[2]Indeed CASE is needed to support the stages through which the controller is developed from its specification into software.

who gives a good account of this development.) In order to put events into proper context, other key influencing factors, chiefly hardware and software developments, are also shown. In this section we briefly describe the background to the emergence of CACSD and CACE tools, starting with technological developments and then moving on to user interface aspects. The aim is to understand the current state-of-the-art by examining the historical context in which these tools have been developed.

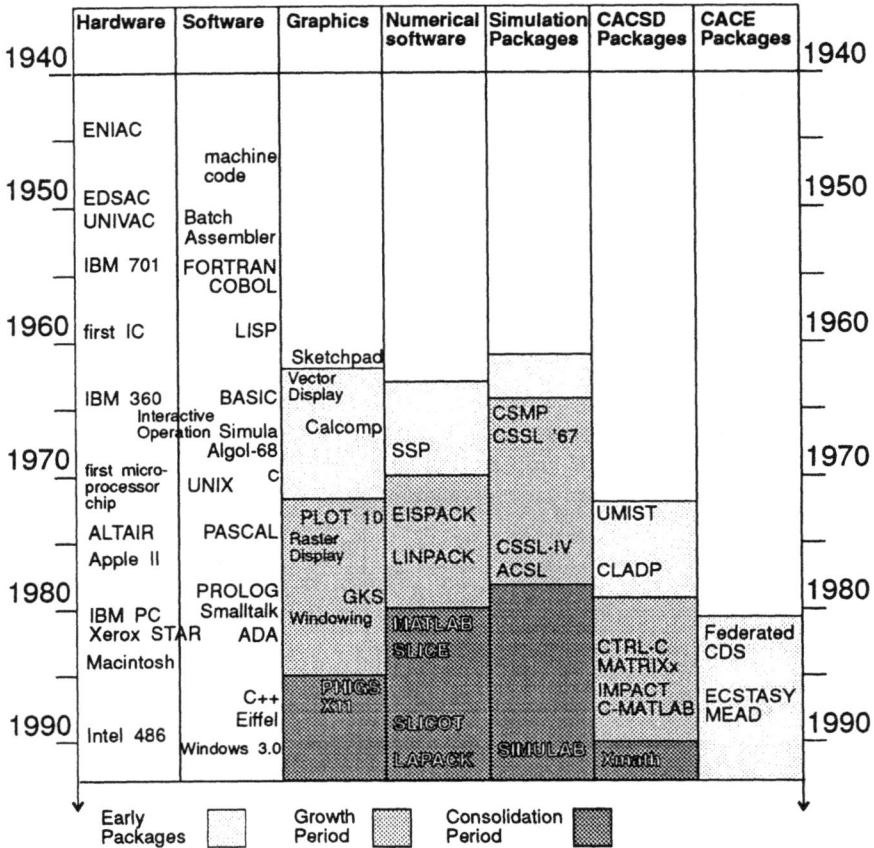

	Hardware	Software	Graphics	Numerical software	Simulation Packages	CACSD Packages	CACE Packages	
1940								1940
	ENIAC							
1950	EDSAC	machine code						1950
	UNIVAC	Batch Assembler						
	IBM 701	FORTRAN COBOL						
1960	first IC	LISP						1960
			Sketchpad					
			Vector Display					
	IBM 360	BASIC			CSMP			
	Interactive Operation	Simula	Calcomp		CSSL '67			
1970	Algol-68 C			SSP				1970
	first micro-processor chip	UNIX						
	ALTAIR	PASCAL	PLOT 10 Raster Display	EISPACK		UMIST		
	Apple II			LINPACK	CSSL-IV ACSL	CLADP		
1980	IBM PC	PROLOG Smalltalk	GKS Windowing	MATLAB SLICE			Federated CDS	1980
	Xerox STAR	ADA				CTRL-C MATRIXx		
	Macintosh							
		C++	PHIGS XII	SLICOT		IMPACT C-MATLAB	ECSTASY MEAD	
1990	Intel 486	Eiffel Windows 3.0		LAPACK	SIMULAB	Xmath		1990

Early Packages □   Growth Period ▨   Consolidation Period ▧

Fig. 5. The historical development of interactive CACSD and CACE tools showing the availability of related hardware and software. Some actual products are included to indicate the state-of-the-art. This figure is updated from that appearing in Rimvall (1987,1988).

Technological Developments

Computing hardware. Since 1953 there has been a phenomenal growth in the capabilities and power of computing hardware. Observers estimate that the power of computing devices (in terms of both execution speed and memory availability) has doubled every second to third year, whereas the size and cost (per computational unit) of the hardware has halved at approximately the same rate (Torrero, 1985). The chief effect of these developments on CACE has been to widen the range of applications for computing and at the same time to make computers, and therefore the applications, widely available to practitioners in all branches of the subject.

System software. The development of system software, such as operating systems, programming languages and program execution environments, has been slower than that of hardware, but is nonetheless impressive. What is less impressive has been the steadily increasing cost of software, estimated at about 80% of the total installation cost of a computing system (Sommerville, 1989), which developments in computer science have been largely unable to reduce.

High level languages. The first electronic programmable computing machine was programmed by physical rearrangement of its circuits using wiring diagrams and a patch panel. The second electronic computer introduced the *stored-program architecture*. In such a machine, which is the model for most computers in use today, both programs and data are stored in *short-term memory*. The configuration of the computing elements can be changed by instructions obtained from memory in the form of binary words. Early programmers had to program such computers by directly loading this 'machine-code' into memory. Programming a computer in this way is a painstaking and error prone process that effectively limits the size and type of applications possible. In the early 1950's shorthand notations called assembly codes were invented which enabled an *assembler*, running on the target computer itself, to translate these assembly codes into machine code and then run the resulting program.

The next big breakthrough in engineering computing was the invention of FORTRAN, a high-level language, and the *compilers* which convert it into machine code. Engineers could then write their own programs in a language that was sufficiently close to mathematical notation so as to be quite natural. Since then numerous high-level languages have been invented. Among these are LISP and Prolog, which have been used extensively in AI research; C, which is the predominant systems programming language; PASCAL, MODULA and ADA which were invented to support software engineering principles; and new object-oriented languages such as Smalltalk-80, C++ and Eiffel.

Graphical displays. Engineers are, in general, more comfortable with pictures than with text as a means of communicating their ideas. Hence the wide availability of graphical displays is of prime importance to many areas of engineering computing. Indeed, the development of computer graphics has been the means by which certain control systems design techniques, such as multivariable control systems analysis, have been made practicable. Computer-graphics has also been instrumental in providing improvements in human-machine interfaces such as schematic systems input and direct manipulation interfaces with windows, icons, pull-down menus and pop-up dialogue boxes. Further improvements in user interfacing techniques such as *hypermedia* will continue to rely on developments in display technology.

Quality numerical software. Following the invention of FORTRAN there was a gradual development of useful general purpose subroutines which could be archived into libraries, distributed and shared. This lead eventually to the development of standard subroutine libraries such as EISPACK, LINPACK, BLAS and LAPACK (for solving eigenvalue problems and sets of linear equations) which have had a direct influence on the development of CACSD.

Simulation languages. For many years before the predominance of digital computers, dynamic system behaviour was simulated using analogue computers. Hybrid computers, which combined the real-time dynamic equation solving of an analogue computer with the sequence control features of the digital computer, provided facilities for parameter optimisation among other things. With the emergence of digital computers, it was discovered that numerical integration could be used to solve systems of nonlinear differential equations numerically. Numerical integration was so successful that *digital simulation* rapidly superseded analogue and hybrid simulation during the mid 1960's.

The first digital simulation programs were written in FORTRAN. Eventually, special purpose languages emerged which allowed statements written in a form close to state equation notation to be translated into FORTRAN, and this in turn enabled the engineer to concentrate on the problem description. In 1967, a standard language called CSSL (Continuous Systems Simulation Language; Augustin *et al.*, 1967) was proposed by the US Simulation Council and this forms the basis of most simulation languages in use today.

Other influences. Other changes in computing hardware, software and user interface 'philosophy' have had, and will continue to have, significant influence on the development of CACE tools.

- The invention of the microprocessor.

- The launch of the IBM personal computer and MS-DOS.

- The launch of the Macintosh *graphical operating system.*

The Development of the User Interface

There have been several identifiable stages in the development of the user interface. These are briefly described in this section. In the very early days, computers had to be communicated with through an intermediary who could speak the machine code. This gradually changed with the invention of assembly and FORTRAN programs and the use of punch cards to prepare programs. Eventually punch cards were replaced by teletype machines and visual display units. The current state-of-the-art is a windowed graphical operating system. Through all the stages, the trend has been to move the control of the machine from the computing centre to the desktop.

Batch mode computation. In the early days, because both short term and long term computer memory was expensive and scarce, it was only economical to load a compiler if there were a reasonable number, or *batch*, of programs waiting to be compiled and run. This batch mode of operation typically operated as follows. An engineer would submit a program on coding sheets to an operator, who would sit at a card punch machine and convert the program text into a set of punched cards. This stack of cards (in the correct order) would then be carefully submitted to the machine operators who would add the stack to the card reader for the computer to get to it when it was ready. Some time later the engineer would get her results back. The first few times there would probably be errors of syntax which halted the compiler, the next few times errors of semantics (bugs) would prevent the machine code from executing correctly, and eventually after many days the program would run properly. Debugging took a long time then, and productivity was very low.

Time-sharing operating systems. A major breakthrough was the invention of time sharing operating systems which allowed several users to access a computing machine at the same time. Early time sharing systems were little different from the old batch systems except that the

310

turn-around time was shorter. Eventually, both the system and the programs it ran became interactive. Programs could be stored in files in long-term memory and the card punch disappeared overnight. Engineers became much more productive. With the advent of "cheap" graphics output devices the stage was set for the next phase in the development of CACSD.

Interactive computing. With the emergence of time sharing computers it became possible for a program to interact with the person using it. Memory became less expensive and so programs could be bigger, and when virtual memory systems became widely available program size was essentially limitless. Programs needed no longer to be single jobs. Many programs were glued together through some kind of "dialogue manager" which enabled the user to easily combine operations to perform a given task.

Command languages. Early interactive computing systems operated using *modes*. For example the user might be in editing mode or compilation mode. This limited the options available at any given time to those commands that were appropriate to the current mode. This was fine for beginners and occasional users since the computer effectively prevented them from doing anything wrong. With familiarity, however, such systems become restrictive and inflexible. With the popularity of operating systems like UNIX, the command language style of interaction became popular. This provided the (experienced) user with a wide range of short (two or three word) commands with which to control the operation of the computer. There were no modes since at any point virtually any command could be issued, and commands could easily be combined to produce new commands. This gave immense flexibility and freedom to the user, but provided little support to the novice. Command language interfaces are nonetheless the state-of-the-art today.

Graphical user interfaces. The most recent advance in user interface technology has been the move to direct manipulation graphical user interfaces. Surprisingly, their development owes much to arcade computer games like tennis and space invaders. The idea in these games is that an object on the screen, say a gun emplacement, moves under the direct influence of a control device which the user manipulates. The speed and direction of movement are related in some tactile way to how the input device is manipulated. This gives the illusion that objects on the screen are under the direct control of the user. Researchers realised that this type of interface could be used in an office information system. Documents, files, printers, editors and other 'office' objects could be represented by pictures on a high resolution bit mapped screen and manipulated by selecting them, moving them and 'activating' them by means of a suitable input device. The first commercial embodiment of these ideas was the Xerox STAR workstation which because of its cost was a commercial failure. However, it was a great influence on the Apple Macintosh. The Macintosh itself has been extremely influential in the development of user interfaces and it is probably only the matter of a few years before all computers will have graphical user interfaces.

Models of User Interaction

Having surveyed the development of user interfaces in general, we can now illustrate how they have been used within the context of CACSD. In the discussion we give examples of how they look to the user, and indicate their complexity in terms of programming. We shall then be in a good position to introduce the benefits of AI as a user interface facility.

Batch mode. The earliest CACSD "tools" were batch programs which did one job on a set of data and returned a set of results. An example might be a program for reading a transfer func-

311

tion and computing the corresponding frequency response. The user prepared the data, submitted it to the program and awaited the result. There was no interaction.

Question and answer. This type of interaction was used in the development of the pioneering UMIST design suite in the early 1970's (Rosenbrock, 1974). Packages like the UMIST suite were essentially collections of control systems analysis tools with a user interface bolted on. The user interface is used in the selection of analysis choices, the definition of model and execution parameters, and to control the display of results. It makes use of what has become known as a Question and Answer (Q&A) dialogue (Fig. 6).

```
***SYSTEM INPUT***

WHAT NEXT? QUIT

***SUPERVISOR***

WHAT NEXT? ?

SYSTEM INPUT (SYS)
ROOT LOCUS DESIGN (RLOC)
BODE DESIGN (BODE)
NYQUIST DESIGN (NYQ)
NICHOLS DESIGN (NICH)

WHAT NEXT?

LEAVE PACKAGE (QUIT)

WHAT NEXT? BODE

***ROOT LOCUS***
```

Fig. 6.   Question and answer dialogue.

The *computer* controls the interaction by asking the user "what shall I do next?" after each action has been completed. In answer to each question there are only a small number of possible replies. The one required by the user is typed at the keyboard. If the user doesn't know what to type, context-sensitive help, in the form of a list of currently available options, is obtained by typing a question mark. On the basis of the user's response, the program branches to the subroutine that implements the required task. This subroutine in turn might begin a further dialogue before executing an analysis procedure. In the UMIST package, the style of interaction, response to queries etc., is strictly defined and all programmers working on it use the same style. The package is therefore consistent and easy to use. However the inflexibility imposed by the program, which controls the order of events, quickly becomes tedious for regular users of the package. For example, after a false numerical input, say a miss-typed polynomial coefficient, the user has to continue a now meaningless conversation until the computer produces some mendacious result, before he can go back to correct the original error.

Menu. A variation of Q&A dialogue is the menu driven dialogue (Fig. 7) which differs only in the details of what the user sees. Each choice is explicitly given at each stage and the user selects the one required by number.

Command language. The current generation of CACSD interfaces use what is called a command line interpreter. This differs from Q&A and menu dialogues in that all commands are available all of the time and the user merely submits a request for data to be processed in a way analogous to making a procedure call. The command line interpreter reads each line of input and breaks it down into calls to primitive functions. Figure 8 illustrates a typical sequence of commands in MATLAB.

```
***SYSTEM INPUT***

1. INPUT PLANT G(S)
2. INPUT CONTROLLER D(S)
3. INPUT FEEDBACK H(S)
4. RETURN TO SUPERVISOR

YOUR CHOICE? 4

***SUPERVISOR***

1. SYSTEM INPUT
2. ROOT LOCUS DESIGN
3. BODE DESIGN
4. NYQUIST DESIGN
5. NICHOLS DESIGN
6. LEAVE PACKAGE

YOUR CHOICE? 3
```

Fig. 7.   Menu driven dialogue.

```
% define transfer function G
num = [1, 4];
den = [1, 2, 2, 1];
% define frequency range
w = logspace(-1,1,50);
% calculate magnitude and phase of G(jw)
[mag, phase] = bode(num, den,  w);
% draw Bode plot
subplot(211)
semilogx(w, 20.*log10(mag))
subplot(212)
semilogx(w, phase)
```

Fig. 8.   Command line interface (MATLAB).

The advantage of this type of interface is that it provides the user with great flexibility without being too much of a burden on the programmer. An example of this flexibility is that it often doesn't matter to the interpreter where the commands originate, i.e. from the user's keyboard or from a file. This means that users can easily extend the functions of a package by grouping collections of commands into a 'macro file' which is then executed when necessary. The disadvantage is that the number of commands and their syntax often makes such systems difficult to master and so they tend to find more favour with experienced users rather than with novice or infrequent users.

Direct manipulation graphics. We expect the next stage in CACSD and CACE package development will be a gradual movement away from keyboard operated interfaces to highly pictorial, direct manipulation graphical interfaces.

Graphical user interfaces (GUIs) give a higher priority to the data than to the programs that operate on the data. Data (e.g. files, output devices; graphs; diagrams etc.) are often represented pictorially as "icons". To do something, a data object is selected and then the operation required is chosen (See Fig. 9). Compare this with text based interfaces where one usually selects the operation and then gives the data as a formal parameter. Such "data centred" rather than "operation centred" interfaces are remarkably intuitive to use and seem to be accepted by

313

both novice and experienced users. For the so called "power user", keyboard shortcuts are often provided.

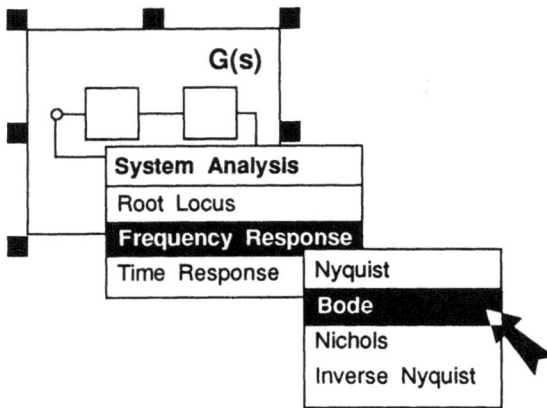

Fig. 9. Graphical user interface. A control *system* data object is selected. The operations available are therefore limited to those which are appropriate to *systems*. This combines the best features of Q&A and command driven interfaces.

The programming effort in user interface development. The criterion by which user interfaces are judged must be the satisfaction of the user. No other yardstick is appropriate (Schneiderman, 1987). Nonetheless, the effort involved in constructing user interfaces will have a bearing on the cost of the tools produced and may limit their adoption for reasons of economy.

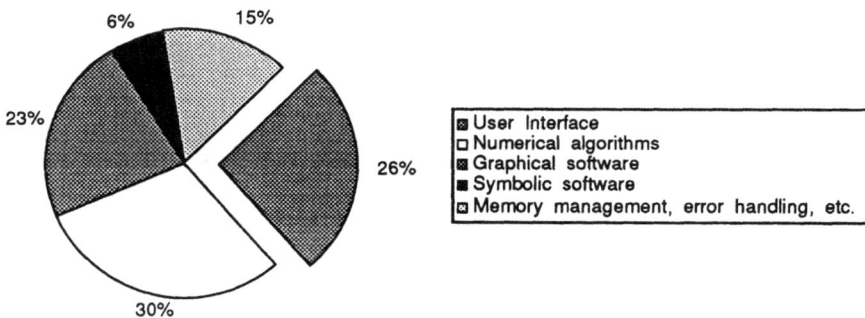

Fig. 10. Relative size of different code segments of IMPACT.

Rimvall (1987) estimates that in early CACSD packages, such as the UMIST suite, as much as 80% of the programming effort went into the numerical algorithms. In his own command-language based CACSD package, IMPACT, the coding effort was distributed as shown in Fig. 10. This distribution of code is probably typical of the current generation of CACSD packages. GUIs, on the other hand may swing the proportion of code to the other extreme, with as much as 40–60% being attributed to the user interface (Cox, 1986). GUIs are also difficult to program since they are "user centred" and not "programmer centred". That is the program must react to what the user wants to do, not what the computer (or rather the original programmer) thinks she wants to do. Because today's developers are predominantly control

314

engineers rather than software engineers, these difficulties may account for the slow adoption of GUIs in CACSD.

THE STATE OF THE ART IN COMPUTER-AIDED CONTROL ENGINEERING

Before we move on to consider the impact of AI in CACE we need to consider the current state-of-the art of CACSD. In this section we shall describe the matrix environments that have come to dominate CACSD. We shall then move on to examine the wider requirements of CACE and see that these are less well served by the current generation of tools. Finally we shall briefly review early attempts at providing integrated CAD tools for control systems design.

Consolidation in CACSD

As can be seen in Fig. 5, the 1980's was a decade of consolidation during which CACSD technology matured. Menu driven and Q&A dialogues were superseded by command languages. The matrix environment became the *de facto* standard for CACSD.

The reasons for this are due to the *simplicity* and the *flexibility* of the interface model. We illustrate these properties using MATLAB (MATrix LABoratory), the original matrix environment: a program giving interactive access to the linear algebra routines EISPACK and LINPACK, which was released into the public domain (Moler, 1980). In MATLAB, matrices and matrix operations are entered in a straightforward fashion as shown in Fig. 11.

```
» a = [1 3 5
7 6 5; 0 0 5];
» [vec, val]= eig(a)

vec =

    -0.7408    -0.3622    -0.1633
     0.6717    -0.9321    -0.8981
          0          0     0.4082


val =
    -1.7202          0          0
          0     8.7202          0
          0          0     5.0000
»
```

Fig. 11. Entering and manipulating matrices in MATLAB. In this example a matrix is defined and its eigenvectors and eigenvalues are determined.

This elegant user interface model, although intended for numerical analysts, quickly caught the eye of control scientists who realised that MATLAB was equally applicable to the solution of 'modern control' problems based on linear state-space models. The flexibility of MATLAB is due to its support of *macro files* which makes it easy to extend the program to fit a particular application domain (Fig. 12).

3. *CACSD packages should support an algorithmic interface.* This principle simply is a recognition of the fact that no package developer can anticipate the requirements of every user. So the package must be extensible by provision of user defined macros and functions.

4. *The transition from basic use to advanced use must be gradual.* The user of a CACSD package is faced with two different types of complexity: the complexity of the user interface and the complexity of the underlying theory and algorithms. In both cases extra guidance is need for novice users. Typically, the designers of CACSD packages do not wish to stand in the way of the expert users, so they provide direct access to the whole package and interfere in the use of the package as little as possible. This creates problems for novice or infrequent users of the package.

5. *The system must be transparent.* This is a fundamental principle of software engineering that simply means that there should be no hidden processing or reliance on side effects on which the package depends for its correct operation.

6. *Small and large systems should be equally treated by the user interface.* This simply means that account should always be taken of the limitations of numerical algorithms. The package should warn the user when limits are reached and the algorithms should thereafter 'degrade gracefully'.

7. *The system must be able to communicate with the outside world.* At the very least it must be possible to save data in a form that can be retrieved by an external program. MATLAB and its cousins provide basic file transfer capabilities, but the ideal CACSD package would have some link to a much more convenient data sharing mechanism such as could be provided by a database.

Other desirable features.

- *Form or menu driven input* is often more useful than a functional command driven interface for certain types of data entry.

- *Graphical input* is useful for defining systems to be analysed.

- *Strong data typing*, is useful for since it provides a robust means of developing extra algorithms within the context of the CACSD package.

- *Data persistence.* Unless explicitly saved, data is not maintained between sessions. Neither can data easily be shared between users. The evolution of models and results over time cannot be recorded. Hence a CACSD package needs a database.

- *Symbolic computation.* Delaying the replacement of symbolic parameters for numerical values for as long as possible can often yield great insight into such properties as stability, sensitivity and robustness.

The Emergence of CACE

During the design of a control system, the designer typically makes use of her skill, intuition and experience to design a controller such that the overall closed-loop control system is stable, robust and has good performance characteristics. The stages that the designer typically goes through in creating a design are tabulated in Table 1 (Taylor and Frederick, 1984). Spang (1984) provides a useful summary of the techniques used in the various stages of the control system design procedure (see Fig. 13). It should also be noted that since design is a feedback process, it is likely that several of these stages will be repeated as more knowledge is gained about the plant and what control options are possible.

Table 1. Control Engineering Activities.

1	Modelling the plant to be controlled.
2.	Determining the characteristics of the plant model.
3.	Modifying the configuration to make the plant more amenable to control (e.g. moving or adding actuators and sensors).
4.	Formulating the elements of the design problem.
5.	Checking to see that the design problem is well posed; especially, determining the realism of the specifications in light of the given plant model and design constraints.
6.	Executing appropriate design procedures.
7.	Performing design trade-offs, if necessary.
8.	Validating the design.
9.	Documenting the design.
10.	Implementing the design.

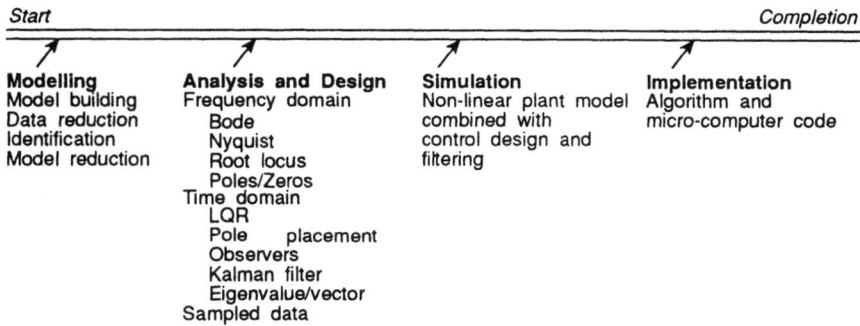

Fig. 13. The stages in control systems design.

Careful review of the design activities and the techniques used, reveal that CACSD, as it is currently implemented, fails to provide support for the whole design process. Although good software packages are available to help the designer during most of the individual stages of the design process, the current state-of-the-art fails to provide comprehensive support for the whole design process. The problems are (Taylor and Frederick, 1984):

1. Most packages are specialised to one aspect of the overall design process, to one "domain" (time or frequency), and often to one particular synthesis or design approach. Therefore, many designers will not be willing or able to restrict themselves to the use of a single package.

317

2. The problem solving environment provided by most current packages is generally power-ful but rather low-level. Combined with the first point, this means that the user may need to know several hundred commands in several different syntaxes, in order to be able to fully exploit a comprehensive package or set of packages.

3. Most packages provide little or no guidance. The user must formulate a meaningful model of the plant, pose the design problem appropriately, know exactly what procedures to exe-cute, and have the experience and the practical feel for the control engineering problem in order to be able to judge the results obtained and to know how to proceed.

4. Most packages provide little or no useful documentation of the design process. The user must keep separate record of the design procedure, trade-off studies and other considera-tions involved in arriving at a final controller design.

5. Most packages provide little or no useful support in validating and implementing the final design. It is up to the user to recognise the need for validation and to know how to ac-complish that vital step, and it is up to the user to decide how to realise the theoretical controller design in hardware.

Package integration. In the early 1980's, workers in the field of CACE realised that package integration was important. The earliest truly integrated CACSD system was the Federated Computer Aided Control Design System (Spang, 1984). This has been followed more recently by ECSTASY (Munro, 1988) and MEAD (Taylor and McKeehan, 1989). The architecture of all these systems is essentially that shown in Fig. 14. We name this type of integration *external integration* since the normal external interfaces of the constituent packages are connected to the environment through command language and data translators. The user communicates with the packages through a single user interface and the supervisor is responsible for converting the user's commands into package commands. The packages pass data, through data convert-ers, to a common database where it may be shared with other packages. Added value is achieved by giving the user direct access to the database where models and results may be manipulated.

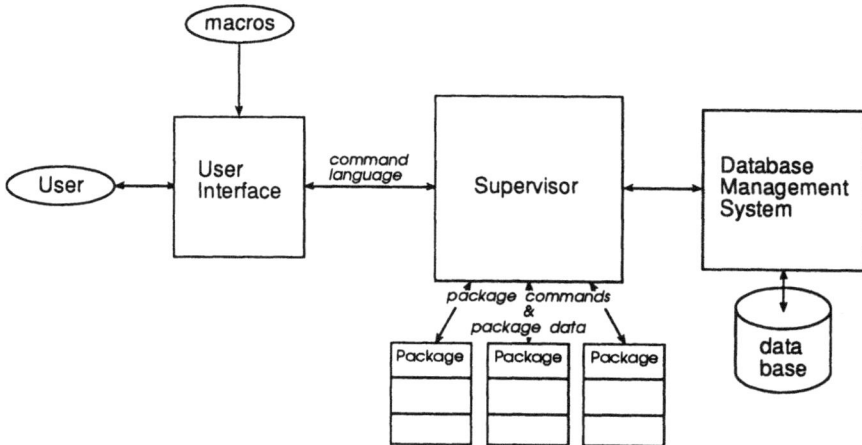

Fig. 14. External package integration.

318

The disadvantage of external integration mechanisms is due to the continual development and enhancement of the constituent packages which makes the effective maintenance of such an environment impossible.

A role for AI. The use of AI can alleviate some of the problems of package integration in CACE. It also has a possible role in reducing complexity, providing help for novice and infrequent users and even automating certain stages of the design process. We shall turn our attention to this in the next section. However, in a real sense, the use of AI to overcome deficiencies in the current generation of CACE packages provides only a short term solution. A better solution is made possible by a fundamental rethink in the way CACE tools are packaged. Some early ideas on this are discussed in *Section "The Fourth Generation of CACE"*.

APPLICATIONS OF ARTIFICIAL INTELLIGENCE IN CACE

The application of AI to CACE is usually justified as follows (Taylor and Frederick, 1984; Pang, 1989):

- Control systems design is a complex problem,

- Second generation CACE packages are complex and difficult to use, packages typically work at too low a level,

- New control techniques and new applications of control are steadily increasing the complexity of the design.

AI is thus seen as a way of reducing the immediate complexity of the control systems design problem by allowing the designer to work at a higher level of abstraction, by providing additional help to novice or inexperienced users, and by tutoring users in the application of novel control systems design techniques. In all cases, the aim is to improve the *interface* between human and machine so as to free the human from the nitty-gritty procedural details best handled by the machine and enable her to concentrate on the solution of the problem. This is precisely the view of interactive design environments advocated by MacFarlane *et al.* (1989) and presented in *Section: "A Model of Computer-Aided Design"* above. It is clear that AI, at least as it has been used so far, is a user interface issue and its treatment in this manner is justified.

Expert System Shells for CACE. Most applications of AI in CACE make use of expert system shells. There are many commercial expert system shells available on the market and its is not our intention here to review any of these packages. However James (1987) and Taylor (1988) both give a list of required or desirable features that have proved useful in the development of expert systems applications in CACE. These features are:

1. The support of production rules with variables which avoids the need to 'hard-wire' numeric and symbolic values into the rule base,

2. The support for both backward and forward chaining,

3. Support for numerical operations such as comparisons, and arithmetic. This will enable comparisons to be made between design values and specifications for example,

4. The ability to revise previously believed facts which is often necessary during design iteration as the engineer realises that a dead-end has been reached and she needs to backtrack,

5. Support for the partitioning of knowledge into separate rule-bases which helps to keep the development of large knowledge-based systems manageable,

6. Comprehensive explanation facilities so that the designer can determine how she came to a particular point and why certain steps in the design had been taken,

7. The ability to customise the expert-system shell to a particular application,

8. The provision of external interfaces. As a bare minimum it should be possible to interact with external files but it is also useful to be able to exchange data with external applications as illustrated in Fig. 15.

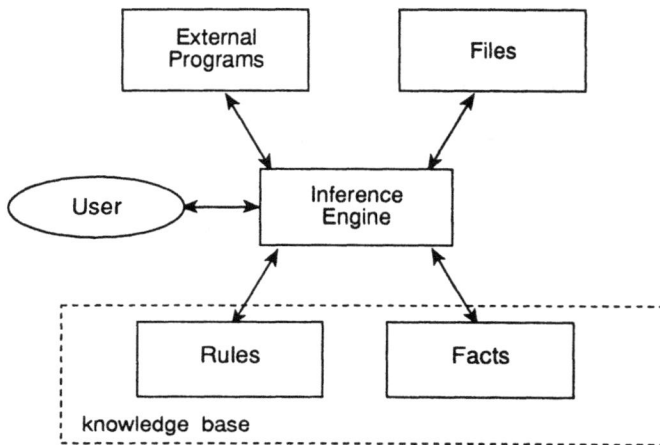

Fig. 15. The architecture of a typical expert system.

Many of the current generation of expert systems shells will provide support for most of the above features. In addition, facilities such as frames, object-orientation, and graphics are now available for use in CACE.

Case Studies

In the sections that follow we shall present case studies of the use of AI in the following key areas:
- As a means of providing symbolic computation,
- As a means of reducing the complexity of the user interface,
- As a CACE package integration mechanism,
- As an intelligent front-end,
- As tutor,
- As progress monitor, and
- As design assistant.

Provision of symbolic computation. Symbolic computation is especially useful during the early stages of control system design that is: setting up a model of a control system, identifying important parameter relationships, etc. At the author's institution, two approaches to symbolic manipulation of control system models have been explored. The first of these is the application of computer algebra systems such as *MACSYMA* (Grant *et al.*, 1988) and *Mathematica* (Shouaib *et al.*, 1992) to the linear analysis of control systems. The second is the application of Prolog to the analysis of linear systems.

Computer algebra packages were among the earliest successes of AI research. Perhaps the oldest package is *MACSYMA*, a large LISP program. More recently, efficient computer-algebra packages, written in C, such as *Mathematica* and *MAPLE* have been made available. These run on a variety of machines, including Macintosh and IBM-PC's and so make computer-algebra systems available to control engineers. A computer algebra system is essentially a sophisticated string manipulation program. It uses techniques such as pattern matching, term re-writing and rules to manipulate strings which represent mathematical objects. An example of a problem that can be solved by such systems is the inverse Laplace transformation. A student, given a problem, typical of an early control course, like this:

$$\mathcal{L}^{-1}\left(\frac{25}{s(s^3 + 3.25s^2 + 27.5s + 50)}\right),$$

would endeavour to solve it by systematically applying well known rules. In this case, the denominator of the rational polynomial would be factorised, the expression would be converted into a sum of simple first and second order terms by partial fraction expansion and then table look-up would be used to convert the terms in *s* into expressions in *t*. The basic operations used in this transformation are usually built-in to computer-algebra systems. For example *pattern matching* is used to recognise forms of sub-expressions and for table look-up, polynomial factorisation and partial fraction expansion are used to *re-write* expansions into forms that are simpler to manipulate further, *rules* and *heuristics* are used to guide the application of the various techniques. The application of some of these techniques are illustrated in Fig. 16. The final stage would be to use pattern matching to recognise the inverse Laplace transforms of the partial terms (Fig. 17).

```
<c1> Expression = 25/(s*(s^3 + 3.25*s^2 + 27.5*s + 50));

<c2> Factorize(Expression,s)

<d2> 25 / s*(s + 2)*(s^2 + 1.25*s + 25)

<c3> PartialFractionExpand(%d2,s)

<d3> 0.5/s - 0.472/(s + 2) - (0.028*s + 0.979)/(s^2 + 3.25*s + 25)
```

Fig. 16. Stages in the solution of an inverse Laplace transformation problem
in a computer-algebra system.

Recognise pattern		Replace with re-write text
`b`	$\Rightarrow$	`b*d(t)`
`b/s`	$\Rightarrow$	`b*u(t)`
`b/(s + a)`	$\Rightarrow$	`b*exp(-a*t)`
`(s + b)/((s+a)^2+b^2)`	$\Rightarrow$	`exp(-b*t)*sin(a*t)`
`b/((s + a)^2 + b^2)`	$\Rightarrow$	`exp(-b*t)*cos(a*t)`

```
<c4> Rewrite(-0.472/(s + 2))
```

```
<d4> -0.472*exp(-2*t)
```

Fig. 17. Pattern matching for term re-writing. Look-up table and example result.

Prolog is a computer implementation language for predicate calculus (Dodd, 1990) and is thus very useful for implementing logic-based inference systems. A logic program in Prolog consists of two parts, a set of *axioms* or facts and a set of *rules of inference* which determine what new facts can be determined given that certain axioms are true (Barr and Feigenbaum, 1981). The simplest of axioms is the *proposition* which is either true or false: "a pole of this transfer function lies in the right-half plane" is a proposition. It could be written in Prolog as

```
?- right_half_plane(pole).
```

More complex propositions can be formed by the use of *logical connectives* and the *existential* and *universal quantifiers*. An example of such a proposition is "a system is unstable if it has a pole in the right-half plane." This could be written in Prolog as

```
unstable(System):-    transfer_function(System,X),
                      has_pole(X,Pole),
                      right_half_plane(Pole).
```

However, the language has a much wider application than this. In particular, it is eminently suitable for many AI problems including symbolic processing and expert systems. Prolog has been used by the author to analyse and manipulate structural models of control systems such as signal flow graphs (Jobling and Grant, 1988) and block diagrams (Grant and Jobling, 1991). Prolog is particularly suited to these tasks because it provides the following programming features:

- *Pattern matching* — which is vital for symbolic processing and eliminates the need for conditionals and assignments. An example of pattern matching in Prolog are the rules for differentiation (Clocksin and Mellish, 1985), some of which are shown in Fig. 18. The pattern matcher is of the form `d(pattern,X,result)` and the right hand side of the clause indicates the extra processing that may be required, which in most cases produces a further differentiation step. Differentiation stops when one of the primitive patterns is matched. Note how close the rules are to the mathematical rules.

```
d(X,X,1)  :- !.                                          dx/dx = 1

d(X,X,0)  :- atomic(C).                                  dc/dx = 0

d(U+V,X,A+B)  :- d(U,X,A), d(V,X,B).          d(u + v)/dx = du/dx + dv/dx

d(C*X,X,C*A)  :- atomic(C), C \= X, d(U,X,A), !        d(cu)/dx = c du/dx

d(U*V,X,B*U+A*V)  :- d(U,X,A), d(V,X,B).      d(uv)/dx = u dv/dx + v du/dx
d(U^C,X,C*U^(C-1)*W)  :- atomic(C),

              C \= X, d(U,X,W).               d(u^c)/dx = cu^(c-1) du/dx


/* an example run */
?- d(x+1,x,X).
X = 1+0
```

Fig. 18. Some of the rules of differentiation: an example of pattern matching in Prolog.

- *Rule-based execution* — the Prolog execution engine works by seeking to satisfy a goal by firing applicable rules. This is similar to backwards chaining in expert systems, indeed it is easy to implement an expert system shell in Prolog (MIKE, 1990). An example of a Prolog rule that is applicable to block diagrams is the series reduction rule illustrated in Fig. 19.

- *Automatic backtracking* — when a solution is found to the current goal, or if a solution cannot be found, the Prolog interpreter will automatically seek another solution. This is useful for problems with many possible solutions.

- *Terms* — Prolog terms can be used to represent mathematical expressions in a natural manner (e.g. the differential terms in Fig. 18). This avoids the necessity to map expressions to some other internal representations.

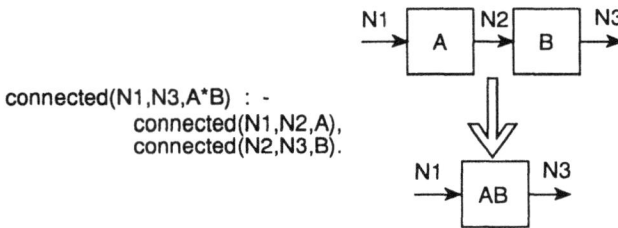

```
connected(N1,N3,A*B) : -
       connected(N1,N2,A),
       connected(N2,N3,B).
```

Fig. 19. Series reduction rule for block diagrams.

So far at in the author's research group, Prolog has been used for the following CACE applications: signal flow graph reduction (Jobling and Grant, 1988), block diagram manipulation (Grant *et al.*, 1989), diagram transformation, block-diagram to signal flow graph and *vice versa,* and automatic diagram layout (Barker *et al.*, 1988), modelling, simulation and analysis of discrete event dynamic systems (Barker *et al.*, 1989a), and automatic generation of programmable-logic controller code from ladder diagrams (Barker *et al.*, 1989b). A good overview of all these applications is given in (Grant *et al.*, 1992). Prolog has also been found to be useful by other researchers in CACE, in particular we refer the reader to Tan and Maciejowski (1989) who have used it as a database language and Gawthrop *et al.* (1991) and Linkens *et al.* (1991a) who have used Prolog for the modelling and analysis of systems represented in bondgraph form.

<u>Reduction of the complexity of the user interface</u>. Using 2nd generation CACSD tools, there is a considerable 'impedance mismatch' between the level at which the designer works and the level at which the package works (Taylor and Frederick, 1984; Pang, 1989). The result is that the user must issue many commands, in the correct sequence, in order to be able to move from one stage in the design to another. This is an inefficient use of her skills.

To illustrate this, consider the following statement of the first stage in the frequency response design of a cascade compensator for a plant with transfer function $G_o(s)$:

> "Adjust the low frequency gain to achieve the required steady-state performance and check the resulting phase margin."

We would argue that this is the correct level of detail at which to be instructing the machine. However, watch how quickly the task is lost in the minutiae of the low-level package commands.

One level of decomposition at least is needed to give us a procedural algorithm for tackling this task:

> 1. calculate the error required
> 2. calculate the steady-state error of the open-loop plant
> 3. determine the open-loop gain required
> 4. calculate the frequency response for the plant
> 5. add the gain (in dB) to the magnitude response (in dB)
> 6. determine the phase-margin of the gain compensated plant.

This is then translated into package commands, in this case in MATLAB:

```
(1)>> sse=0.02;          % steady state error
(2)>> numGo=[5];         % plant numerator
(3)>> denGo=[1,5,0];     % plant denominator type 1

% plant is type 1 so use Kv = sGo(s)|s->0 to determine
% velocity error constant
(4)>> Kv = polyval(numGo,0)/polyval(denGo(1:length(denGo)-1),0);
(5)>> K = Kv/sse;        % open loop gain required

% get frequency response
(6)>> w = logspace(-2,2,100);
(7)>> [magGo,phGo]=bode(numGo,denGo,w);

% convert magnitude to decibels
(8)>> magGodB = 20.*log10(magGo);

% convert gain to decibels
(9)>> KdB = 20*log10(K);
% add gain to open-loop magnitude and determine phase margin
(10)>> [Gm,Pm,Wcg,Wcp]=margin(magGodB+ones(w)'.*KdB,phGo,w);
```

Note how a simple statement of a procedure becomes at least 10 package statements by the time the description is at a level that the package can understand. The obvious response is to put the whole procedure into a script or macro, but the problem with this solution is apparent when one considers a slight change to the requirements such as:

> "Adjust the phase margin to achieve the required transient performance and determine the resulting steady-state error."

324

Clearly the macro that performs the steps needed for the first task would be useless for the second. Trying to extend the macro to cover both steps would just add extra complexity to the code and writing a new macro for each possible case increases the complexity of the package.

A better solution is provided by making use of the rule-based features of expert system shells. A possible approach is to create rules by which some parameter may be obtained from other parameters, so we might have rules like

```
rule open-loop-gain forward
     if
             the steady-state-error of system is Kv &
             the error-constant of system is Ess
     then
             execute_in_matlab(K = Kv/ess) &
             note the open-loop-gain of system is K.
```

A sufficient number of well chosen rules ought to be enough to allow high level-goals like those given above to be satisfied automatically by an inference engine. In essence we 'declare' what we want to do and the computer determines how to do it. This is known as the *declarative computing* paradigm. An interesting branch of research is being pursued by Pang (1989,1991,1992), who has embedded a rule-based inference engine within a MATLAB-like matrix environment and therefore has the capability of providing just such high level facilities directly within the CACSD package. Most other attempts e.g. CCAE-III (Taylor and Frederick, 1984; James *et al.*, 1987; Frederick *et al.*, 1991), have made use of links to external packages as illustrated in Fig. 15.

Integration of CACE packages. The use of AI for integrating CACE packages was one of the aims of the CACE-III project (Taylor and Frederick, 1984). In this project, the expert system was seen to take a central role in the environment similar to that of the supervisor in Fig. 14. The user would submit high level commands to the expert system, which would then use its rules to determine how the packages should be driven to give the desired solution. The requirements for such a role are of course the ability of the expert system to communicate with external packages and also with files.

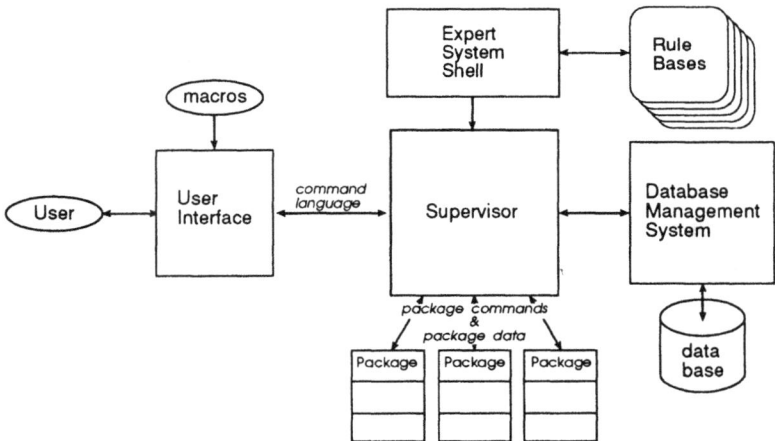

Fig. 20. Use of expert advisor in MEAD.

325

Unfortunately, this role proved to be too difficult, since only rules for a few design techniques were encapsulated. A more sensible role for an expert system is to provide help on particular design techniques, so the architecture shown in Fig. 20 was developed in the MEAD project (Taylor and McKeehan, 1989; Hummel and Taylor, 1989; Taylor *et al.*, 1990). The expert system can be called to assist with particular design problems, but is not the chief intermediary.

Intelligent front end. A common dilemma in CACE package design is how to provide the expert user with a highly flexible and responsive design environment, yet at the same time not leave novice or occasional users floundering. The solution has typically, if regrettably, to ignore the novice completely!

Luckily, expert system techniques can also be used in such cases to provide assistance by means of *intelligent front ends* (IFE). One of the earliest examples of such a system was the "command spy" developed by Larsson and Persson (1986). This system kept a complete history of the user's dialogue, the intention being to build-up knowledge of the user's use of the CACE package so that if he made a mistake, or asked for assistance, the command spy could give advice based on the current context. The knowledge representation scheme used in this system was *scripts* or sequences of typical commands that would be used in particular cases. A pattern matcher, written in LISP, was used to select the correct script according to the history of the user's interaction. In this way it was hoped that the system would, at any point in the interaction, know what the user was trying to do; find out what, if anything, went wrong; and offer advice or corrective measures suitable to the current context. The use of automatically generated scripts also provided the potential for the system to learn by 'watching' an expert user going through a design. Since a design requirement was that the command spy be unobtrusive, its ability to provide meaningful assistance to novice users was probably limited.

A much more general purpose facility is the IFE proposed by Pang (1988). The three level architecture is shown in Figs. 21, 22 and 23. In Fig. 21, the set-up intended for novice users, the IFE acts as an intermediary between the user and the package. In Fig. 22, the IFE acts as an assistant providing help when needed. In Fig. 23, the IFE acts as a command spy and only intervenes when the user, an expert himself, makes a typing error or some other mistake.

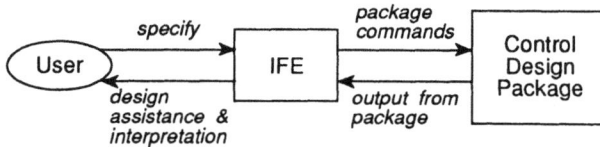

Fig. 21. Intelligent front-end for a novice user.

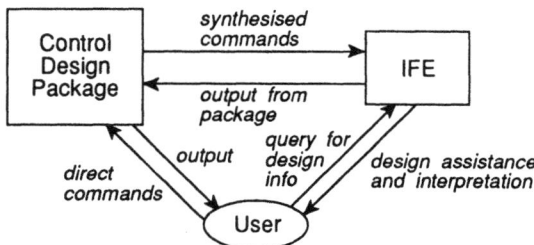

Fig. 22. Intermediate intelligent front-end (advisor).

326

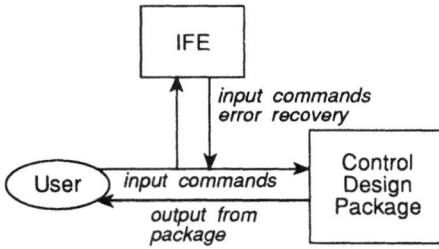

Fig. 23. Intelligent front-end for an expert user (command spy).

Such a layered IFE has the potential of adapting itself by changing its architecture and behaviour according to some internal model of typical users. Other advantages are that, with the addition of suitable rules, it can be used to provide tutorial introductions to a package.

Tutor. Tutorial assistance is useful in at least two cases: to help the user to learn a CACE package; and to help the user to learn a new design technique or algorithm. The first use of tutor systems is best left to an IFE as discussed above. The second use, which becomes ever more critical as control techniques and algorithms become increasingly complex, is the topic of this section.

To illustrate how an expert system can operate as a tutor, we shall use as a case study a program called "Tutorial Assistant" (TA), developed at the University of Zimbabwe (Appiah *et al.*, 1991). TA is a simple forward chaining expert-system shell written in C. It is embedded within PC-Matlab using the MEX file facility (Mathworks, 1989). TA is intended to teach students control systems analysis techniques by self-study, but the principles are applicable to any tutoring system.

In addition to the inference engine, the TA system consists of three components:

1. a set of script files used to describe the problem to the user and to present its solution. M-files[3] are used for this although some other method such as hypertext might be more appropriate.

2. A set of facts which define the problem and set the initial conditions. In TA, facts are either MATLAB workspace variables or functions which are executed at the start of the tutorial session. As the tutorial runs, additional facts are made available.

3. A set of rules whose pre-conditions and post-conditions are all MATLAB functions. The preconditions typically run pages of the script, elicit variables from the fact store or request input from the tutee. The post-conditions are further functions that set variables in the workspace indicating the completion of some stage in the tutorial process.

Although the inference engine selects which rules to fire at any given stage in the process, it is MATLAB itself that controls the calculation of variables, the processing of user inputs and the performance of any simulations. TA is thus a nice symbiosis of a CACE package and an expert system shell, similar to that found in the MAID (Pang, 1989) and MEDAL (Pang, 1992) systems.

---

[3] An M-file is the MATLAB name for a macro file, or external script of MATLAB instructions.

327

```
Rule: 1
Determination of cross-over freq, w1, given phase margin
IF
      ~exist('w1')
AND
      exist('phasem')
THEN
      w1 = omega1(mag,phasem,re,im)

Rule: 2
Determination of phase margin, given cross-over frequency
IF
~exist('phasem')
AND
exist('w1')
THEN
phasem = phase_in(w1)
      :
      another 7 more rules
      :
```

Fig. 24. Partial TA rule-base for gain and phase margin problems.

Figure 24 shows the first two rules in a TA rule-base for solving gain and phase margin problems. Such problems can be broadly described as being one of the following three cases:

"given an open-loop transfer function $KG_o(s)$ use frequency response methods to find either: (i) a value of K and the gain margin for which the phase margin is ϕ_m; (ii) a value of K and the phase margin for which the gain margin is γ_m; or (iii) the gain and phase margin when K is given."

Which case is in use depends on the initial facts loaded into TA before its starts. Figure 25 shows the facts that would start a problem of type (i).

```
Facts
[num,den] = preamble
[mag, re, im] = get_parm(num,den)
wpi = omegapi(re,im)
modG = modGwpi(wpi)
phasem = 40
```

Fig. 25. Facts for a gain and gain margin problem.

In operation, TA calculates the actual values for each of the parameters, and at the same time requests that the tutee makes calculations of her own and inputs her values for all the missing parameters. At the end of the run the user's results are compared with the calculated parameters and the tutee may then review the process or backtrack if she has made any mistakes. A more complex problem is to have the expert system itself determine the source of any errors. This is in essence possible by keeping a complete record of the tutee's actions and using a backwards chaining diagnostic system to find the point at which an error was introduced. Additional, tailored help could then be given to try to overcome the student's misconceptions.

Progress monitor. There is a dual role for this type of facility. The first is to provide explanations of the current state of the design, and the second is to provide a complete record of the design for the purposes of documentation. Many commercial expert systems provide some form of explanation and diary or log facility which could be adapted for this purpose. A further use for progress monitors might be to train the expert system itself, although this is possibly not within the scope of today's systems.

Design assistant. This is potentially the most important application for AI technology in CACE. It is one that has, not surprisingly, received the most attention from researchers in control systems design. A typical application (Taylor and Frederick, 1984) is one that has been used to provide assistance in the design of lag-lead compensators by frequency response methods, described in some detail by James *et al.* (1987). The architecture of this expert system is shown in Fig. 26.

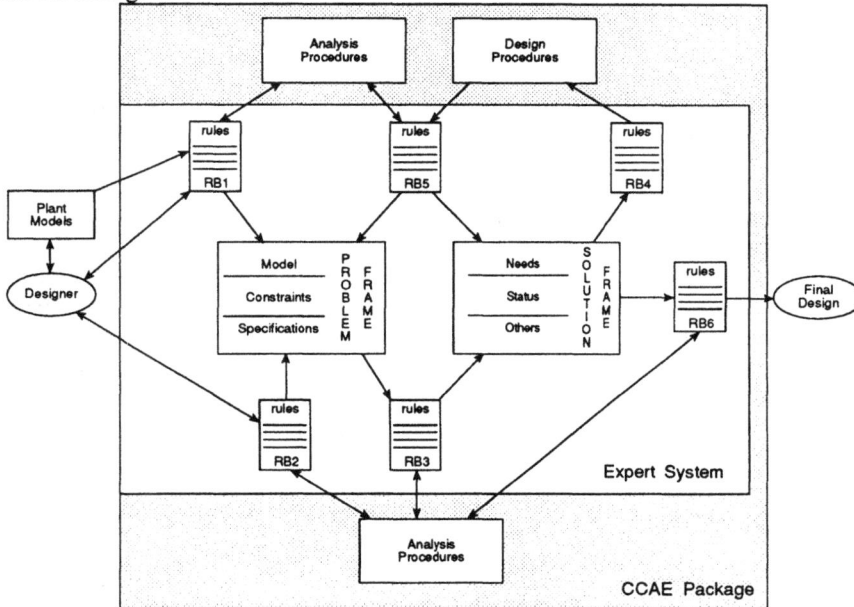

Fig. 26.   The architecture of the CACE-III design assistant.

It consists of two frames and six rule-bases. The *problem frame* contains details of the model, the constraints, and the specifications to be satisfied. This frame is populated by the user at the start of the session although some of the parameters may change during the design. The *solution frame* is populated as the design proceeds. It contains a *needs slot* that lists the parameters still to be found, a *status slot* which keeps track of what the design assistant is doing and an *other slot* that is used to record such things as what design methods have been tried. The six rule bases are used as follows:

- *RB1* — governs interactions among the design engineer, plant models and the model and constraint components of the problem frame,

- *RB2* — governs interactions between the designer and the specification component of the problem frame,

- *RB3* — governs interactions between the problem frame and the solution frame,

- *RB4* — governs interactions between the solution frame and the available (external) design procedures,

329

- *RB5* — governs interactions between the design procedures and the problem and solution frames,

- *RB6* — governs the final control system design validation process and conversion from idealised controller design to practical implementation.

Frames and rule-bases are separated for reasons of ease of program development and maintenance. Technical details of some of the design heuristics used in CACE-III are given in (James *et al.*, 1987). A recent paper describes the update of CACE-III to make use of the extra features of expert system shells that were not available originally (Frederick *et al.*, 1991).

Other applications of design assistant expert systems are given in (Trankel and Markosian, 1985; Birdwell and Cockett, 1985; Trankel *et al.*, 1986; Pang and MacFarlane, 1986; Basu *et al.*, 1988; Basu *et al.*, 1991; Tebbutt, 1990; Muha, 1991 and Linkens *et al.*, 1991b).

THE FOURTH GENERATION OF CACE

Before control engineers can properly benefit from the power of modern computing technology, the problem of package integration must be solved. As we have seen, modern CACSD packages provide support for only a small subset of these functions required for CACE so a wide range of different packages is needed to support a large subset. To satisfy all requirements a large number of packages need to be integrated into a single environment that provides the full range of features that computing technology has made available. In this section we shall gaze into our crystal ball to find out what the fourth generation may have in store.

The Open CACE Environment

The most successful architectures for integration have been those which have been based on well publicised interfaces or on published standards. The facilities we have for CACE are in some respects standardised, but the standards are at the package level rather than the environment level. If one starts from the view that integration is *necessary*, a different sort of architecture emerges. This is an architecture that provides a *framework* for integration, that is a sort of 'software bus' into which tools can be slotted. The basis for this architecture is the recognition that the most basic data entity in control engineering is a model of a dynamic system. Once this is recognised then the tools-based architecture discussed in the remainder of this section results.

A tools based approach. Integration cannot be achieved without a fundamental re-assessment of how CACE tools should be implemented. Andersson *et al.* (1990) have advocated a tools based architecture which has been further refined by Barker *et al.* (1992a,1992b). This approach is based on the dual recognition that dynamic system models are the fundamental data object in CACSD and that user interfaces need to be made consistent across all tools. If a common modelling language could be provided, then all tools could be made to adopt it, eliminating the need for separate operating system interfaces or data translators. One such modelling language is Omola developed by Andersson (1989,1990) from ideas on modelling expressed by Mattsson (1988) and Cellier and Elmqvist (1992).

A common user interface, preferably making use of direct-manipulation graphics for commonly used commands would ensure that the environment operated on a single paradigm, was consistent and easy to use. Within such an architecture, illustrated in Fig. 27, the basic tools need no more than a simple control protocol, eliminating the need for individual user interfaces. Thus the tools become much simpler and in principle interchangeable.

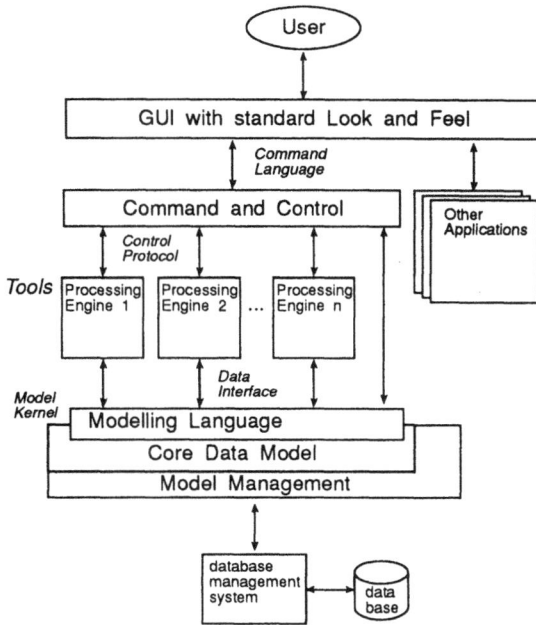

Fig. 27. A tools based architecture.

Graphical interfaces. The next generation of CACSD systems will certainly make use of graphical user interfaces, for both systems data entry, of all kinds of dynamic system representations, and for package control. A front end, called eXCeS, (Fig. 28) is being developed (Barker *et al.*, 1991) that has potential to be the basis of the front end shown in Fig. 27. Other research, such as that of Hara *et al.* (1988,1992), may yield CACE environments with levels of interaction much closer to those enjoyed in desktop publishing packages on the Macintosh and under Windows 3.0.

Symbolic processing. Symbolic processing tools will readily fit within a tools based architecture and there is every reason to suppose that there will be continue to be roles for them in model translation, linearization, etc.

Model development. Block diagrams are only one possible way of describing systems. In fact they are a special case of a rather more general technique which Cellier (1991) calls *stylised block diagrams* of which circuit diagrams, process flow diagrams and mechanical mobility diagrams are concrete examples. The support for more general model descriptions and the provision of (graphical) tools for their input and manipulation will continue to be important. In fact, as control engineering becomes just one part of systems engineering, it will become a requirement to be able to import models defined by other specialists, for example mechanical, electrical or process engineers, into the CACE environment (Grübel, 1991).

331

BlockEdit – lag–lead

File ▽ View ▽ Edit ▽ Props ▽ Mode ▽ Tool ▽

Library Manager

File ▽ Edit ▽
Categories... Block ▽

Category: common
Comments: Commonly-used

P.   Gain   Lag   Lead   Plant   C

Mode: None

OperationEdit: 0.1 – lag–lead

Done Operation ▽ View ▽ Props ▽

$$\frac{s + 0.1}{s + 0.047}$$

Current Operation: Dig

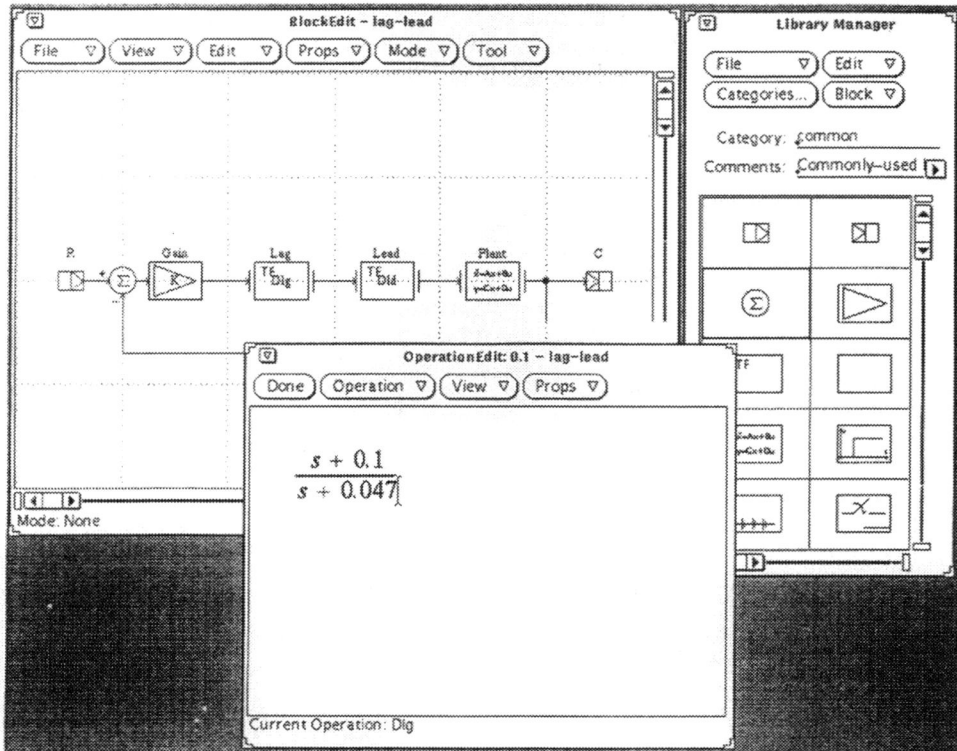

Fig. 28. eXCeS: a graphical front-end for CACE

Model management. The maintenance of models, results, reports etc., right through the life-time of a control project from the creation of the mathematical model to the implementation of the digital controller, is also an important future area for research. Some preliminary work has already been done, for example through the MEAD project (Taylor *et al.*, 1988; Mroz *et al.*, 1988; Taylor *et al.*, 1989; Rimvall and Taylor, 1991) and by Joos (1991) . Future work will need to make use of computer science techniques such as object-oriented programming and object-oriented databases (Hope *et al.*, 1991) as well as utilising computer graphics (Jobling, 1991) to a much greater extent.

On-line documentation. It is desirable to provide computer support for the maintenance of de-sign documentation during the life-cycle. This is a more subtle requirement than just the stor-age and retrieval of data (although even this is given poor support in packages for the com-puter-aided design of control systems). Rather, it is a requirement for computer implementa-tion of the engineer's log book in which all the stages of the design process are recorded: the assumptions and decisions made; the specifications given and modified; the designs tried and rejected. As yet, there is very little systematic computer-based support for recording such documentary information for any engineering activity.

Expert systems. Within a tools-based open CACE environment, expert systems will no-longer be needed to provide support for integration and so their use as adviser systems to help in design, to shield the user from the complexities of the environment or to explain an analysis technique, can be extended. Because the data interface is standardised, expert systems for one particular task or role will be portable and so are likely to be used more. The day may come when advice, intelligent help and tutorial rule-bases come with all new tools.

Interactive Computer-Aided Design of Control Systems.

As we have seen in these lectures, the user interface is a vitally important part of good quality CACE. Much can be done to improve things for the human designer by providing expert systems support for design. However, I also believe that the full potential of computer graphics in the interactive design environment has yet to be realised. Indeed, this may prove to be even more important than AI.

The statement that "design is an interactive process" is one that cannot be over-emphasised when considering CACE. One needs to be able to "try things out to see what happens" quickly and safely. By the inventive use of interactive graphics we should better be able to support this aspect of design and also make it more fun. To illustrate this, we present an example of an "ideal" frequency response design facility, see Fig. 29.

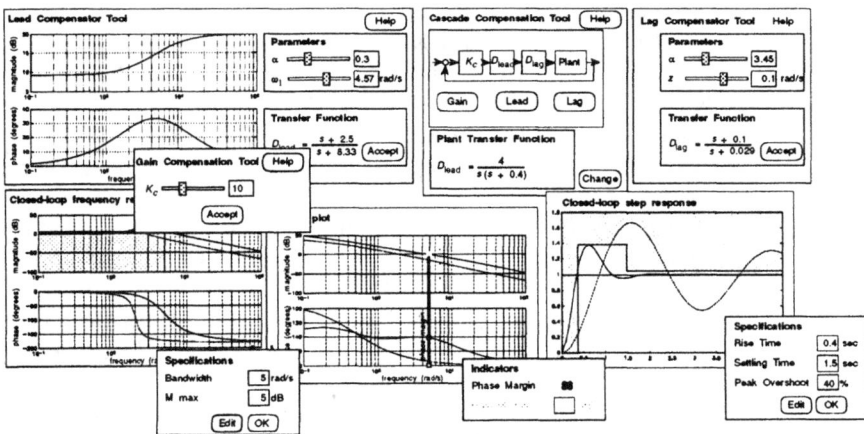

Fig. 29. True CACE – an interactive frequency response design system.

The designer has several windows each of which provides feedback on some aspect of the current status of the design (in this case that of a lag-lead compensator). In the main window the uncompensated open-loop frequency response curve is displayed. The open-loop gain can easily be changed by moving the slider. The display automatically adjusts to indicate this change but *so do all the other displays*: the time response, closed-loop frequency response and compensator time response. All the performance measures also change instantaneously. As the phase lead compensator is adjusted by modifying the centre frequency or the value of the lag fraction the effect can be instantly observed. If the designer gets stuck, she presses the help button, and an expert system shell gives her some advice. Finally, when all the design constraints have been satisfied, the designer presses 'OK' and the software for the digital implementation is automatically generated. It is obvious that it wouldn't take long to satisfy a given set of constraints with this system.

We can imagine you saying that this type of design facility is not achievable with current technologies, but you would be wrong. There already exists a Japanese prototype of such a design environment which is described in (Hara *et al.*, 1988,1992). The potential is enormous!

CONCLUSIONS

In these lecture notes we have discussed the user interface for computer-aided control engineering. From a model of the design process we have developed requirements and seen that the current generation fails to meet them. The use of AI to strengthen the user interface is a useful technique, but of more importance is the problem of integration which must be solved. A possible tools-based environment has been presented. Finally an application of powerful interactive computer graphics in design has been illustrated.

REFERENCES

Andersson, M. (1989). An object-oriented language for model representation. *Pages 8–15 of: Procs. 1989 IEEE Control Systems Society Workshop on Computer-Aided Control System Design*. Tampa, Florida, USA.

Andersson, M, S. E. Mattsson, and B. Nilsson. (1991). On the architecture of CACE environments. *Pages 63–68 of: Preprints of the 5th IFAC Symp. on Computer Aided Design in Control Systems — CADCS '91*. Swansea, UK.

Andersson, M. (1990). *Omola — An Object-Oriented Language for Model Representation*. Tech. rept. CODEN:LUFTD2/(TFRT-3208)/1-102/(1990). Department of Automatic Control, Lund Institute of Technology, Sweden.

Appiah, R. K, H. Gabe, and S. Utete. (1991). Expert-system assistant for engineering tutorial problem solving. *In: Proc. Computer Aided Learning Conf.* Lausanne, Switzerland.

Augustin, D. C, J. C. Strauss, M. S. Fineberg, B. B. Johnson, R. N. Linebarger, and F. J. Sansom. (1967). The Sci Continuous System Simulation Language (CSSL). *Simulation*, 2, 281–303.

Barker, H. A, M. Chen, and P. Townsend. (1988). Algorithms for transformations between block diagrams and signal flow graphs. *Pages 231–236 of: Proc. 4th IFAC Symp. on Computer-Aided Design in Control Systems*. Beijing, PRC.

Barker, H. A, M. Chen, P. W. Grant, I. T. Harvey, C. P. Jobling, A. P. Parkman, and P. Townsend. (1991). The Making of eXCeS — A software engineeirng perspective. *Pages 27–33 of: Preprints of the 5th IFAC Symp. on Computer Aided Design in Control Systems — CADCS '91*. Swansea, UK.

Barker, H. A., P. W. Grant, C. P. Jobling, J. Song, and P. Townsend. (1989a). A Graphical man-machine inteface for discrete-event dynamic systems. *Pages 356–365 of: Procs. 3rd European Simulation Conf. — ESC 89*. Edinburgh, Scotland.

Barker, H. A., J. Song, and P. Townsend. (1989b). A Rule-Based Procedure for Generating Programmable Logic Controller Code from Graphical Input in the form of Ladder Diagrams. *Engineering Applications of Artificial Intelligence*, 2(Dec.), 300–306.

Barker, H. A., M. Chen, P. W. Grant, C. P. Jobling, and P. Townsend. (1992a). An open architecture for computer-assisted control engineering. *In: Proc. IEEE-CSS Symp. on CACSD*. Napa, CA, USA.

Barker, H. A., M. Chen, P. W. Grant, C.P. Jobling, and P. Townsend. (1992b). Modern environments for dynamic system modelling. *In: Proc. IASTED Int. Conf. on Control, Modelling and Simulation*. Innsbruck, Austria.

Barr, A, and E. A. Feigenbaum. (1981). *The Handbook of Artificial Intelligence*. Vol. 1. Los Altos, CA, USA: William Kaufmann, Inc.

Basu, A, A. K. Majumdar, and S. Sinha. (1988). An Expert system approach to control system design and analysis. *IEEE Trans. Systems, Man and Cybernetics*, SMC-18(5), 685–694.

Basu, A, A. K. Majumdar, and S. Sinha. (1991). Design of industrial regulators using expert systems approach. *Int. J. Systems Science*, 22(3), 551–577.

Birdwell, J. D, and J. R. Cockett. (1985). Expert system techniques in a computer-based control system analysis and design environment. *Pages 1–8 of: Proc. 3rd IFAC Symp. on CADCE Systems*.

Cellier, F, B. Zeigler, and A. Cutler. (1991). Object-oriented modelling: tools and techniques for capturing properties of physical systems in computer code. *Pages 1–10 of: Preprints of the 5th IFAC Symp. on Computer Aided Design in Control Systems — CADCS '91*. Swansea, UK.

Cellier, F and H. Elmqvist (1992). The need for formula manipulation in object-oriented continuous system modelling. *In: Proc. IEEE-CSS Symp. on CACSD*. Napa, CA, USA.

Clocksin, W. F, and C. S. Mellish. (1985). *Programming in Prolog*. 2nd edn. Berlin, Germany: Springer Verlag.

Cox, B. (1986). *Object Oriented Programming — An Evolutionary Approach*. Reading, Mass., USA: Addison Wesley.

Dodd, T. (1990). *Prolog: A Logical Approach*. Oxford, UK: Oxford University Press.

Floyd, M. A, P. J. Dawes, and U. Milletti. (1991). Xmath: a new generation of object-oriented CACSD tools. *Pages 2232–2237 of: Proc. European Control Conf.*, vol. 3.

Frederick, D. K, J. R. James, A. Automotti, and H. Nitta. (1991). A Second-generation expert system for CACSD. *Pages 223–228 of: Preprints of the 5th IFAC Symp. on Computer-Aided Design in Control Systems*. Swansea, UK.

Gawthrop, P. J, N. A. Marrison, and L. Smith. (1991). MTT: A bond graph tool box. *Pages 274–279 of: Preprints of the 5th IFAC Symp. on Computer Aided Design in Control Systems — CADCS '91*. Swansea, UK.

Grant, P. W, and C. P. Jobling. (1991). The Application of Prolog to the Symbolic Manipulation of Control Systems in a Third Generation CACSD Environment. *In: Proc. European Control Conf.* Grenoble, France.

Grant, P. W, C. P. Jobling, and Y. W. Ko. (1988). A Comparison of rule-based and algebraic approaches to computer-aided manipulation of linear dynamic system models. *In: Proc. IEE Int. Conf. Control '88*. Oxford, UK.

Grant, P. W, C. P. Jobling, and C. Rezvani. (1992). Some control engineering applications of Prolog. *Pages 136–158 of:* Brough, D (Ed), *Logic Programming: New Frontiers*. Oxford: Intellect.

Grant, P, C. Jobling, and C. Rezvani. (1989). The Role of Prolog-Based Symbolic Analysis in Future CACSD Environments. *In: Proc. 19th IEEE Conf. on Decision and Control*. Tampa, Florida, USA.

Grübel, G. (1991). Concurrent Control Engineering (special invited session). *Pages 101–142 of: Preprints of the 5th IFAC Symp. on Computer Aided Design in Control Systems — CADCS '91*. Swansea, UK.

Hara, S, I. Akahori, and R. Kondo. (1988). Computer aided control system analysis and design based on the concept of object-orientation. *Pages 220–225 of: Preprints of the 4th IFAC Symp. on Computer Aided Design in Control Systems — CADCS '88*. Beijing, PRC.

Hara, S, G. Yamamoto, M. Oshima and R. Kondo. (1992). A distributed computing environment for control system analysis and design. *Pages 47–54 of: Proc. IEEE Symp. on CACSD*. Napa, CA, USA.

335

Hope, S, P. J. Fleming, and J. G. Wolff. (1991). Object-oriented database support for computer-aided control system design. *Pages 200–204 of: Preprints of the 5th IFAC Symp. on Computer Aided Design in Control Systems — CADCS '91*. Swansea, UK.

Hummel, T. C, and J. H. Taylor. (1989). Multidisciplinary expert-aided analysis and design (MEAD). *In: Proc. 3rd Annual Conf. on Aerospace Computational Control*.

James, J. R. (1987). A survey of knowledge-based systems for CACSD. *Pages 2156–2161 of: Proc. Am. Control Conf.*

James, J. R, D. K. Frederick, and J. H. Taylor. (1987). Use of expert system programming techniques for the design of lead-lag compensators. *IEE Proceedings Part D*, 134(3), 137–144.

Jobling, C. P. (1991). Building better graphical user interfaces for CACSD — the case for object-oriented programming. *Pages 143–147 of: Preprints of the 5th IFAC Symp. on Computer Aided Design in Control Systems — CADCS '91*. Swansea, UK.

Jobling, C. P, and P. W. Grant. (1988). A Rule-Based Program for Signal Flow Graph Reduction. *Engineering Applications of Artificial Intelligence*, 1(1), 22–33.

Joos, H. D. (1991). Automatic evolution of a decision-supporting design project database in concurrent control engineering. *Pages 113–117 of: Preprints of the 5th IFAC Symp. on Computer Aided Design in Control Systems — CADCS '91*. Swansea, UK.

Larsson, J. E, and P. Persson. (1986). Knowledge representation by scripts in an expert interface. *Pages 1159–1162 of: Proc. Am. Control Conf.*

Linkens, D. A, S. Bennett, E. Tanyi, M. Rahbar, and M. Smith. (1991a). Experiences from a prototype environment for intelligent modelling and simulation. *Pages 217–222 of: Preprints 5th IFAC Symp. Computer Aided Design in Control Systems — CADCS '91*. Swansea, UK.

Linkens, D. A, S. Bennett, E. Tanyi, M. Rahbar, and M. Smith. (1991b). Experiences from a prototype environment for intelligent modelling and Simulation. *Pages 217–221 of: Preprints 5th IFAC Symp. on Computer-Aided Design in Control Systems*. Swansea, UK.

MacFarlane, A. G. J, G. Grübel, and J. Ackermann. (1989). Future design environments for control engineering. *Automatica*, 25(2), 165–176.

Mathworks. (1989). *MATLAB — User's Guide*. MathWorks, Inc., 21, Eliot Street, South Nantwick, MA, USA.

Mattsson, S. E. (1988). On Model Structuring Concepts. *Pages 269–274 of: Preprints 4th IFAC Symp. Computer Aided Design in Control Systems — CADCS '88*. Beijing, PRC. Pergamon Press.

MIKE. (1990). *MIKE Reference Manual*. Open University, UK.

Moler, C. (1980). *MATLAB — User's Guide*. Department of Computer Science, University of New Mexico, Alberquerque, USA.

Mroz, P. A, P. McKeehen, and J. H. Taylor. (1988). An Interface for computer-aided control engineering based on an engineering data-base manager. *Pages 725–730 of: Proc. Am. Control Conf.*

Muha, P. A. (1991). Expert System for SROOT. *Proc. IEE (Part D): Control Theory and Applications*, 138(4), 381–387.

Munro, N. (1988). ECSTASY - a control system CAD environment. *In: Proc. IEE Int. Conf.: Control '88. 13-15 April 1988. Oxford, UK*. IEE.

OU. (1990). *Knowledge Engineering*. P.O. Box 188, Milton Keynes, MK7 6DH, UK: Open University. Study Pack No. PD624.

Pang, G. K. H, and A. G. J. MacFarlane. (1986). An Expert Systems Approach to Computer-Aided Design of Multivariable Systems. *Page 325 pp:* Balakrishnam, A, and M. Thoma

(Eds), *Lecture Notes in Control and Information Sciences*, vol. 89. New York: Springer Verlag.

Pang, G. K. H. (1988). An Intelligent front end for control system design and analysis package. *Pages 505–510 of: Proc. 4th IFAC Symp. on Computer-Aided Design in Control Systems*. Beijing, PRC.

Pang, G. K. H. (1989). Knowledge engineering in the computer-aided design of control systems. *Expert Systems*, 6(4), 250–262.

Pang, G. K. H. (1991). The implementation of an expert system for control system design in the knowledge environment of SFPACK. *Pages 229–234 of: Preprints of the 5th IFAC Symp. on Computer-Aided Design in Control Systems*. Swansea, UK.

Pang, G. K. H. (1992). A matrix and expert system development aid language. *Pages 148–154 of: Proc. IEEE Symp. on CACSD*: Napa, CA, USA.

Rimvall, M. (1987). CACSD software and man-machine interfaces of modern control environments. *Transactions of the Institute of Measurment and Control*, 9(2).

Rimvall, M, and L. Bomholt. (1985). A flexible man-machine interface of CACSD applications. *In: Proc. 3rd IFAC Symp. on Computer Aided Design in Control and Engineering.*

Rimvall, M, and J. H. Taylor. (1991). Data-driven supervisor design for CACE package integration. *Pages 33–38 of: Preprints of the 5th IFAC Symp. on Computer Aided Design in Control Systems — CADCS '91*. Swansea, UK.

Rimvall, M. (1988). Interactive Environments for CACSD Software. *Pages 17–26 of: Preprints of 4th IFAC Symp. on Computer Aided Design in Control Systems – CADCS '88*. Beijing, PRC. Pergamon Press.

Rosenbrock, H. H. (1974). *Computer-Aided Control System Design*. New York: Academic Press.

Schneiderman, B. (1987). *Designing the User Interface*. Addison-Wesley.

Shouaib, A. M. K, P. W. Grant, and C. P. Jobling. (1992). A Control Engineering Toolkit for *Mathematica* on the Macintosh. *Pages 1–4 of: Computer-Aided Control Systems Design, Algorithms, Packages and Environments*. IEE, London, UK. Colloquium Digest No. 1992/004.

Sommerville, I. (1989). *Software Engineering*. 3rd edn. Reading, MA: Addison Wesley.

Spang, H. A., I. (1984). The Federated Computer-Aided Control Design System. *Proc. IEEE*, 72(12), 1724–1731.

Tan, C. Y, and J. M. Maciejowski. (1989). DB-Prolog : A Database programming environment for CACSD. *Pages 72–77 of: Proceedings 1989 IEEE Control Systems Society Workshop on Computer-Aided Control System Design*. Tampa, Florida, USA.

Taylor, J. H. (1988). Expert-aided environments for the computer-aided engineering of control systems. *In: Proc. 4th IFAC Symp. on Computer-Aided Design in Control Systems*. Beijing, PRC.

Taylor, J. H, and D. K. Frederick. (1984). An Expert system architecture for computer-aided control engineering. *Proc. IEEE*, 72(Dec.), 1795–1805.

Taylor, J. H, and P. D. McKeehan. (1989). A Computer-Aided Control Engineering Environment for Multidisciplinary Expert-Aided Analysis and Design (MEAD). *In: Proc. National Aerospace and Electronics Conf. (NAECON).*

Taylor, J. H, D. K. Frederick, C. M. Rimvall, and H. Sutherland. (1989). The GE MEAD Computer-aided control engineering environment. *Pages 16–23 of: Proc. 1989 IEEE Control Systems Society Workshop on Computer-Aided Control System Design.*

Taylor, J. H, D. K. Frederick, and M. Rimvall. (1990). Computer-Aided Control Engineering Environments: Architecture, User Interface, Data-base Management, and Expert-Aiding. *In: Proc. 11th IFAC World Congress*. Tallin, USSR.

Taylor, J. H, K.-H. Nieh, and P. A. Mroz. (1988). A Data-base management scheme for computer-aided control engineering. *Pages 719–724 of: Proc. 1988 Am. Control Conf.*

Tebbutt, C. D. (1990). An Expert systems approach to controller design. *Proc. IEE Part D.,* 137(6), 367–373.

Torrero, E. A. (1985). *Next-generation computers*. New York: IEEE Press.

Trankel, T. L, and L. Z. Markosian. (1985). An expert system for control system design. *In: Proc. IEE Int. Conf.— Control '85*. Cambridge, UK.

Trankel, T. L, I. Z. Markosian, and U. H. Rabin. (1986). Expert systems architecture for control system design. *In: Proc. Am. Control Conf.*

338

PART V

INTELLIGENT CONTROL

Road Map.

The fifth part of the course deals with *Intelligent Control*. Basic knowledge on control engineering is expected and will not be introduced here. For an introduction on the artificial intelligence techniques used in these chapters, the reader is referred to part I.

This part consists of four chapters:

Fuzzy Controllers. Direct control using fuzzy logic is introduced in this chapter. After a quick review of the basic concepts of fuzzy logic and approximated reasoning, the fuzzy controller structure is shown. Finally, the analysis, design and supervision of such fuzzy controllers is discussed.

Adaptive Fuzzy Control A second approach to using fuzzy logic in control is presented in this chapter. Here self- organizing fuzzy control is discussed and the fuzzy expert system shell DICE is introduced. Finally, an insight is given in the adaptation mechanisms and using some examples the working of the adaptive fuzzy controller is shown.

Neural Networks for Control. Another approach for intelligent control can be found in the use of neural networks. In this chapter an introduction is given to this domain. The use of neural networks for system representations, system identification and actual control is shown. A comparison between adaptive control and neural network control is given, which leads to a discussion on some possible adaptive control structures with neural networks.

Associative Memories: The CMAC approach. A special type of neural network is the Cerebellar Model Articulation Controller (CMAC). In this chapter the structure of CMAC, its working and the application to control is discussed.

This road map should enable the reader to make up his own choice in this basic course, depending on his skills and future use.

FUZZY CONTROLLERS

P. Albertos

Departamento de Ingeniería de sistemas, Computadores y Automática.
Universidad Politécnica de Valencia. Spain

1 Intelligent control

The purpose of this chapter is to present the use of Fuzzy Logic theory, already discussed in previous sessions, [see therein the references to this theory], to develop a kind of Intelligent Controllers, (IC), the so called Fuzzy Controllers, (FC). The structure, main components, its use, and the different methodologies to design such a controller will be discussed. Also, the fuzzy logic approach will be shown suitable to represent the approximated knowledge we have about the dynamic behavior of the plant to be controlled.

1.1 Introduction

Let us start by reviewing the general control problem, its different ways of approaching, and its solution by the use of AI techniques.

a. The controller role.

When dealing with control problems, a model of the process to be controlled and some control specifications must be given. The solution, following a control design methodology, will lead to the controller. It should be considered as a system, a dynamical one, with a set of input/output variables, and a mathematical model expressed by the control algorithms or control laws, with some tunable parameters.

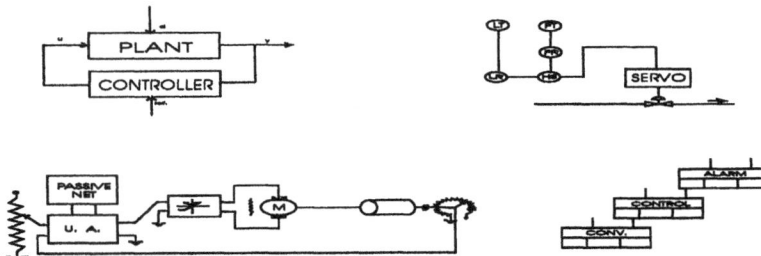

Fig. 1.1. The controller device: control vs. instrument viewpoint.

But the way the controller is considered depends of the user involved. From a control engineer point of view, the controller is just a subprocess which is connected to the plant, fig. 1.1.a. Its purpose being the control of the plant, it is conceived in relation with it, as a full dependant system. Sometimes it is combined with other system components, like filters, power amplifiers, and so on, fig. 1.1.b. On the other hand, from an instrumentation engineer point of view, the controller is a device processing the measurement signals and providing a control action. This control action is mainly related to the actions the plant operator would take under manual control, fig. 1.1.c. In the case of a digital control implementation, it appears as a small part of the global code, fig. 1.1.d.

We will see in this chapter, that Fuzzy Controllers must be analysed from both points of view.

b. Control System

The control design methodology largely depends on both process model and goals. And, of course, they both are conditioned to the kind of signals the measurement or processing systems allow to use.

The signals, also called variables, attached to the process may be modelled as:

- Logical.

   Only two possible values are considered: on-off, (1,0), true-false. In most cases, it is a rough way to discriminate between possible signal value options. The signal domain is mapped into the set: $U \to \{0,1\}$.

- Multivalue

   The physical variable is partitioned into various ranges or zones. Compared to the previous one, it is a finer discrimination allowing a better understanding of the current signal value. Now, its domain may be expressed by: $U \to D \in \mathcal{Z}$, where $\mathcal{Z}$ is the integer number set.

- Discrete

   Although in digital control any number is discretized in magnitude, if the word length is large enough, the signal model is considered as a sequence of values in $\mathcal{R}$.

- Continuous

   Conceptually, the signals are time functions the argument being $t \in \mathcal{R}$.

- Stochastic

   The signal magnitude is a random variable with some stochastic properties. This representation allows the handling of partially known signals. Nevertheless, some information related to the uncertainty is required or must be assumed.

- Fuzzy

   As previously discussed, a reduced number of linguistic variables, X_i, is attached to each physical variable, x. The current physical variable value, $x(t)$, is expressed by the set of membership function values, $\mu_{X_i}(x)$, related to all the corresponding linguistic variables. It is a way to express the approximated knowledge we have about its current value. Probabilistic knowledge may be expressed as a particular case.

344

According to the kind of signals the controller is handling, it will be represented by means of either, a quantitative model (any of the signals above, except the fuzzy ones) implemented by an algorithm, or a qualitative model (in particular for the fuzzy signals) implemented by a reasoning (intelligent) process.

c. Control Design Methodologies

As stated above, the model of the process and the control goals determine the control design methodology.

- Logical controllers [1]

  Dealing with binary or multivalued signals, the control goal is a known finite set of performance values. The underlying theory is that of sequential and combinatorial digital system. The logical controller, usually a Programmable Logic one, (PLC), is expressed by logical functions. It is clearly the case of starting up, shutting down, or transfering between stable operating conditions, in most process control applications.

- Continuous/discrete time controllers.

  According to the control especifications, and the available process model, different control methodologies are used:

  - Time/frecuency domain techniques, [2].
  - Optimization techniques, [3].
  - Adaptive/robust control techniques, [4].

- Supervisory controllers

  Dealing with either complex systems or systems with different modes of operation, the control solution must be also variable and some kind of supervision and/or coordination is required. Then, hierarchical/decentralised control techniques, [5], are applied. Many times, some of these techniques are used together, and a lot of user interaction is involved.

- Intelligent Controllers

  If the control goals are stated unprecisely (good quality, not too far from the ideal profile, and so on) or they are combined in a multiobjective goal, if the process model is basically a qualitative model, or if the controlled process scenario is variable and not well defined, the above techniques use to fail or to become extremely complex, unflexible and difficult to update. The use of AI techniques [6] will lead to IC.

d. Basic Techniques for IC

A common feature in the classical AI techniques is the handling of approximated knowledge, combined with some capability of reasoning and learning. But this is a very broad definition leading to a great variety of control solutions when applied to solve a control problem.

The different intelligent techniques may be characterized by:

- Symbolic and problem solving approach, like Expert Systems. Data are expressed by lists of characters and some semantic meaning may be involved.

- Models with implicit approximated knowledge, such as fuzzy systems. The problem solving approach involves reasoning and, in general, the solution is not unique. Nevertheless, if applied in control framework, a merged solution should be obtained.

- Learning capability, like Artificial Neural Networks (ANN). Based on a basic distributed and tunable structure, the learning capability allows to implement fast non-linear computations and pattern recognition.

Any of the IC concrete implementations involve some of these features. Sometimes it is just a *desideratum*. As we will see, learning capability is one feature to be provided in Fuzzy Controllers. But, it is still in a very experimental stage.

In this framework, let us review the key issues in fuzzy logic and reasoning, to further go on the FC.

1.2 Fuzzy Logic basic concepts

In order to fix the already known concepts, [12], we will refer to a steam generator, fig. 1.2, as an example.

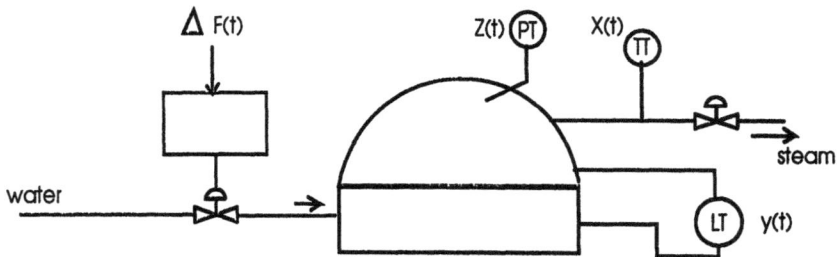

Fig. 1.2 Steam generator example

Given a physical variable, $x(t)$,[i.e. outlet temperature], we are interested in its current value if and only if it belongs to some precise interval named **universe of discourse**, $\mathcal{X}$, [i.e. $\mathcal{X} = (80, 150)$]. Otherwise, the variable will be out of range and its value may be associated to the corresponding extreme.

In this universe of discourse, some **linguistic variables** are defined, [Cold = X_1, OK = X_2, Hot = X_3]. Each one defines a fuzzy subset. Attached to each fuzzy subset is a **membership function**, (MF), $0 \le \mu_{X_i}(x) \le 1$, for all $x \in X_i$, being $= 0, if x \notin X_i$. That is, $\mu_{X_i} : \mathcal{X} \rightarrow [0,1]$.

For any linguistic variable, X_i, we name:

- Support

$$\mathcal{X}_i = \{x_i\} \quad \text{such as} \quad \mu_{X_i}(x_i) > 0$$

For instance, Cold support may be (80,110), (95,125) for the OK support, and (110,150) for Hot support.

- Crossover

$$\eta_i, \quad \text{such as} \quad \mu_{X_i}(\eta_i) = 0.5$$

- Bandwith

$$w_i, \quad \text{such as} \quad \forall x \in w_i, \mu_{X_i}(x) \geq .5$$

- Peakvalue

$$p_i, \quad \text{such as} \quad \mu_{X_i}(p_i) = 1$$

Fig. 1.3 shows these concepts for the OK outlet temperature fuzzy subset.

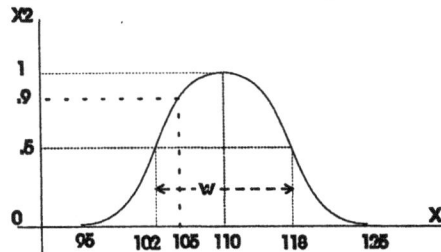

Fig. 1.3. Fuzzy subset and attached parameters

If the MF bandwith is narrow, the MF will be a discriminant one. In an heuristic sense it means a clear significance is given to the linguistic variable. On the other hand, a flat MF will be non-discriminant, that is, almost any variable value is strongly belonging to the fuzzy subset. In some cases, for instance if we have a measurement without error, the support of a fuzzy subset may be reduced to a single value s_i, taht is, the bandwith is zero, and the crossover and the peakvalue are the same. This fuzzy subset is call a **singleton**.

For the steam generator example, let us define two more variables: the drum water level, [Y_1 = Low; Y_2 = OK; Y_3 = High; $\mathcal{Y} = (-10, 10)$], and the steam pressure [Z_1 = Low; Z_2 = OK; Z_3 = High; $\mathcal{Z} = (1.2, 2.8)$]. Membership fuctions, similar to that in fig. 3 may be defined for any of these fuzzy subsets. Some operations have been defined between fuzzy subsets.

We know that in Fuzzy algebra we can define, with a particular interpretation, some basic operations similar to the basic boolean algebra operations: Equality, Inclusion, Complementation, Intersection, and Union. In particular, these operations are valid for fuzzy sets with common universe of discourse. Also, classical properties like Commutativity, Associativity, Idempotency, Distributivity, Absorption, Identity, and the Morgan's law are appliable. Moreover, from the point of view of FC, very useful operations can be defined when different physical variables are considered. For instance:

- Fuzzy relation

347

Between n fuzzy subsets, a joint fuzzy relation, F may be defined. The joint membership function is

$$\mu_F = \mu_{X_1, X_2, \ldots, X_n}(x_1, x_2, \ldots x_n).$$

The fuzzy relation, temperature OK - drum level High, that is, $\mu_F = \mu_{X,Y}(x, y)$, will express the membership function of a given pair (temperature,level) to this combined fuzzy subset.

- Cartesian Product

It is a particular case of a fuzzy relation, where the joint membership function is either the minimum of the components ones, or their product. That is:

$$\mu_{X_1, X_2, \ldots, X_n}(x_1, x_2, \ldots x_n) = min(\mu_{X_i}(x_i)), \qquad \text{or}$$

$$= \prod_{i=1}^{n} \mu_{X_i}(x_i)$$

- Sum-star Composition

If F_1 and F_2 are fuzzy relations in $X \times Y$ and $Y \times Z$, respectively, it is possible to compose both relations, the result being denoted by $F_1 \circ F_2$. The sum-star composition of F_1 and F_2, is defined by the membership function

$$\mu_{X,Z}(x, z) = sum(\mu_{F_1}(x, y) * \mu_{F_2}(y, z))$$

where $*$ stands for any operator in the class of triangular norms (minimum, algebraic product,...), and sum stands for any operator in the class of triangular conorms (maximum, algebraic sum, bounded sum,...).

Let us assume a fuzzy relation F_1 between Low level, Y_1, and High temperature, X_3, and a second fuzzy relation F_2 between the same temperature and High pressure, Z_3. If we apply the max-min composition operation, that is, the Sum-star composition one taken the maximum as triangular conorm and the minimum as triangular norm, the resulting fuzzy relation will be:

$$\mu_F = \mu_{Y_1, Z_3}(y, z) = \max_x(min(\mu_{X_3, Y_1}(x, y), \mu_{X_3, Z_3}(x, z))$$

1.3 Approximated Reasoning

Again, in order to fix the basic concepts already discussed, [12, section 4], let us review them when applied to the same example we introduced in the last section.

In classical reasoning there are two basic inference rules, the Modus Ponens (MP) and the Modus Tollens (MT), that can be generalized for the case of approximated knowledge:

- Generalized modus ponens (GMP)

348

```
premise 1:   IF x is A THEN y is B
premise 2:   x is A'

conclusion:  y is B'
```

• Generalized modus tollens (GMT)

```
premise 1:   IF x is A THEN y is B
premise 2:   y is not B'

conclusion:  x is not A'
```

If $A' = A$, and $B' = B$, then MP and MT inference rules are obtained.
For instance, let us assume the following premise:

```
premise:   IF temperature is  OK          THEN  pressure   is  OK
```

If such a rule is applied, because, in some sense, the temperature is OK, the conclusion will be:

$$\mu_Z(z) = \mu_{X'}(x) \circ \mu_{XZ}(x, z)$$

where $\mu_{X'}(x)$ is the MF of the input data, and $\mu_{XZ}(x, z)$ is the rule composition fuzzy MF.

When dealing with data from physical processes, they may be crisp data. If $x = 105$ then, fig.1.3, $\mu_{X_2}(x) = .9$. Then, if we apply the cartesian product to this fuzzy relation we obtain:

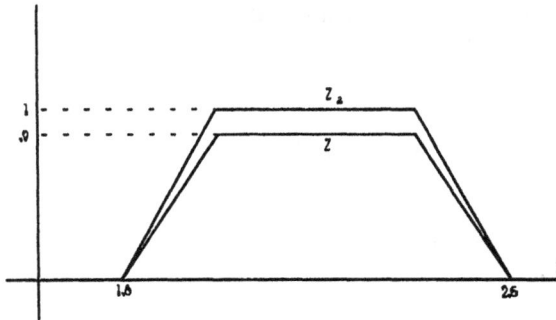

Fig. 1.4. Basic fuzzy inference

$$\mu_Z(z) = \mu_{X_2}(x).\mu_{Z_2}(z) = 0.9\mu_{Z_2}(z)$$

as a conclusion about the fuzzy value of the pressure.

But, with such a type of premise, it is possible to have a different knowledge, more or less confident. It would be the case if, for instance, the temperature was not a direct measurement but the result of some fuzzy implication (the result of pevious rules). Then, it is possible to have, for instance:

349

```
premise 1:    IF temperature is  High          THEN  pressure    is  High
premise 2:       temperature is  Sure High

conclusion:      pressure    is  Sure High
```

By Sure High we are declaring a more confidence in High temperature peakvalue. That means, for instance, $\mu_{X_3'}(x) = \mu_{X_3}^2(x)$, where X_3' stands for the Sure High temperature fuzzy subset. A similar expression will be used for Sure High pressure.

Among the many freedoms we have in selecting operations and compositions to express a fuzzy inference, the correspondence between A' and B' fuzzy subsets in GMP or GMT inferences is one of the most experience oriented selections. Even the way we express this approximated knowledge. On a fuzzy subset A, we can define, for instance, the following variations:

- A' := very A, or sure A. We point out our bigger confidence in the peakvalue of A.

- A' ;= almost A. The corresponding membership function will be smoother.

- A' := few A. The MF will be almost that of "not A".

- A' := not A. $\mu_{A'}(x) = (1 - \mu_A(x))$.

Similar approximated fuzzy sets should be considered at the conclusion B'. Possible criteria to relate the linguistic variables A' and B' are intuitive but not unique.

But this is not the only way we have to compose rules, [7]. Let us just mention some other options:

- Fuzzy conjunction

  It is the general case of the one we applied in the example above (fig. 4.). If $A \rightarrow B$, is the rule R, then $\mu_R(x.y) = \mu_A(x) * \mu_B(y)$.

- Fuzzy disjunction

  Under the same rule of inference, $\mu_R(x.y) = \mu_A(x) + \mu_B(y)$

- Fuzzy implication

  Many other ways to relate A and B have been used in the literature [7], again using some intuitive relations, such as $(not A) + B, (not A) + (A * B), (not A) \times (not B) + B$, and so on.

2 Fuzzy Controller Structure

In this section, the general structure of a FC will be discussed. It is assumed that the FC is directly connected to the process, that is, the process/controller interface involves A/D and D/A converters. On the other hand, the FC also interacts with the operator, exchanging both numeric and heuristic data. A linguistic interface should be also available. This scheme is called Direc FC, [10].

Fig 2.1 shows the general scheme of such a controller. Measurements from the process may be previously filtered or processed, in order to get trends, compensations, integrals and so on. Then, the FC input from the process is a set of discrete variables, such as the error and functions of it, state variables, and/or references. This information, together with the numeric one introduced by the operator, if any, must be converted into fuzzy data in order to be processed by the inference engine, a simple one.

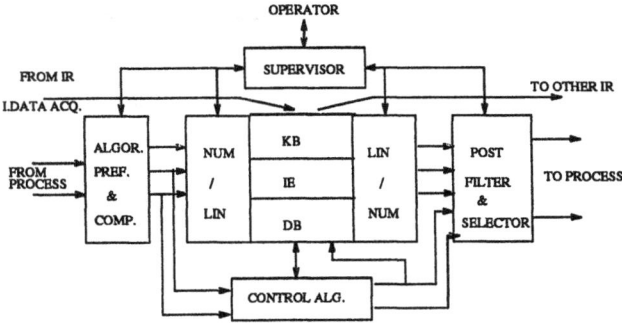

Figure 1: Fig. 2.1 Fuzzy Controller Structure

The output of the FC is twofold. That to the process must be numeric or logic, asking for some control action such as a manipulated variable variation or a binary actuator switch on/off. The output to the operator should be both, graphical or numeric, to show the process variables evolution, and linguistic or simbolic, to give some explanation about the reasoning process, if required.

As shown in fig. 2.1, supervision is a natural feature of FC even when applied in Direct control. The availability of controller and process behavior data allows some kind of supervision, autotuning or, in the more desirable case, learning.

Other than the digital pre- y post- processor modules, the FC includes: a fuzzification module, to convert from numeric to linguistic data, a knowledge-based system composed by a data base, a knowledge base and an inference engine, a defuzzyfier module converting data from fuzzy to numeric, and a supervisor taking care of some tunable parameters and selective options, also dealing with the operator interface.

Let us go in detail through the different FC modules.

2.1 Knowledge Base (KB)

As in the Expert Systems structure, the KB includes data (objects) and rules (relation between data objects). Information about the variables and fixed known facts are included in the first block. Parametric and heuristic relations between variables are expressed by rules.

351

2.1.1 Data Base (DB)

In the DB, linguistic and physical variables may be defined. Attached to each physical variable, the following information must be given:

- The universe of discourse

- The linguistic variables, LV,

- The membership functions, MF, parameters.

In general, parametrized fuzzy sets are predefined, and at this level, a set of instantiations is attached to each physical variable. So, the number of LV and their support, and other MF parameters are fixed.

Different basic MF are used. The rectangular one converts the fuzzy set into a multivalue logical variable. But many other geometrical functions are used: triangular, trapezoidal, polynomial, and so on. In order to easily modify the MF, those including a reduced number of parameters are preferable. Functions like the exponential or the sigmoidal one are used.

The exponential MF is defined by,

$$\mu_X(x) = e^{\frac{-(x-m_X)^2}{2\sigma_X^2}}, \quad \text{for} \quad \forall x \in \mathcal{X}$$

where m is the peakvalue and σ determines the bandwidth. The sigmoidal is:

$$S(x,a,c) = 0, \quad x \leq a$$

$$= 2[\frac{x-a}{c-a}]^2, \quad a \leq x \leq \frac{a+c}{2}$$

$$= 1 - 2[\frac{c-x}{c-a}]^2, \quad \frac{a+c}{2} \leq x \leq c$$

$$= 1, \quad x \geq c$$

or in symetrical shape,

$$\Pi(x,p,w) = S(x,p-w,p), x \leq p$$

$$= 1 - S(x,p,p+w), x \geq p$$

where w is directly the bandwidth. Fig 2.2 shows these MF.

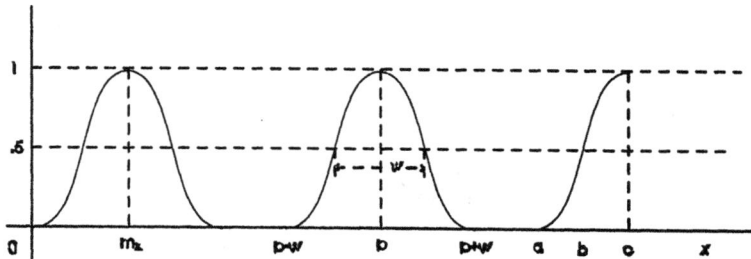

Fig. 2.2. MF : a) Exponential, b) Symetric sigmoidal, c) Sigmoidal.

352

In order to reduce the number of parameters, discretized fuzzy subset may be used. For instance, if the universe of discourse of the error variable, e is $\mathcal{E} = (-10, 10)$, and a set of seven linguistic variables is used (NB, NM, NS, Z, PS, PM, PB, where N := negative, P := positive, S := small, M := medium, B := big, and Z := zero), $\mathcal{E}$ may be divided into a number of intervals, say (-10,-7), (-7,-4), (-4,-1), (-1,1), (1,4), (4,7), and (7,10). For any value of the error, the range it belongs to it is selected and the most appropriated linguistic variables with also discretized membership function are assigned to them. Table 2.1 is an example of such a discretization.

Membership functions values									
		Linguistic Variables							
Int.	Range	NB	NM	NS	Z	PS	PM	PB
1	$e < -7$	1.	.7	.3	0	0	0	0
2	$-4 > e \geq -7$	.7	1	.7	.3	0	0	0
3	$-1 > e \geq -4$	.3	.7	1	.7	.3	0	0
4	$1 > e \geq -1$	0	.3	.7	1	.7	.3	0
5	$4 > e \geq 1$	0	0	.3	.7	1	.7	.3
6	$7 > e \geq 4$	0	0	0	.3	.7	1	.7
7	$e \geq 7$	0	0	0	0	.3	.7	1

Table 2.1 Discretized MF

If the number of intervals is larger than that of linguistic variables, the MF values may be expressed by $[0, \ldots, .3, .7, 1, .7, .3, 0, \ldots]$, the unity value being assigned to the interval characterizing the linguistic variable.

Another option giving more resolution to the physical variable current value is to normalize the universe of discourse, to divide it in intervals, as before, and to assign an easely parametrized MF to each linguistic variable, such as the sigmoidal or the exponential one.

2.1.2 Rule Base

Rules have the following classical structure:

```
rule R : IF  fact1  AND  fact2 ....    THEN       facth
```

where fact1, fact2, ..., are the *antecedents*, and facth is the *consequent*.

At this moment, only "AND" connective logical operation in the antecedent side and one consequent are considered. Later on, this constraint will be removed.

Rules are based on:
- Human experience based relations
- Generalization of algorithmic non fully satisfactory control laws.
- Inverse relations to those obtained by a fuzzy or qualitative modeling of the process.
- Training and learning.

It is clear that the greater the number of rules is the more complex the reasoning process, leading to more difficult to predict needed computer time. If the FC handles n physical variables, with m attached linguistic variables each, the full set of rules will contain n^m rules.

Some kind of hierarchy should be stablished, in such a way that a set of rules is only fired if a given conclusion has been previously proved as a valid one. On the other hand, this approach may lead to rules nesting and recursion. Again, time constraints must be fulfilled.

The Rule Base must have some properties:

- Completeness. For any input pattern an output must be generated. No-action is considered as one of the possible outputs, of course.

- Consistency. Same input must conclude the same action.

- Interactivity. Any of the possible output values must be the result of, at least, one input pattern.

2.1.3 Types of Rules

The typical rules are the so called *state evaluation* rules, such as:

$$\text{rule R1 :} \quad \text{IF x is X1 and} \quad \text{y is} \quad \text{Y1} \quad \text{THEN} \quad \text{z is Z1}$$

Using the fuzzy composition notation, it means: $z = (x \circ y) \circ R_1$. The conclusion MF will be

$$\mu_{R_1} \doteq sum(\mu_{Z1}(z) * sum(\mu_{X1}(x) * \mu_{Y1}(y)))$$

Also, rules may have as antecedent a performance evaluation. For instance, the rule

$$\text{rule R2 :} \quad \text{IF [(u is C1 ---> } \quad \text{p is P1) p is P1]} \quad \text{THEN} \quad \text{u is C1}$$

is read as follows: If the control C1 was applied, the performance p would be P1. P1 is the desired performance value. Then, make $u \in C1$.

This kind of rules will provide something like a "predictive fuzzy control".

In FC, input use to be a set of physical variables. As described above, these values are converted to LV, loosing any further interest. Another kind of rules are the so called **linear combination conclusion** rules. In them, the conclusion is no more a fuzzy subset, but a linear combination of the input variable values. That is, the conclusion is a numeric output value. For instance,

```
Rule R1 :  IF x is High   AND   y is Low      THEN   z = 2 x + .3 y
Rule R2 :  IF x is High   AND   y is Medium   THEN   z = 1.8 x + .5 y
```

This kind of rules will simplify the output generation procedure, as described below.

354

2.2 Fuzzifier

The digital preprocessor provides all the physical variables the FC may deal with. The supervisor module, if there is one, will determine the current used variables, for instance, the error and its increment (PD-like control), the error and the accumulated error (PI-like control) or any other available option. The fuzzyfier role is to convert this information into fuzzy data, accessing to the fuzzy variables parameters already defined in the DB.

Another option is to directly handle the measured data, converting them into singletons, attaching a certainty coefficient to them, or using their stochastic properties, if they are random variables, to create a fuzzy subset. In that case, the KB must be able to add new facts and/or rules from a predefined class ot them, or the fuzzy implication will use the generalized reasoning, as later shown.

2.3 Rules Selection

Let us consider a simple FC with two inputs, the error $e(t)$ and the change of the error, $c(t)$, and also to simplify the notation, only three LV attached to each one, i.e., P positive, N negative, and Z zero. For each one of the nine input combination an output will be assigned. Table 2.2 shows this set of rules, where, as before, B, M, and S stand for big, medium and small, respectively.

		error change: c		
control:u		N	Z	P
	N	NB	NM	PS
error:	Z	NM	Z	PB
e	P	NS	PM	PB

Table 2.2 Rule table for the control

The same table may be used to compute the change in the control, instead of the control itself.

For instance:

```
IF error is Z  AND  change-in-error is P   THEN  change-in-control PM
```

The fuzzifier provides the fuzzy values of the inputs. Let us see how the rule selection is performed.

Inference Matrix

Assume a set of MF, $\mu_i(e)$ for the error LV, and $\mu_j(ce)$ for the change-in-error variables. For a given pair of current data, (e, ce) the following matrix, named *inference matrix*, may be formed

$$m_{ij} = [\mu_i(e) * \mu_j(ce)]$$

where, again, * stands for any triangular norm (min, product,...).

The entries of this matrix allow the rule selection

Selection Criteria
Some of these entries may be zero. That means unactivation of the corresponding rule. In order to elaborate the output, the rules to be fired, and in which order, must be decided. Usually, a threshold discard the less significant rules. We can dispose to evaluate only a fixed number of rules, those with higher m_{ij} value.

Let us assume a selective control in the steam generator example. It should be a rule structure, allowing to derive the control, for instance, from either the error in the temperature or that in the pressure, in an independent way. A first set of rules will decide how to compute the control, and then, the selected rule table will be evaluated and the next rules to be applied selected.

2.4 Rules Evaluation

Each rule expresses a relation between LV, and a MF has been previously defined for each LV.

In our previous example, let us assume there are rules such as:

| rule R1: | IF | drum-level | is OK | THEN | water-flow | No-Change |
| rule R2: | IF | pressure | is High | THEN | water-flow | Decrease |

If our knowledge about the antecedents is just fuzzy, the conclusion of each rule will be the referenced linguistic variable, with the predefined MF.

If our knowledge is not the fuzzy subset previously defined, but something else, for instance, the drum-level is almost OK, or the pressure is quite sure High, the generalized modus ponens reasoning will provide the conclusion with a modified MF, as discussed in section 1.3.

If the controller input is a physical measurement, the fuzzy subset it generates must be treated by the GMP reasoning to get the conclusion MF. The simplest way to handle physical data, if they are assumed to be deterministic, i.e. without any added random noise, is to consider the conclusion MF as an hometetic function of the predefined one, as follows:

$$\mu'_C(z) = \mu_A(x_0).\mu_C(z)$$

wher x_0 is the current value of the physical variable in the rule antecedent, $\mu_C(z)$ is the predefined MF of the LV C, attached to the variable z, and $\mu_A(x)$ is the MF of the antecedent LV.

But, if the measurement is a random variable, a similar to singleton MF may be attached to the input variable. Fig. 2.3 shows two of these functions.

356

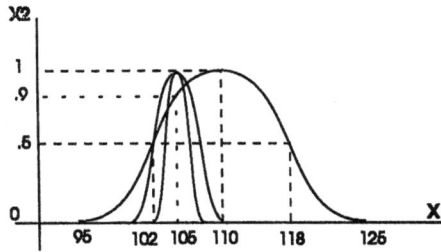

Fig. 2.3. Input fuzzy subsets

These MF may correspond to a basic temperature measurement of 105 degrees, with different accuracy meters. If the rule antecedent applies for "temperature OK", this MF should combine with the predefined one, similar to that shown in fig. 1.3.

The resulting MF may be evaluated in different ways:

- Taking the maximum value of the MF covered by the input, that is,

$$\mu'_C(z) = \max_x\{\min(\mu_{A'}(x), \mu_A(x))\}.\mu_C(z)$$

- Taking the maximum combined MF value, as for instance,

$$\mu'_C(z) = \max_x\{\mu_{A'}(x) \times \mu_A(x)\}.\mu_C(z)$$

- Evaluating a MF averaged product, like

$$\mu'_C(z) = \frac{\int_X \mu_{A'}(x) \times \mu_A(x)dx}{\int_X \mu_{A'}(x)dx}\mu_C(z).$$

In a rule, if there are two or more antecedents connected by ÁND', the antecedent fuzzy evaluation is ussually performed by fuzzy conjunction, taking as result either, the minimum or the product of the antecedents MF. Once the antecedent MF is computed, same as before apply to obtain the conclusion MF.

2.5 Fuzzy Conclusion

In the previous paragraph, as a result of the evaluation of a rule, the consequent MF is obtained. We face now two further problems: 1) if two or more rules have the same LV consequent, how to evaluate its final MF, and 2) once the rules have been combined in a fuzzy way, how to compute the numeric value of the physical output variable, if there is such a variable. This is always the case in Fuzzy Controllers. Some conclusions about the control input are derived and, finally, a control action must be taken.

To solve the first problem, the **also** connective should be considered. The second one will be treated in the defuzzification module.

If the same LV is obtained as a conclusion of different rules, the total MF of this LV may be evaluated before the defuzzification process, in order to get one MF per LV. But also, the conclusions may be treated as a set of LV attached to a unique physical variable, and then apply the defuzzifier strategy to the ensemble.

357

2.5.1 The "Also" Connection

Rules, as defined in the rule base, consider neither, the connective "OR" in the antecedent nor the connective "AND" in the consequent. The second case is easely decomposed in as many rules as consequents are, owing to some sort of linearity in the output. All the rules will have the same antecedents.

To deal with the first case, and in order to allow a possible different fuzzy implication strategy for each antecedent AND-connected set, the rule base incorporates the also connective, that is, rules with different antecedents having the same LV in the consequent.

To combine conclusions referred to the same LV any of the triangular conorm or sum operations, $(+)$, may be used: union (maximum value), algebraic sum, bounded sum and so on. The result is bounded between 0 and 1.

Again, total freedom to choose the also connecting operation is allowed to the designer. Based on his/her reasoning experience, if any, the final LV MF is evaluated.

2.6 Defuzzification

In the FC structure, fig. 2.1, there is a module to convert linguistic to numeric information. The difuzzifier will combine the reasoning process conclusions into a final control action. So, the problem is: given a set of LV, X_i, with attached MF, $\mu_{X_i}(x)$, all defined in the universe of discourse $\mathcal{X}$ of a physical variable, x, what is the most appropriated value $\bar{x}$ for this variable.

Again, different models may be applied. They have different computational complexity. The more complex they are the more reasoning conclusion try to keep in.

Let us assume the following conclusions pattern, with reference to fig. 2.4, :

- Rule R102: water-flow Decrease, $\mu_1(\Delta F)$

- Rule R205: water-flow Decrease, $\mu_2(\Delta F)$

- Rule R209: water-flow No-change, $\mu_3(\Delta F)$

- Rule R301: water-flow Increase, $\mu_4(\Delta F)$

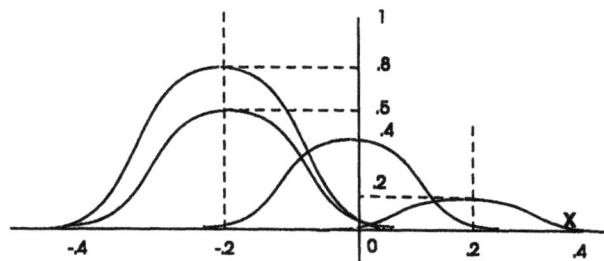

Fig. 2.4. Water-flow rate conclusion

It is assumed that other output LV, such as Highly decrease or Highly increase, have not been concluded.

358

2.6.1 Most Significant

It is the simplest approach. The output is the MF peakvalue having the greatest MF value. The remaining conclusions are not considered. In the fig 2.4 example, it will be: $\Delta F = -0.2$, which corresponds to $\mu_2(-0.2) = 0.8$, the highest MF value.

2.6.2 Mean of Maxima

Also trying to reduce the computation burden, but taking into account all the conclusions, the final output is computed averaging the MF peakvalues.

$$\bar{u} = \frac{\sum_i p_i \times \mu_i(p_i)}{\sum_i \mu_i(p_i)}$$

In the example above, it will be:

$$\Delta F = \frac{-0.2 \times 0.8 - 0.2 \times 0.5 + 0 \times 0.4 + 0.2 \times 0.2}{0.8 + 0.5 + 0.4 + 0.2} = .158$$

2.6.3 Mass Centroide

In this case, not only the peakvalues but the full MF are considered. The computed output is:

$$\bar{u} = \frac{\sum_i \int_{X_i} x \times \mu_i(x)dx}{\sum_i \int_{X_i} \mu_i(x)dx}$$

The term mass centroide refers to the weighted average of all the concluded MF.
For the example above, the output would be $\Delta F = 0.164$. In fact, the computation time is much higher.

2.6.4 Linear Combination

If the rules are of the linear combination type, each one will provide an output value:

$$u_i = \sum_j a_{i,j} x_{i,j}$$

where $x_{i,j}$ is each one of the input variables of the i-rule.
The total conclusion about the output variable u will be:

$$\bar{u} = \frac{\sum_i u_i \prod_j \mu(x_{i,j})}{\sum_i \mu_j(x_{i,j})}$$

If the MF are also linear functions, the computation is quite simple.
Example
Let us assume a very simple situation. There are two input variables, x_1 and x_2, and the output u. The data base is as follows, fig 2.5:
* For x_1, LV: Small S_1 , $S_1 = [0, 16]$, $\mu_{S_1}(x_1) = 1 - x_1/16$,
and Big, B_1 $B_1 = [10, 20]$, $\mu_{B_1}(x_1) = -1 + 10x_1$.

* For x_2, LV: Small S_2 , $S_2 = [0, 8]$, $\mu_{S_2}(x_2) = 1 - x_2/8$,
and Big, B_2, $B_2 = [2, 10]$, $\mu_{B_2}(x_2) = -.25 + x_2/8$ . .

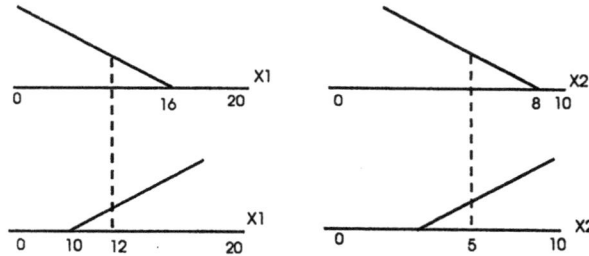

Fig. 2.5. Linear MF

The rule base is:

```
R1     IF   x1 = S1     AND     x2 = S2     THEN     u = x1 + x2

R2     IF   x1 = B1                         THEN     u = 2.x1

R3     IF   x2 = B2                         THEN     u = 3.x2
```

Let the input variables take the value: $x_1 = 12; x_2 = 5$.
The output will be:

$$u = \frac{.25 \times 17 + .2 \times 24 + .375 \times 15}{.25 + .2 + .375} = 17.79$$

3   FC Analysis and Design

3.1   FC Analysis

The FC, as described in the last section, may be considered as a dynamical system.
The most interesting feature being its user oriented configuration and description, the
mathematical analysis is rather complex.

This controller may be considered as an experimentally tuned **non linear controller**,
[8]. For instant, in a Proportional-like FC, the control action is computed from the
measurement error. In order to simplify the notation, let us assume the following set of
rules, with numerical conclusions:

- If error is PB $u = 20$

- If error is PM $u = 10$

- If error is Z $u = 0$

- If error is NM $u = -10$

- If error is NB $u = -20$.

360

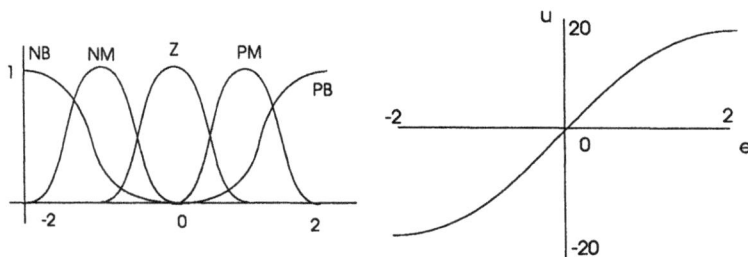

fig. 3.1. a) error MF . b) I/O function

If the MF are as shown in fig 3.1a, for any error value, e, the output may be computed by, for instance, addition of the $u_i \times \mu_i(e)$. The controller input/output relation is plotted in the fig. 3.1b. Note that the gain $\Delta u / \Delta e$ is variable, leading to a non-linear controller. The nice property being the user option to change the shape of this non-linear function according to his/her own experience.

On the other hand, as previously discussed, the FC is linear in the output. This allows us to design a FC for each control input, the process input interaction being taken into account in the controller input.

The FC complexity is also related to the number of inputs and how many LV each input has. In general, simple FC only consider two inputs, with a number of 7 or 9 LV. But, following the user oriented approach, the FC can change the set of inputs and outputs it is dealing with. For instance, the input vector may be formed by the error and the change in the error, leading to a PD-like controller, or it may be changed to deal with the error and the accumulated error, leading to a PI-like controller. According to the operating conditions, the same FC may perform as decided. Another simple option is to change the output variable. Dealing with the error and the change in the error as input, the output may be decided to be the control or its change.

Changing the universe of discourse in both input and output, the same structure may perform as a coarse or a fine controller, without any additional requirement.

Some techniques to reduce the computation time have been mentioned in the last section. Also, a FC may be implemented as a **look-up the table** controller. Let us assume a discretized universe of discourse of the input variables, if the output for any possible input combination is precomputed, the output value can be easely computed by a trapezoidal interpolation:

$$u(e, c) = \begin{bmatrix} c_{i+1} - c & c - c_i \end{bmatrix} \cdot \begin{bmatrix} u(e_i, c_i) & u(e_{i+1}, c_i) \\ u(e_i, c_{i+1}) & u(e_{i+1}, c_{i+1}) \end{bmatrix} \cdot \begin{bmatrix} e_{i+1} - e \\ e - e_i \end{bmatrix}$$

$$c_i < c < c_{i+1}; \qquad e_i < e < e_{i+1}.$$

The control variable will be expressed by the non-linear mapping the fig. 3.2 shows.

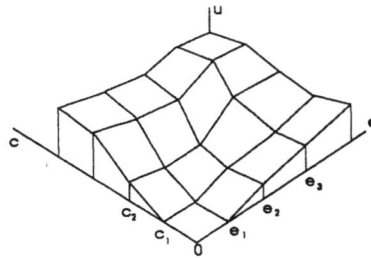

Fig. 3.2. Non-linear interpolated I/O mapping

3.2 FC Design

The design of a FC always follows a trial and error procedure. If the plant to be controlled has been running and some data records are available, the FC design and parameters tuning may be done off-line. If not, an iterative approach will lead to the most convenient controller. One of the desirable features of an intelligent controller is learning. If the supervisory module does not include this facility, the user should update the FC parameters as a result of the experience.

Let us review the FC parameters, the different approaches to design the FC, and its on-plant tuning.

3.2.1 FC Parameters

When designing a FC some parameters must be decided based on either the designer experience or the control experimentation. For a given process, some of them are fixed upon normal operating conditions and others must be updated from time to time. Among the first ones, we can enumerate:

- **Structural parameters**
  - Number of output variables.
  - Number of input variables.
  - Number of input trends, accumulations, ...
  - Linguistic Variables.
  - Parametrized MF.
  - Discretization and Normalization ranges and intervals.
  - Rule base structure.
  - Basic rule set.

Parameters to be on-line updated are, for instance:

- **Tuning parameters**
  - Universe of discourse of the dealt with variables.
  - MF parameters, such as peakvalue, bandwidth, or support.
  - Input and/or output static gains.

362

One of the main problems in the FC Design is to capture the expert Knowledge and to translate it into LV, MF, rules, and so on. Later on we will discuss the tuning approach.

The aforementioned properties of the Rule base must be tested: completeness, consistency, interaction, and robustness. The last one is related to the sensitivity of the control to some unmodelled behavior or disturbance. A possible way of robustness measurement is to introduce a random input x_i characterized by its average $\bar{x}_i$ and width w_i and to get the same values for the output.

Many other options must be decided. Among them: the type of operation to implement the fuzzy implication, the fuzzy subsets matching in approximated reasoning, the also connective implementation, and the defuzzification method.

3.2.2 Experimental FC Design

Let us now consider an experimental design based on linear combination conclusion rules.

From a rough process behavior knowledge:

- Define LV and Simple MF
- Write some control rules are stated:

```
Rule R1:   IF   (x1 is   *)   AND   (x2 is   *)   AND   (x3 is   *)       THEN
```

$$u_1 = a_{1,0} + a_{1,1}x_1 + a_{1,2}x_2 + a_{1,3}x_3 + b_{1,1}y_1$$

where y_1 is a variable not appearing in the antecedents. Assume there are m such a rules.

- Form an experimental data table with columns $x_1, x_2, x_3, y_1, \bar{u}_1$ for any of the rules above, where $\bar{u}_1$ is the control action usually applied by a skilled operator. Let N be the number of measurements.

- For any measurement table, express:

$$\bar{u}^j = \left(a_{i,0} + a_{i,1}x_1^j + a_{i,2}x_2^j + a_{i,3}x_3^j + b_{i,1}y_1^j\right)$$

- Apply Weighted Least Squares, (WLS), to get $\hat{a}_{i,k}$. For any of the rows above, the weight will be β_j:

$$\beta_j = \frac{\mu_*^j(x_1)\mu_*^j(x_2)\mu_*^j(x_3)}{\sum_{j=1}^{N} \mu_*^j(x_1)\mu_*^j(x_2)\mu_*^j(x_3)}$$

These are the best parameters fitting the experimentally recommended data. Of course, the FC will operate in conditions which are different to those considered in the design step.

3.2.3 FC Tuning

As it is always true, there is nothing for nothing. If you are able to design a FC, from a weak process knowledge and not well defined control specifications, and a good flexibility to update the controller is required, in some stage of the design procedure, a great design effort should be involved. And this happens in the FC training phase.

As previously mentioned, one desirable feature of the FC is learning. But learning is not yet a well established issue in FC, and it is the operator or designer job to train and tune the controller via trial and error approach. We are faced to an AI typical activity. Much effort is also devoted to this issue in classical control theory. Recently, in [11], a new control design strategy, the so called windsurfing adaptive control, involving a learning capability, has been proposed. According to the controlled process knowledge you have, the controller design technique may be more or less ambitious. With poor process knowledge, a conservative control design must be implemented to avoid unexpected bad behavior.

We can say that the tuning approach complexity depends on both, the process knowledge and the precise control objectives statement. If the plant is well modeled and the goals are easily understood, probably the best is to forget the FC techniques and control the process by some classical control algorithm. The weaker is the knowledge the more appropriated the FC.

The methodology we propose is in the way of the classical experimental tuning approaches.

- Start with the simplest control, such as a proportional-like controller, with

  - reduced gain.
  - more relevant variables
  - low number of linguistic variables
  - non-discriminant membership functions

  Increase the gain until some undesirable response appears.

- Add knowledge according to the experience

  - Looking for new linguistic or physical variables, to cover with the new control difficulties
  - Tuning the membership functions, controllers parameters
  - Adding new components or changing the control structure

- Check the validity and coherence of the incorporated new knowledge under operating conditions others than those initially tried.

If some past experience over the controlled process is available, in the form of variables and events records, and a controller structure has been decided as convenient, try to estimate the controller parameters better fitting these records.

Anyway, in control of complex processes, the controlled process is frequently changing and control updating and tuning is a must. What that requires is a powerful MMI, allowing an easy operator interaction. The MMI must include specific options, such as editors for rules, objects, and procedures, graphical capabilities, and a good explanatory subsystem, to show the involved reasoning in getting a result.

3.3 Process Fuzzy Modeling

The Fuzzy knowledge representation may be also used to model a plant from its qualitative knowledge, [9]. By the term plant we include the proper controller, if there is

364

a previous system controlling the process (human operator, conventional decision based software, or a complex combined system). Following the approach of linear combination conclusion rules, the modeling procedure may be supported by algorithmic computations, enabling a faster model building.

Let us outline the procedure for a process modeling. As previously stated, some *a priori* knowledge of the process behavior must be available. Otherwise, on plant experiments should provide this information. For instance, we must know which are the input/output physical variables to be considered. Assuming that (if not, we can come back to this assumption and repeat the procedure including/removing some variables), we must decide:

1. Number of LV.

2. Parameters in the MF of the antecedents.

3. Consequent parameters.

The iterative procedure will be:

- **a.-** If 1 and 2 are fixed, compute the optimal consequent parameters as in section 3.2.2.

  A record of experimental data is required, or the results must be checked by the expert.

- **b.-** If 1 is fixed, assuming linear MF, derive the optimal LV support (which is the only MF parameter), by nonlinear programming techniques. A set of Weighted Least Squares, problems must be solved.

- **c.-** Start with one variable in the antecedent side, assuming two LV attached to it. Writte the simple rules and go to the following sequence, applying **b**.

  - Combine one input variable more.

  - Double the number of LV of the more significant input.

  - .

  - .

  - Stop if: i) too much rules, or ii) the WLS index in not longer improved.

The resulting model may be experimentally improved if more flexible MF and/or conclusions are allowed. The advantage of the outlined procedure is its possibility to be automatized.

4 FC Supervision

In a further chapter the use of FC or any other kind of Intelligent Control strategy to implement a supervisory control will be discussed. But, as we have seen in the previous section, the rule-base structure of a FC allows an easy implementation of its basic supervision. The FC supervision will cover the basic ideas of the tuning procedure detailled above. Let us briefly comment the most common options.

With reference to fig. 2.1, the supervisor module will compute some extra performance indices. For instance, if a step change in an input is applied, the ISE, or any other process performance, such as the output overshoot, settling time or steady-state error may be evaluated.

Based on a long term sampling scheme, it may detect oscillations, slow drifts or a more unusual type of performances, such as unexpected behavior, unclear response or complex reasoning.

The kind of expected actions from the supervisory module are:

- Changes in parameters.

  In that case, a smoothed or progressive change is recomended.

- Deletion or addition of rules.

  Based on lake of use, rules may be removed. Some indicator may recommend a new rule. For instance, to reduce static errors, rules for narrower universe of discourse of the error may be added, incrementing the static gain.

  In that case, a test of the actions effect will confirm them or recommend to go back to the previous rule set.

- Changes in variables.

  Nondiscriminant MF may lead to neglection of some antecedents in prestablished rules. Unsatisfactory behavior may ask for the consideration of new variables.

  As before, trial and error procedure will confirm the goodness of the action.

5 REFERENCES

[1] Warnock, I. G.,*Programmable controllers.Operation and application*, Prentice Hall 1988.

[2] Kailath, T. *Linear Systems.* Prentice Hall. 1980.

[3] Anderson, B.D.O. and J. B. Moore *Linear Optimal Control* Prentice Hall 1971.

[4] Goodwin G.C.G., and Sin K.S. *Adaptive Filtering, Prediction, and Control* Prentice Hall, 1984.

[5] Sing, M.G. *Decentralized Control* North Holland. 1981.

[6] Tzafestas, S. "AI Techniques in control: an Overview" Int. Symp. on *AI, Expert Systems and Languages in Modeling and Simulation* Brussels. 1987.

[7] Lee, C.C. "Fuzzy Logic in Control Systems: Fuzzy Logic Controller" *IEEE Trans on Systems, Man, and Cybernetics*, Vol SMC-20, Num 2. pp 404-435. 1990.

[8] Ollero A., A. Garcia-Cerezo, and J. Aracil. "Design of Rule-based Expert Controllers". *ECC91 European Control Conference* ,Grenoble, France pp 578-583. 1991.

[9] T. Takagi and M. Sugeno. "Fuzzy Identification of Systems and its Application to Modeling and Control" *IEEE Trans on Systems, Man, and Cybernetics*, Vol SMC 15, Num 1, pp 116-132. 1985.

[10] Albertos, P., A. Crespo, F. Morant , and J.L. Navarro. "Intelligent Controllers Issues". *IFAC Symp. on Intelligent Components and Instruments for Control Applications*. Malaga. 1992

[11] B.D.O. Anderson, and R.L. Kosut. "Adaptive Robust Control. On-line Learning" *Proc. IEEE Conference on Decision and Control*, Brighton. England. 1991.

[12] R. Vepa. "Introduction to Fuzzy Logic and Fuzzy Sets". This book, Previous chapter.

Adaptive Fuzzy Control

René Jager

Delft University of Technology
Department of Electrical Engineering, Control Laboratory
P.O. Box 5031, 2600 GA Delft, The Netherlands

Abstract

In this paper adaptive fuzzy control is addressed. In the application of fuzzy control one can distinguish a theoretical and a practical approach. The theoretical approach is relation-based, the practical approach rule-based. More memory and calculation time is needed when the theoretical approach is applied. By an extension of the fuzzy rules of which the fuzzy controllers exists an additional degree of freedom is introduced which does, however, not result in extra memory and calculation time needed. The extension of the fuzzy rules consists of allowing the fuzzy rules to have more than one weighted conclusion. By varying the step size of the adaptation of the fuzzy rules, it is possible to start with 'rough adaptation' and end with 'fine tuning' of the fuzzy controller. Fuzzy expert systems provide a way to combine the advantages of fuzzy logic and converntional expert systems techniques. A fuzzy expert system for process control with minimal process knowledge is developed and tested in an intelligent real-time control environment.

Keywords

adaptive, control, fuzzy, rule-based, self-organizing, expert systems, real-time

1 Introduction

An increasing number of applications of fuzzy control is reported in literature. The fuzzy controllers used, often replace tasks executed by a human operator or a conventional controller (for example PID-controller). It can be shown that fuzzy controllers are analytically equivalent to (non-linear) PID-controllers and it is possible to construct linear fuzzy controllers that have the same performance as PID-controllers [Buckley and Ying 1990, Ying et al. 1990]. In table 1 the type of commonly used fuzzy rules and the corresponding PID-control configuration is given.

368

fuzzy rule	P	I	D
if e is ... then u is ...	×		
if e is ... then Δu is ...		×	
if e is ... and Δe is ... then u is ...	×		×
if e is ... and Δe is ... then Δu is ...	×	×	
if e is ... and Δe is ... and $\Delta^2 e$ is ... then Δu is ...	×	×	×

Table 1: *Type of rule and corresponding PID-controller type.*

However, in situations where the fuzzy rules, with which the fuzzy controllers are build, are not (a priori) completely known or have a standard initial form in order to control a large number of different processes, an adaptation mechanism has to be provided. Such an adaptation mechanism should be able to 'shape' the (initial) basic and global knowledge to more application specific knowledge. In the field of fuzzy control this comes down on changing the fuzzy rules. In order to change those fuzzy rules some algorithm must be defined which decides when and how fuzzy rules have to be adapted.

In the next section a short description of fuzzy control will be given. In section 3 self-organizing control will be adressed. The extension of fuzzy rules is described in section 4. A description of a real-time fuzzy expert control environment will be discussed in section 5. The section thereafter will describe an adaptive fuzzy expert system developed using this control environment. Experiments and result will be given in section 7. Conclusions and remarks are given in the last section.

2 Fuzzy control

Fuzzy controllers provide a relative simple way to control systems which have a considerable nonlinear behaviour [Aoki et al. 1990, Huang and Tomizuka 1990, Yu et al. 1990]. Usually the fuzzy rules of which a fuzzy controller consists, represent knowledge or experience of an operator. In figure 1 the general scheme of a fuzzy controller is depicted. In case the inputs of the fuzzy controller are quantized the fuzzy controller can be reduced to a look-up table.

In the field of fuzzy control one can distinguish two approaches: a theoretical approach

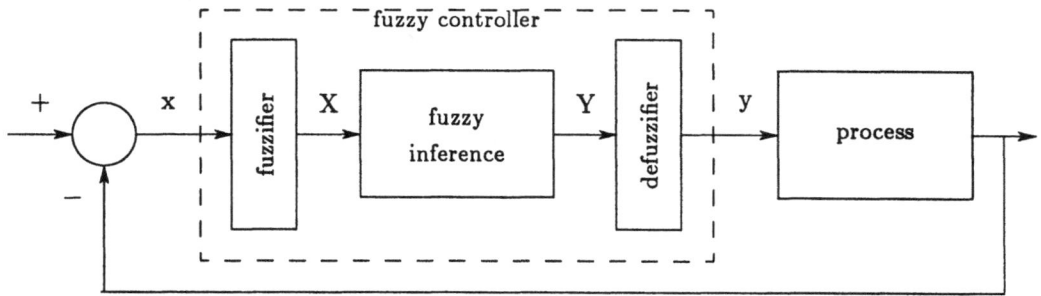

Figure 1: *Fuzzy controller in control loop.*

and a more practical approach. In theory a fuzzy controller is represented by a $(N+1)$-dimensional relation R (N inputs + 1 output). This relation R can be seen as a function, restricted between 0 and 1, of $(N+1)$ variables. R is a combination of several N-dimensional relations R_i each representing a rule

$$r_i : \text{IF } x_1 \text{ is } X_1^i$$
$$\text{AND } \dots$$
$$\text{AND } x_N \text{ is } X_N^i$$
$$\text{THEN } y \text{ is } Y^i$$

The inputs x_j of the controller, for example the error signal and first difference of the error signal, are fuzzified into a relation X, which is in case of M.I.S.O[1]-control (N)-dimensional:

$$\mu_X(x_1, \dots, x_N) = \bigvee_{j=1}^{N} \mu_{X^i}(x_j) \tag{1}$$

By applying Zadeh's 'compositional rule of inference' the fuzzy output Y is calculated:

$$Y = X \circ R \tag{2}$$

Using the multi-dimensional membership functions, this can be represented by:

$$\mu_Y(y) = \mu_X(x_1, \dots, x_N) \circ \mu_R(x_1, \dots, x_N, y) \tag{3}$$

A crisp, numerical output y is obtained by defuzzification of the fuzzy output $\mu_Y(y)$.

Because in this theoretical approach multi-dimensional functions are used to describe the relations R, X and Y, a great amount of memory is needed when implementing a

---

[1]Multi-Input-Single-Output.

370

discrete approximation. This is why in applications normally a more practical way of fuzzy inference is applied.

Most practical applications of fuzzy control use a simplified method to determine the fuzzy output of the controller, by determining which rules could participate in the final fuzzy output, given the measured input(s). In this approach the task of the 'fuzzifier' consists of determining for each input x_j the grade of membership for each linguistic qualification $X_{j,k}$ of that input. For each rule r_i the membership function $\mu_{Y_o^i}(y)$ of the fuzzy output is determined by two intersections:

$$\mu_{Y_o^i}(y) = \{\bigwedge_{j=1}^{N} \mu_{X_j^i}(x_j(t))\} \bigwedge \mu_{Y^i}(y) \tag{4}$$

The complete membership function of the fuzzy output is a combination of the fuzzy output of the individual rules r_i:

$$\mu_Y(y) = \bigvee_i \mu_{Y_o^i}(y) \tag{5}$$

In case of one input x, two overlapping membership functions and two fuzzy rules r_1 and r_2, a schematic representation of the inference strategy is given in figure 2. In figure 2 the $min(.)$- and $max(.)$-operations are used for the $\wedge$- and $\vee$-operands, respectivily. Defuzzification of the fuzzy output results in a crisp, numerical output y.

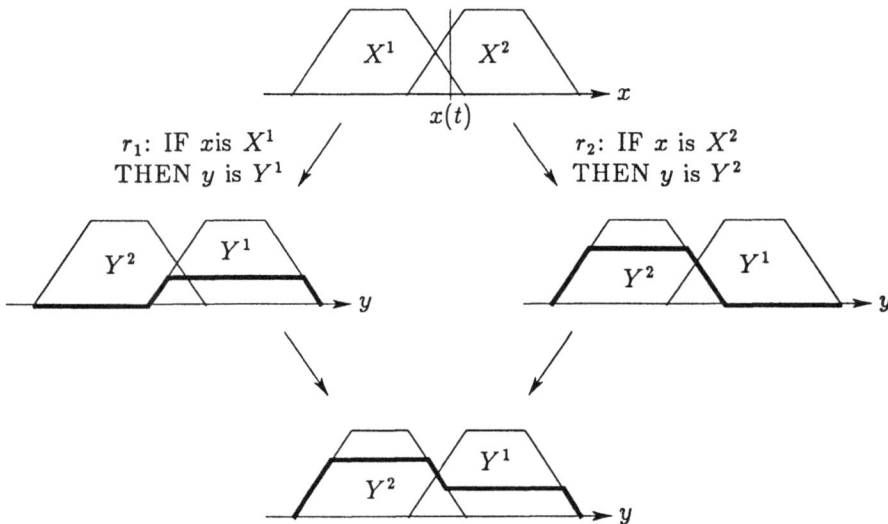

Figure 2: *Practical inference in fuzzy control.*

This approach is usually applied in fuzzy control applications. In those applications rules are used as for example:

if *error is positive big*
and if *error change is negative small*
then *change control signal positive medium*

The defuzzification of the fuzzy output of the controller can be performed by the two basic defuzzification methods [Harris and Moore 1989] or modified versions of those. The most commonly method used is the *Centre-of-Gravity* defuzzification method:

$$y = \frac{\int y \cdot \mu_Y(y)}{\int \mu_Y(y)} \tag{6}$$

Another defuzzification method, which is much faster in terms of calculation times, is the *Mean-of-Maxima* defuzzification method. The result of this defuzzification method corresponds with the mean of all (local) maxima of the fuzzy output. More detailed information about defuzzification methods can be found in [Jager et al. 1992a].

The next section will adress self-organizing control in which a fuzzy controller as described in this section is adapted to the process to be controlled.

3   Self-organizing fuzzy control

Self-organizing fuzzy control [Procyk and Mamdani 1979] is a well known way of adapting the fuzzy controllers. Applications of and studies into the practical as well as the theoretical approach can be found in literature, [Brown et al. 1991, Neyer et al. 1990, Shao 1988, Sugiyama 1988] and [Linkens and Hasnain 1990, Wakileh and Gill 1990]. It consists in fact of two systems: a fuzzy controller and an adaptation system. In figure 3 the schematic representation of self-organizing control is depicted.

The primary loop controller is a fuzzy controller as described in the previous section (2). The second fuzzy system is based on the same kind of rules, except for the fact that the conclusions of the rules represent a linguistic performance measure, for example:

if *error is positive big*
and if *error change is negative small*
then *performance is very bad*

These linguistic performance measures represent in fact a linguistic reference model. The **performance measure** possibly results in modifications of the fuzzy rules. The **rule modifier** modifies the fuzzy rules of the fuzzy controller in the primary control loop according to the result of the performance measure, using a **minimal model** of the process. The most minimal model necessary is the sign of the DC-gain of the process to be controlled.

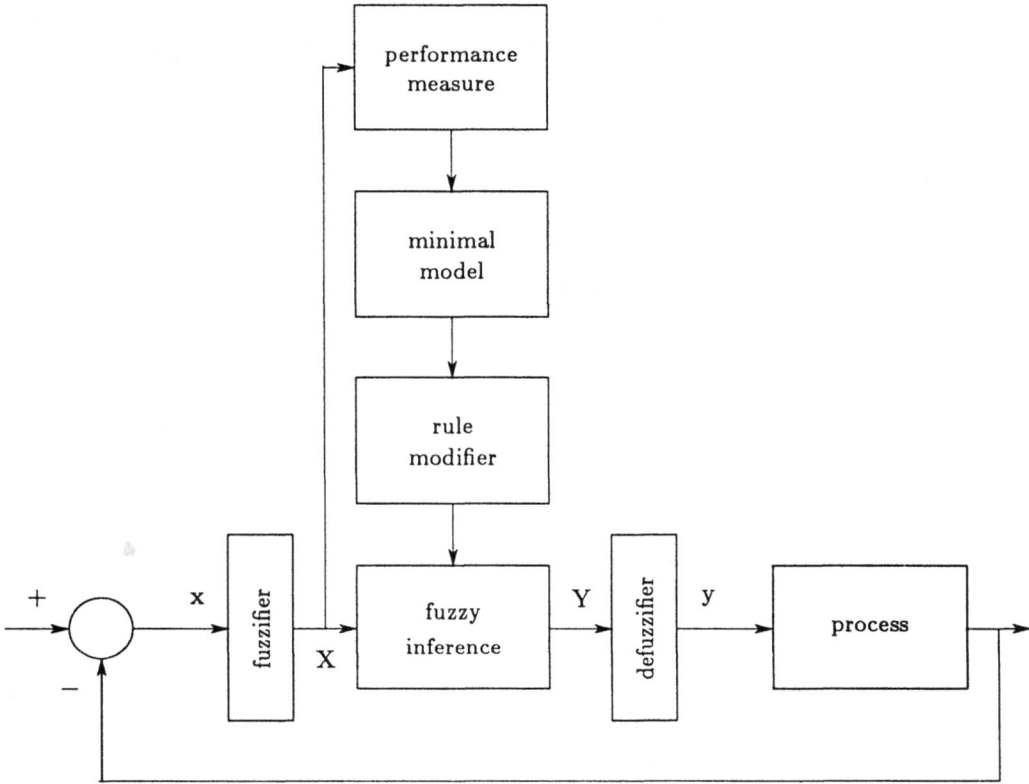

Figure 3: *Self-organizing control.*

In self-organizing control also a theoretical and a practical approach can be distinguished. The theoretical approach of self-organizing fuzzy control consist of modifying the relation R, which represents the fuzzy controller in the primary control loop (see section 2). This is done by applying the learning rule [Harris and Moore 1989, Pedrycz 1989]:

$$R := (R \wedge \neg R_{old}) \vee R_{new} \tag{7}$$

In which R_{old} and R_{new} are the relations describing the old 'bad' and the new 'better' situation. The relation R_{old}, constructed by

$$R_{old} = \{\bigwedge_{i=1}^{N} F(x_i[kT - mT])\} \bigwedge F(y[kT - mT]) \tag{8}$$

is 'removed' from the relation R and the new relation R_{new}, represented by

$$R_{new} = \{\bigwedge_{i=1}^{N} F(x_i[kT - mT])\} \bigwedge F(y[kT - mT] + r[kT]) \tag{9}$$

373

is 'inserted'. In (8) and (9) the $F()$-function is a fuzzifier which constructs a fuzzy membership function around a numerical value. The adjustment of $y[kT - mT]$ is determined by

$$r[kT] = M^{-1}p[kT] \qquad (10)$$

in which $p[kT]$ is a value from the performance table, representing a sort of reference trajectory, and M^{-1} is the 'inverse model' of the process to be controlled, in the most simple case the sign of the DC-gain of the process.

The theoretical approach allows the fuzzy controller to be modified by new rules containing new classifications of inputs and output. So during each adaptation of the fuzzy controller, new membership functions can be generated using the numerical measurements of the inputs and a new rule, in the form of a multi-dimensional relation, can be 'added' to the fuzzy controller after removing the old rule(s). Due to the fact that this approach in comparison with the practical approach needs much more memory for storage and calculation time, it is not applicable in real-time control problems, although the use of transputers can decrease the calculation time considerably [Linkens and Hasnain 1990].

In the practical approach the most simple form of self-organizing control consist of the modification of the conclusions of the fuzzy rules in the rule base. This modification is done by 'changing' the conclusion of rule r_{old}

$$r_{old} : \text{IF } x_1 \text{ is } X_1^{old}$$
$$\text{AND } \dots$$
$$\text{AND } x_N \text{ is } X_N^{old}$$
$$\text{THEN } y \text{ is } Y^{old}$$

which results in a new rule r_{new}

$$r_{new} : \text{IF } x_1 \text{ is } X_1^{old}$$
$$\text{AND } \dots$$
$$\text{AND } x_N \text{ is } X_N^{old}$$
$$\text{THEN } y \text{ is } Y^{new}$$

in the rule base.

In figure 4 an example of the restrictions of the numerical result for the output of the fuzzy controller in case of one input and one output is shown. In this figure the solid dots mark the numerical value of the output y of the controller in case the numerical input (x) matches only one of its possible classifications. This comes down on just one fuzzy rule being fired at that time.

374

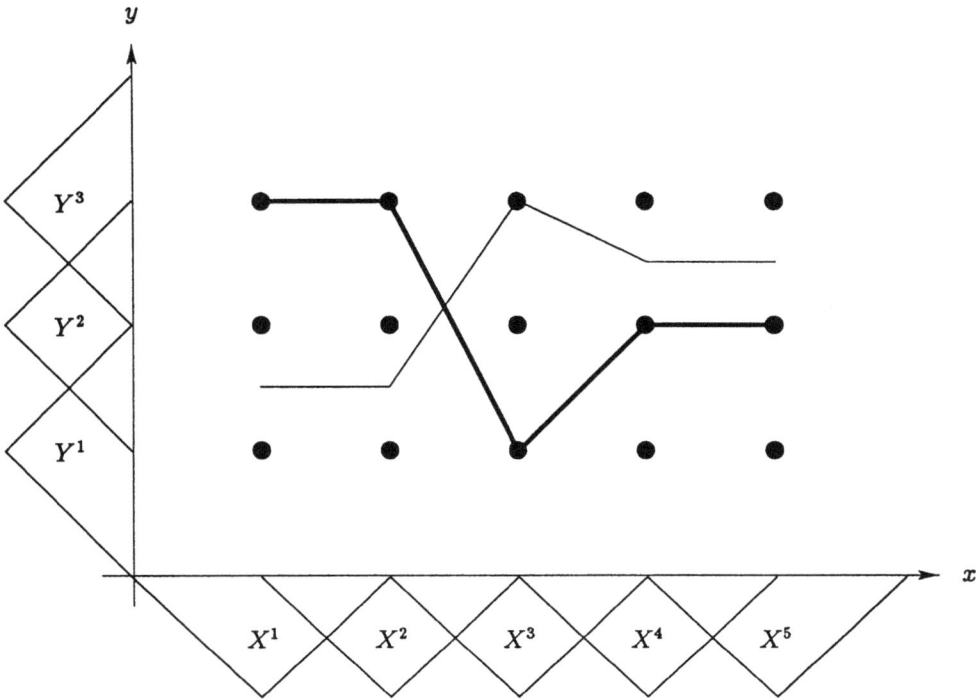

Figure 4: *Restrictions on relation between numerical inputs and output in practical approach of self-organizing control (thick line: possible relation, thin line: impossible relation).*

How the numerical output is related to the numerical inputs in between those points depends on the membership functions chosen for the classifications of inputs and output as well as on the defuzzification method used to defuzzify the fuzzy output. Assuming no run-time modifications of membership functions or defuzzification methods, the self-organizing controller as in figure 4 can represent one of the 3^5 possible relations. A relation in figure 4 is represented by a curve connecting five dots, from each column one. This curve is in fact the numerical value of output y as function of input x.

A way to overcome this drawback is to create a new rule according to the current inputs and the decision of the performance index just as explained above when discussing the theoretical approach. Such a new rule is then inserted in the rule base. To limit the size of the rule base redundant and/or resembling rules have to be deleted. To do so a mechanism has to applied which compares the fuzzy rules and deletes the most alike ones. Applying such a mechanism will be expensive in time and therefor not suited for real-time applications. In the next section a simple way of adaptive fuzzy control is described by allowing the fuzzy rules to have more than one conclusion about the same output of the controller.

4 Extension of fuzzy rules

The fuzzy control and self-organizing control as described in the previous sections can be extended by allowing fuzzy rules to have more than one conclusion for the same variable. One could say that this is in fact allowing contradicting knowledge in the controller. On the other hand one could say that allowing this extension of the fuzzy rules is more or less the way humans tend to apply their qualitative knowledge about quantities: weighting numerical reference points. This idea is also related to the defuzzification of the fuzzy result of a fuzzy controller [Jager et al. 1992a]. The extended fuzzy rules in the rule base have the following form:

$$r_i : \text{IF } x_1 \text{ is } X_1^i$$
$$\text{AND } \ldots$$
$$\text{AND } x_N \text{ is } X_N^i$$
$$\text{THEN } y \text{ is } Y_1^i \, [\alpha_1^i]$$
$$\text{THEN } \ldots$$
$$\text{THEN } y \text{ is } Y_M^i \, [\alpha_M^i]$$

All M conclusions in the rules are weighted. The relations between the weights α_k^i of the conclusions in rule r_i are

$$\sum_{k=1}^{m-1} \alpha_k^i = 0 \tag{11}$$

$$\sum_{k=m}^{n-1} \alpha_k^i = 1 \tag{12}$$

$$\sum_{k=n}^{M} \alpha_k^i = 0 \tag{13}$$

in which $1 \leq m < M$ and $m < n \leq (m + 2)$. From (12) it can be concluded that no more than two successive weights can have a value greater than 0 and that their sum equals 1. For example, fuzzy rules like

if *error is positive big*
and if *error change is negative small*
then *change control signal positive medium* $[\frac{1}{3}]$
and then *change control signal positive small* $[\frac{2}{3}]$

are possible within the system. Each fuzzy rule in the rule base of the fuzzy controller can have one or two successive conclusions out of the M possible conclusions. Using this

376

extension a better fine tuning of the primary controller can be achieved in comparison with the practical approach of self-organizing control, because an extra degree of freedom to adjust is introduced in the system. Because of the restrictions defined for α_j^i all those weights can be represented by one number, which does not cost any additional memory at implementation level.

Allowing the fuzzy rules to have multiple, weighted conclusions an interpolation between the possible fuzzy relations in case of self-organizing control, as described in the previous section, can be made. This introduces the possibility to avoid the restrictions of the possible relations as in figure 4, to which the fuzzy controller can be adapted by applying the simple approach of self-organizing control. Even in cases only one fuzzy rule is fired it is possible that the numerical result of that fuzzy rule is 'any' numerical value. There is no longer a restriction to a finite number of numerical values (as many as classifications are defined for the output).

A new fuzzy rule no longer has a different conclusion in comparison with the old rule, but has another division of weights of the conclusions. In comparison with the practical approach of self-organizing control it can be stated, that allowing fuzzy rules to have more conclusions, provides a way to obtain a better estimation of the 'ideal' (in relation with the process to be controlled) fuzzy relation R as in the theoretical approach.

In figure 5 an example of three subsequent adaptations of a fuzzy rule is shown. Adaptation is performed with a certain step size. In case this step size is chosen equal to one, the adaptive fuzzy controller acts exactly like the simple practical version of the self-organizing controller discussed in the previous section. By allowing a variable step size, it is for example possible to initially perform a 'rough' way of adaptation and to perform 'fine' tuning in case of convergence of the fuzzy rules by decreasing the step size. One could compare this principle with the one used in search algorithms with variable step size.

In order to deal with the system dynamics the adaptation of the weights of the conclusions of the fuzzy rules can be performed according to a certain reward mechanism. This mechanism adapts a rule using an adaptation factor according to the number of samples ago the rule was fired. In most cases the adaptation factor will be a monotonic decreasing function of the history in samples.

5 Real-time fuzzy expert control environment

The practical approach of fuzzy inference in a fuzzy controller results normally in a static form of the rules and a matrix representing the rule base. When the in- and outputs are quantised the output can be determined preliminary for each combination of the inputs. In that case the complete fuzzy controller can be, or in fact is, reduced to a simple look-up table. To avoid the static form of the rules and to increase the flexibilty of the applications, a fuzzy inference engine is build. This inference engine

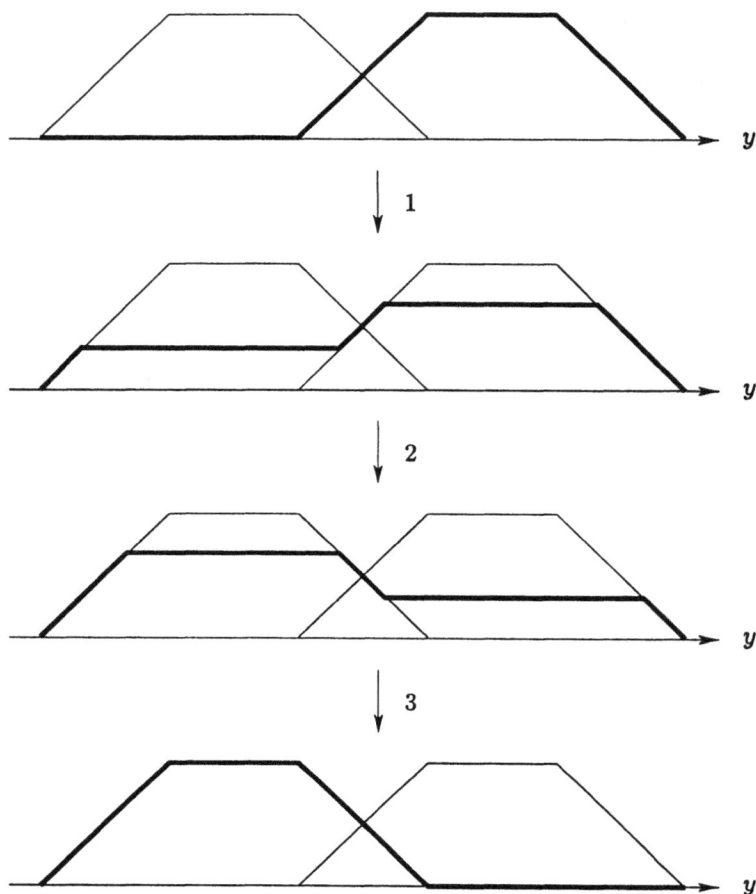

Figure 5: *Example of the result of three subsequent adaptations of a fuzzy rule.*

has 'facts' (for example: *error is big*') as in- and outputs, all having a 'grade of truth'. The inference engine propagates these grades from the input facts to the output facts. In a simple application as in figure 2 the inference engine replaces formula 2. More detailed information about fuzzy inference in real-time expert systems can be found in [Jager et al. 1992b].

The fuzzy inference engine uses search and reasoning methods as available in other (boolean) expert system shells. Because of the fuzziness, operands which are unique in a boolean inference engine can have multiple interpretations in a fuzzy inference engine. The intersection '$p \wedge q$', in a rule like '*if p and q then* ...', can be determined by, for example: $min(p,q)$, $max(p+q-1,0)$ or pq. In a rule like '*if p or q then* ...', the union '$p \vee q$' can be determined by, for example, one of the following operations: $max(p,q)$, $min(p+q,1)$ or $p+q-pq$.

The user can choose an appropriate interpretation. For example, in case of a complete correlation between p and q, the Lukasiewicz-operations $(\max(p + q - 1, 0)$ and $\min(p + q, 1))$ can be used, in case of no correlation at all between p and q the Zadeh-operations $(\min(p, q)$ and $\max(p, q))$ can be used. Rules are not necessarily completely certain. Therefor each conclusion in the consequent of a rule can have a 'grade of truth' representing the strength of the rule [Heatwole and Zang 1990]. This allows the extension of fuzzy rules as discussed in section 4.

The fuzzy expert system shell is a part of DICE[2], which consist of three major parts: a set of expert system kernels (ES_1, ES_2 and ES_3), a blackboard for data storage and a server which controls the streams of information between tasks, optionally running on different computers in a network. In figure 6 the overall structure of DICE is shown.

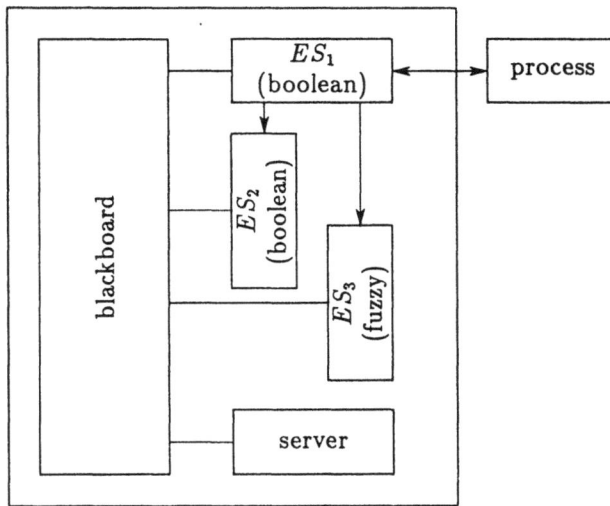

Figure 6: *Overall structure of DICE.*

Expert system kernel ES_1 is used for direct communication with the process and contains several I/O-routines, to interface the shell to the outside world. The kernel also contains a small rule base which controls the timing aspects of the other kernels (ES_2 and ES_3) and can be extended with (user-defined) demons. Expert system kernel ES_3 is used for the implementation of a fuzzy rule-based controller. Like expert kernel system ES_2, its structure is more complex than the structure of ES_1 and is capable of backward, forward, progressive [Lattimer et al. 1986] and temporal reasoning. The difference between the second and the third expert system kernel is the fact that the second only deals with boolean logic and the third deals with fuzzy logic.

---

[2]Delft Intelligent Control Environment.

6 An adaptive fuzzy expert controller

This section will describe the use of fuzzy inference and adaptation within an fuzzy expert control system. One can see this as an extension of the more basic use of fuzzy set theory in contol as discussed in the previous sections.

The rule-based controller is designed to control various processes with the following minimal a priori process knowledge and criteria: range of reference and control signal, sampling time and general performance criteria like maximum overshoot, undershoot and settling time [Jager et al. 1991b]. The task of the rule-based controller is to determine a new control signal increment every sampling time. To increase the performance of the system the controller adapts itself to the process to be controlled.

Two control strategies are used within the rule-based controller: *direct expert control* (d.e.c.) and *direct reference expert control* (d.r.e.c.).

In **d.e.c.** the control signal depends directly on the classification of the measured state into one or more symbolic states. This state classification is a translation of a measured numerical state (error $e[kT_s]$ and error change $\Delta e[kT_s]$) into a symbolic state, which can be used within a rule base. Using a fuzzy state classification one, two or at most four symbolic states are possibly to be (partly) 'true', when assuming that no more than two fuzzy sets per input (error or error change) overlap. Rules in this part of the controller are like:

> **IF** *error is big positive*
> **AND** *error change is small negative*
> **THEN** *change control signal medium positive*

This d.e.c.-part is a 'standard' fuzzy controller (see section 2).

In **d.r.e.c**, the control signal depends on a symbolic clasification of a predicted behaviour, compared with a desired reference behaviour. Rules in this part are of the form:

> **IF** *predicted behaviour is too fast*
> **THEN** *change control signal big negative*

This part is also a fuzzy controller. The input consists of the angle between the reference and the predicted behaviour or the process. This predicted behaviour is determined by a curve, which is in the phase plane represented by a line through the current and the previous state. The reference behaviour is determined by the behaviour of a first order reference model, which in the phase plane is represented by a straight line through the current state and the origin of the phase plane, so each sampling time a new reference behaviour is determined. The angle between these two lines is taken as a measure of the deviation between the predicted and the desired behaviour. In figure 7 is shown that the classification of the predicted behaviour is determined by two angles, φ_{fast} and φ_{slow}, representing the maximum allowed 'faster' and 'slower' behaviour, and a radius, δ_{alert}, representing the 'alertness' of the controller. The vector $\underline{e}_k$ stands for the measured state $(e[kT_s], \Delta e[kT_s])$.

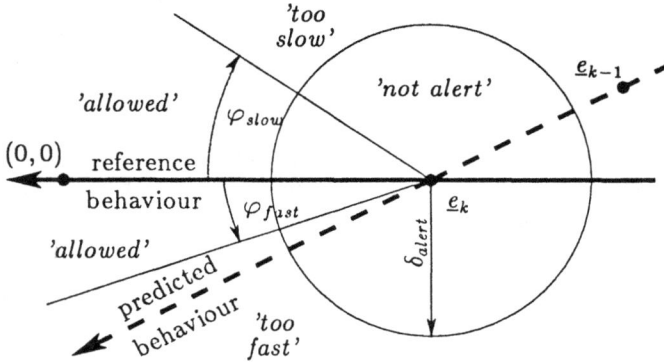

Figure 7: *Classification of behaviour.*

The change in the control signal depends on wether the behaviour is classified as 'too fast', 'allowed' or 'too slow'. In figure 7, for example, the behaviour would be classified as 'too fast'. All these classifications use membership functions and therefore all statements concerning the behaviour of the process can be partly true In this way a far reaching decision which can occur when only boolean (crisp) classifications are used, is smoothened in case of fuzzy classifications. The new control signal is determined in the same way as determined by the direct expert control strategy.

For improvement of the performance of the controller two types of adaptation are performed: *in-line adaptation* and *on-line adaptation*.

In-line adaptations can occur each sample. It consists of a zoommechanism, used to improve steady state control, of mechanisms to adapt the margins (see figure 7) of the direct reference expert control part of the rule-based controller, and of a mechanism to achieve a constant steady state value for the control signal.

The zoommechanism projects the phase plane on a smaller area around the origin. Because the relative error change is classified instead of the error change, the zooming mechanism does only effect the classification of the error. Using a zoommechanism the same set of rules designed for the complete phase plane can be used for a smaller area after scaling of the range of the control signal change. The zoommechanism is primarily used around the origin of the phase plane and therefore rules concerning direct reference expert control are not applied when zoomed in. Only rules of the direct expert control part (see subsection are used in this case.

381

The adaptation of the margins concerning the direct reference expert control strategy consists of changing the maximum allowed deviation(s) between the predicted and the reference behaviour (in figure 7 represented by φ_f and φ_s). As the error decreases the maximum allowed 'faster' behaviour decreases and the maximum allowed 'slower' behaviour increases.

To achieve a constant steady state value for the control signal a scaling of the range of the control signal change is performed: every time the setpoint is crossed (for example during an undesired limit cycle) the scaling will be decreased. Using this mechanism a limit cycle will be damped. When a setpoint change occurs, the range of the control signal change is set to the initial range.

The knowledge layer in which the on-line adaptations is implemented is in fact a supervisory layer. Decisions of the knowledge base made in this layer are based upon the overall behaviour of the system and therefore should extend over a time related to several times the slowest time constant of the system. The on-line adaptations are performed to improve the overall performance of the controller. Therefore detections and classifications are made of the following features of the system:

- overshoot

- undershoot

- rise time $(10\% - 90\%)$

- settling time

These classifications are used to decide how the margins of the direct reference expert control part of the controller or the scaling of the control changes should be changed. For example, 'too much overshoot' and a 'small rise time' will lead to a reduction of the changes in the control signal next time a similar setpoint control is desired.

The knowledge concerning on-line adaptations is implemented to reason about the overall performance of the system, so this knowledge layer is not initialized every sampling time like the 'lower' knowledge layers. The detections necessary for this knowledge layer have to be available before the inference of this knowledge layer can be even started. The results of this knowledge layer will effect the rule-based controller over a longer period, in contrast with the lower knowledge layer (for in-line adaptations), of which the decisions result in temporary adaptations and in principle lasting one sampling time.

7   Experiments and results

In this section some examples of adaptive fuzzy control will be shown. The experiments will consist of control of linear as well as non-linear processes. Also some experiments with the fuzzy expert controller, as described in the previous section, will be discussed.

Now an example of a simple fuzzy self-organising controller will de described. The fuzzy controller uses the error and the error change as inputs and a new control signal as output. The rules have the form *'if error is ... and the error change is ... then control change is ...'*. The rule base used is shown in table 2, where AZ means 'about zero', PM 'positive medium', etcetera.

PB	AZ	PS	PM	PB	PB	PB	PB
PM	NS	AZ	PS	PM	PB	PB	PB
PS	NM	NS	AZ	PS	PM	PB	PB
AZ	NB	NM	NS	AZ	PS	PM	PB
NS	NB	NB	NM	NS	AZ	PS	PM
NM	NB	NB	NB	NM	NS	AZ	PS
NB	NB	NB	NB	NB	NM	NS	AZ
	NB	NM	NS	AZ	PS	PM	PB

Table 2: *Rulebase of fuzzy controller, control change as 'function' of error (horizontal) and error change (vertical).*

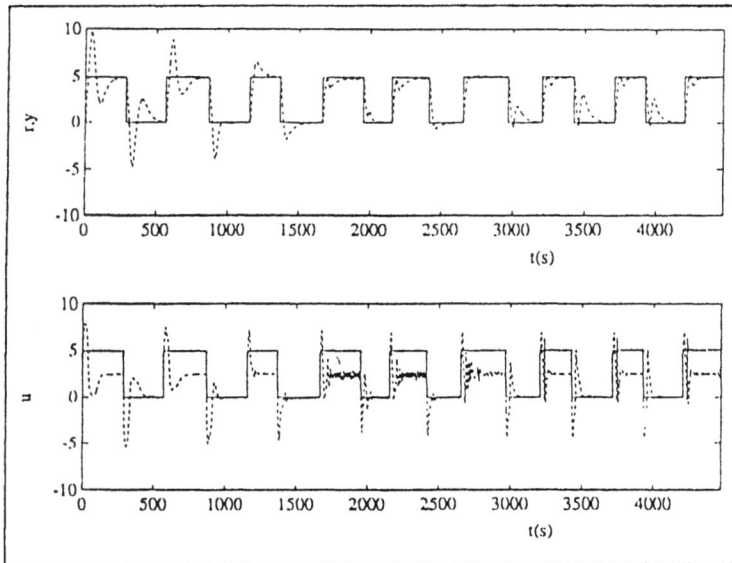

Figure 8: *Adaptive fuzzy control of linear second-order process.*

From the results shown in figures 8 one can see the lack of global convergence.

When treating the process different for several intervals of the set-point a better convergence for more set-points can be obtained, as shown in figure 9. In fact one can regard this extension as having several adaptive fuzzy controllers: for each chosen set-point interval one.

383

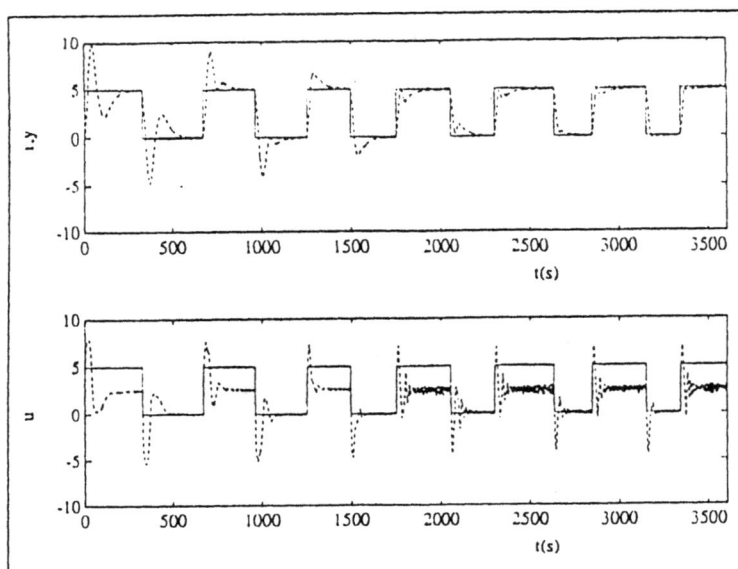

Figure 9: *Adaptive fuzzy control of linear second-order process, distinguishing different set-point intervals.*

The adaptive fuzzy expert system for the process control with minimal process knowledge is a more advanced control system than the above given example. A detailed description is given in section 6. To test the robustness of the rule-based controllers, several simulations were done. They showed that various processes can be controlled. The on-line adaptations adapt the controller to the process in question and increase performance.

Real-time tests on a μVAXII[3] were done to determine the minimal sampling time of the rule-based controller. In table 3 the calculation times of the several parts of the controller are given.

task	rules	time (ms)
I/O, graphics	-	24
direct expert control	57	112
direct reference expert control	12	14
in-line adaptation	5	2
on-line adaptation	22	50
total	96	202

Table 3: *Calculation times of several parts of controller.*

It appears that the use of fuzzy logic within the inference mechanism results in a minimal

---

[3]VAX is a trademark of the Digital Equipment Corporation.

sampling time that is about four times the sampling time needed by a comparable boolean logic [Jager et al. 1992b] inference engine.

As can be noticed from the previous section, the complete fuzzy rule-based controller consists of several fuzzy sub-controllers, as shown schematically in figure 10.

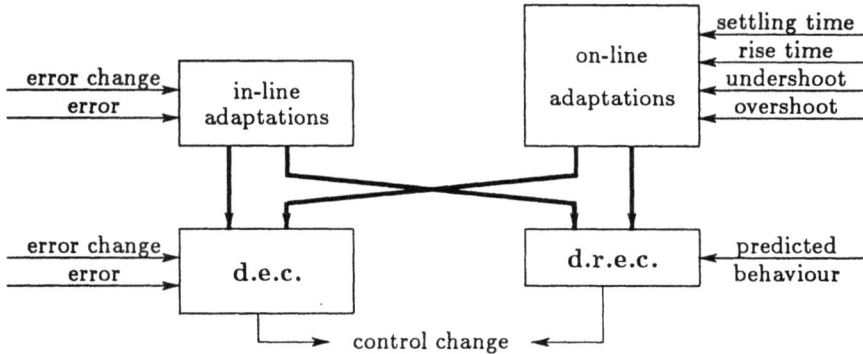

Figure 10: *Schematic representation of rule-based controller.*

8   Discussion and conslusions

It can be concluded that of the types of fuzzy control discussed in this paper the theoretical approaches (relation-based) have the disadvantage of needing more memory and calculation time than the practical approaches (rule-based).

Introduction of the extra degree of freedom, by allowing more than one conclusion, restricts the number of extreme values (local minima and maxima) of the relation to the maximum number of fuzzy rules. This restriction is due to the fact that in between the 'columns' of the raster of points in figure 4 y is a monotonic (not strict) function of x (assuming consistent choices of membership functions). The extension of the fuzzy rules yields a way to apply the practical approach of (self-organizing) fuzzy control and still provide an extra degree of freedom. A smooth adaptation is provided which results in a control system behavior which is more related to the requirements of the process operator.

In most applications found in literature, PI- or PD-control like fuzzy controllers are used. In case of a non-linear process to be controlled the non-linearity is normally not a function of the error and/or error change. When, for example, the process exhibits a dead zone, the non-linearity is a function of the control signal. Because of this, a fuzzy controller with rules like 'if error is big and error change is small then make control signal change big' can not 'capture' such a non-linearity. In case of convergence of the rule base within the fuzzy controller a setpoint change will emerge the need for a number of adaptation cycles: convergence is only local, not global. To overcome these problems in case of non-linearities one should introduce extra 'dimensions'. This results in, for example, using rules like 'if process output is big and process output change is small and desired process output is medium then make control signal small'.

The fuzzy expert control system for the adaptive control of process with minimal process knowledge consist of various interacting fuzzy sub-controllers. These controllers perform on a time scale equal to the sampling time (d.e.c., d.r.e.c. and in-line adaptations) or on a larger time scale (on-line adaptations). The inputs as well as the outputs can differ for each sub-controller. The presented rule-based controller is able to control various processes even though almost no process model is required. Delay time or non-minimum phase behaviour can be dealt with.

References

[Aoki et al. 1990] S. Aoki, S. Kawachi and M. Sugeno. *Application of fuzzy control logic for dead-time processes in a glass melting furnace.* Fuzzy Sets and Systems, vol. 38, pp. 251-265, 1990.

[Brown et al. 1991] M. Brown, R. Fraser, C.J. Harris and C.G. Moore. *Intelligent self-organizing controllers for autonomous guided vehicles.* Proceedings of IEE Control 91, pp. 134-139, Edinburgh, U.K., March 1991.

[Buckley and Ying 1990] J. Buckley and H. Ying. *Linear fuzzy controller: it is a linear nonfuzzy controller.* Information Sciences, vol. 51, pp. 183-192, Elsevier, 1990.

[Harris and Moore 1989] C.J. Harris and C.G. Moore. *Intelligent identification and control for autonomous guided vehicles using adaptive fuzzy-based algorithms.* Engineering Applications of Artificial Intelligence, vol. 2, pp. 267-285, December 1989.

[Heatwole and Zang 1990] C.D. Heatwole and T.L. Zang. *Representing uncertainty in knowledge-based systems: confirmation, probability, and fuzzy sets theories.* Transactions in Agriculture, vol. 33, pp. 314-323, January/February 1990.

[Huang and Tomizuka 1990] L. Huang and M. Tomizuka. *A self-paced fuzzy tracking controller for two-dimensional motion control.* IEEE Transactions on Systems, Man, and Cybernetics, vol. 20, no. 5, pp. 1115-1124, September/October 1990.

[Jager et al. 1991a] R. Jager, H.B. Verbruggen, P.M. Bruijn and A.J. Krijgsman. *Real-time fuzzy expert control.* Proceedings of IEE Control 91, Edinburgh, U.K., March 1991.

[Jager et al. 1991b] R. Jager, A.J. Krijgsman, H.B. Verbruggen and P.M. Bruijn. *Rule-based controller using fuzzy logic.* Proceedings EURISCON '91, Kanoni (Corfu), Greece, July 1991.

[Jager et al. 1992a] R. Jager, H.B. Verbruggen and P.M. Bruijn. *The role of defuzzification methods in the application of fuzzy control*. Proceedings IFAC SICICA '92, Málaga, Spain, May 1992.

[Jager et al. 1992b] R. Jager, H.B. Verbruggen and P.M. Bruijn. *Fuzzy inference in rule-based real-time control*. Proceedings IFAC AIRTC '92, Delft, The Netherlands, June 1992.

[Kong and Kosko] S.G. Kong and B. Kosko. *Adaptive fuzzy systems for backing up a truck-and-trailer*. IEEE Transactions on Neural Networks, vol. 3, no. 2, March 1992.

[Krijgsman et al. 1991] A.J. Krijgsman, R. Jager, H.B. Verbruggen and P.M. Bruijn. *DICE: a framework for intelligent real-time control*. Proceedings 3th IFAC Workshop AIRTC '91, Napa (Ca), U.S.A., September 1991.

[Lattimer et al. 1986] M. Lattimer Wright, M.W. Green, G. Fiegl and P.F. Cross. *An expert system for real-time control*. IEEE Software, pp. 16-24, March 1986.

[Linkens and Hasnain 1990] D.A. Linkens and S.B. Hasnain. *Self-organizing fuzzy logic control and application to muscle relaxant anaesthesia*. IEE Proceedings, vol. 138, no. 3, pp. 274-284, May 1990.

[Neyer et al. 1990] M. De Neyer, D. Stipaniev and R. Gorez. *Intelligent self-organizing controllers and their application to the control of dynamic systems*. IMACS Annals on Computing and Applied Mathematics, Brussels, Belgium, 1990.

[Pedrycz 1989] W. Pedrycz. *Fuzzy Control and Fuzzy Systems*. Research Studies Press Ltd., Taunton, Somerset, England, 1989.

[Procyk and Mamdani 1979] T.J. Procyk and E.H. Mamdani. *A linguistic self-organizing process controller*. Automatica, vol. 15, pp. 15-30, 1979.

[Shao 1988] S. Shao. *Fuzzy self-organizing controller and its application for dynamic processes*. Fuzzy sets and systems, vol. 26, pp. 151-164, 1988.

[Sugiyama 1988] K. Sugiyama. *Rule-based self-organizing controller*. Fuzzy Computing, Theory, Hardware, and Applications, M.M. Gupta and T. Yamakawa (editors), North-Holland, 1988.

[Wakileh and Gill 1990] B.A.M. Wakileh and K.F. Gill. *Robot control using self-organizing fuzzy logic*. Computers in Industry, vol. 15, pp. 175-186, Elsevier, 1990.

[Ying et al. 1990] H. Ying, W. Siler and J.J. Buckley. *Fuzzy control theory: a nonlinear case*. Automatica, vol. 26, no. 3, pp. 513-520, Pergamon Press, 1990.

[Yu et al. 1990] C. Yu, Z. Cao and A. Kandel. *Application of fuzzy reasoning to the control of an activated sludge plant*. Fuzzy Sets and Systems, no. 38, pp. 1-14, 1990.

NEURAL NETWORKS
FOR CONTROL

Gabriela Cembrano and Gordon Wells

Instituto de Cibernética
Univ. Politécnica de Cataluña - Consejo Superior de Investigaciones Científicas
Diagonal, 647, 08028 Barcelona, Spain.

1 Introduction

The design of control systems to modify the performance of dynamic systems in order to achieve certain desired dynamic behaviours has been thoroughly studied for linear systems. Classical control techniques have mostly dealt with systems where a complete knowledge is assumed of the process to be controlled and its environment. In this case, a solid theoretical basis exists for studying the observability, controllability and stability properties of systems, as well as for controller design.

Similarly, during the last decades important results have been achieved in adaptive identification and control of time-varying and partially unknown systems. This is particularly so for linear systems with invariant structure and unknown parameters, for which it is possible to establish adaptive control laws which ensure the global stability of the controlled system [Astrom and Wittenmark 89], [Narendra and Anawasamy 89]. This is not the case, however, for non-linear and otherwise complex systems. Since no generally applicable principles or techniques exist for these systems, the study of their observability, controllability and stability must be performed on a case-by-case basis. Therefore, the controller design methods as well as the adaptive control laws derived for linear systems cannot readily be extended to the nonlinear case.

From this standpoint, the use of connectionist techniques appears to be a promising approach to the adaptive identification and control of certain classes of nonlinear systems. On the one hand, they provide a good representation of complex mappings between inputs and outputs of dynamic systems, as will be shown later, and secondly, they have been proven capable of efficiently *approximating broad classes of non-linear functions*, as reported in [Funahashi 89], and [Girosi 89]. One of the most relevant features of neural networks is their *adaptability* to a changing environment, which allows them to emulate time-varying behaviours of nonlinear plants and of the adaptive controllers required to effectively control them. Additionally, they have shown great potential in terms of their *robustness* with respect to noise, and *fault-tolerance* with respect to local disablement of neurons, much like the qualities observed in their biological counterparts which to some extent they were designed to imitate.

2 Representation of Dynamic Systems

Neural networks are useful as tools for dynamic system representation. This is especially clear for the case of discrete-time, state-space models (although they can also provide good representation for continuous-time and transfer function models), where feedforward and recurrent networks have been used successfully.

A common way to express the behaviour of a dynamic system in discrete time is the state-space representation, which uses recursive equations such as:

$$
\begin{aligned}
x^{k+1} &= f(x^k, u^k) \\
y^k &= g(x^k)
\end{aligned}
$$

where:

x^k represents the state of the system at time k

u^k represents the input applied to the system at time k and

y^k is the output of the system observed at time k

This type of recursive expression may formally be expanded into another expression involving *all* the past observations of the inputs and outputs, by writing x^k in terms of x^{k-1} and u^{k-1} and so on. However, in most practical cases, it is possible to reconstruct the state of the system with only a few past measurements of the input and the output, so that the output at time k is expressed in terms of:

- the input at time k

- inputs at previous times $k, k-1, ..., k-p,\ p < k$

- previous values of the output at times $k-1, k-2, ..., k-q\ ,\ q < k$

2.1 System Representation using Feedforward Networks

A straightforward way to implement this type of model with neural networks is to include tapped delays of the inputs and outputs of the process as inputs to a feedforward neural network, which will then produce the output of the process in its output layer. The nonlinearities of the mappings between inputs and ouputs of the neural networks can be

389

efficiently handled by one or more hidden layers between the inputs and the outputs. Figure 1 shows a representation of a SISO (single-input single-output) model with a feedforward network.

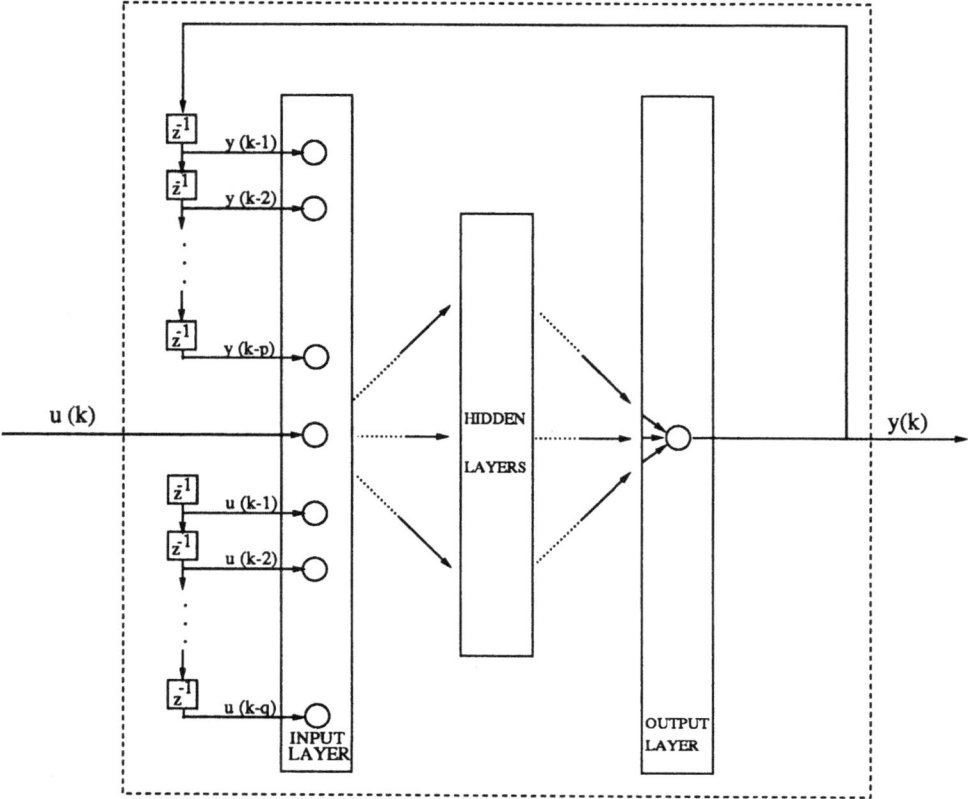

Figure 1: Representation of a dynamic system using a feedforward network.

2.2 System Representation using Recurrent Networks

Another option for the representation of dynamic systems is the use of recurrent networks, which allow the designer to implement any sequential or recursive dynamic behaviour explicitly. It also provides a straightforward way to represent internal states of the processes. Figure 2 shows a recurrent neural network implementing the representation of a SISO dynamic model.

The recurrent network implementation provides a more general framework for the identification of an unknown system, since no previous estimation is required of the number of delayed values which significantly affect the output and it allows the inclusion of a potentially infinite information on the past history of the process. Conversely, the implementation with tapped delays requires a deeper knowledge of the dynamics of the

390

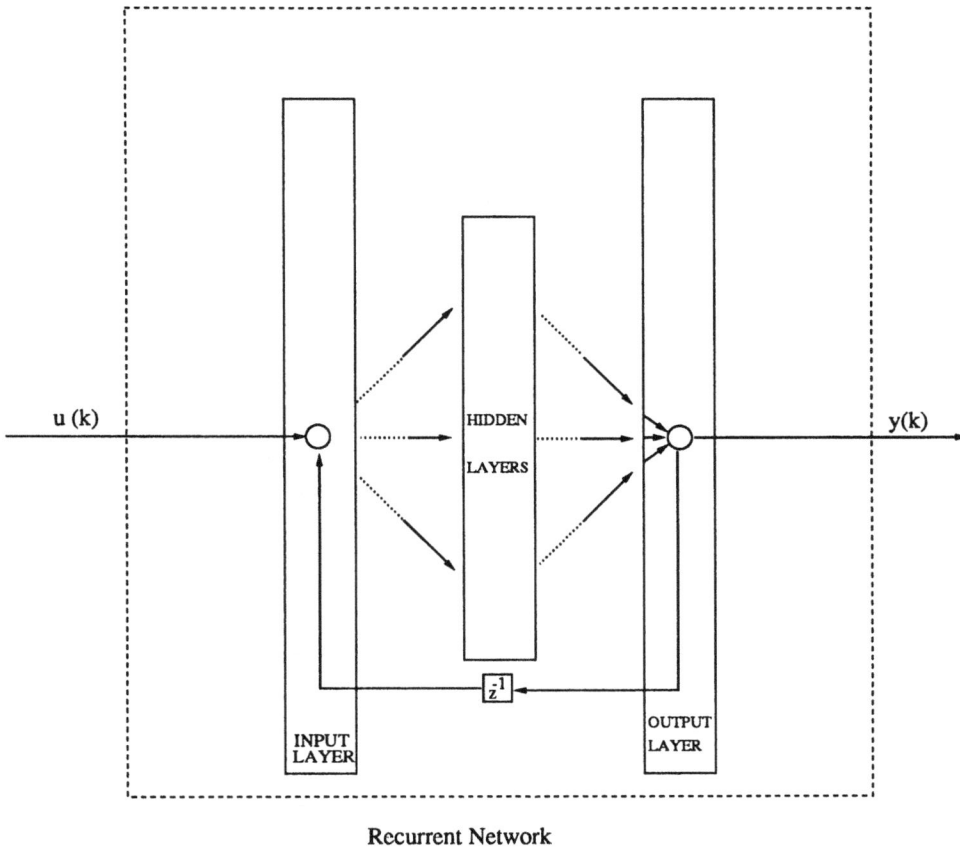

Recurrent Network

Figure 2: Representation of a Dyanmic System using recurrent network

process and usually a trial-and-error process to define the number and placement of tapped delays. However, since feedforward networks have so far been much more widely understood and used, more theoretical developments and tools exist which make their implementation easier.

3   System Identification

System identification, one of the fundamental issues in systems theory, is concerned with the mathematical representation of the behaviour of a system. More specifically, system identification seeks to derive the mathematical expressions which relate some sets of observed input and output data of the process. Identification may be carried out off-line, with finite sets of input/output data of the process or on-line, with a continuous monitoring of input/output data during operation. The latter is also referred to as adaptive identification.

Strictly speaking, system identification is a recursive process containing several iterations of the following steps:

- **structural identification** which seeks to assign the process to a particular class of parameterized mathematical models

- **parameter estimation** which derives suitable values for the parameters in these models, so as to provide the best possible approximation of the actual behaviour of the real system

- **validation** of the mathematical model with new sets of input/output data of the real system.

However, the term identification is often used in Control literature to refer merely to the second step, the parameter estimation, while the structure of the model is assumed to be defined, based on some knowledge on the temporal and frequential behaviour of the sytems. This is especially the case in *adaptive identification*, where the structure of the model is fixed and the parameters are recursively estimated by attempting to minimize some error measure of the output of the model as compared to the actual output of the plant. While the properties these recursive algorithms have been thoroughly studied and the conditions for their convergence and stability have been clearly established for linear systems, no similar results are available for nonlinear ones.

3.1 Identifier Structures

In an attempt to extend the constructive procedures of linear identifiers, the identification structures which have proved successful for linear systems are reformulated for neural nonlinear identifiers. This is the case of the *parallel* and the *series-parallel* identification structures proposed in [Narendra and Pathasarathy 90]. These two structures are constructive models of dynamic systems, which produce an estimation of the output at time $k + 1$, in terms of past observed values of inputs and outputs. In the case of linear systems, this is a linear combination of past values of inputs an outputs. In the case of nonlinear systems, nonlinear functions of these past values of inputs and outputs may be efficiently represented by neural networks.

While the parallel structure uses the past values of the input and of the *model output*, in the series-parallel structure, the output of the model is computed by means of the past values of the inputs and the *real-system outputs*. Better stability and convergence results are achieved with the series-parallel model in linear system, so that a similar condition may be assumed to hold for nonlinear systems.

3.2 Parameter Estimation and Neural Network Training

Once an identification structure has been chosen, the neural networks involved may be designed as feedforward or recurrent networks, using the system representations described in the previous section. Then, the remaining problem of adaptively estimating the weights or parameters of the model to emulate the real system is translated into *training* the neural network(s) to behave like the real system. The well-known concepts of backpropagation may be successfully used in both feedforward and recurrent networks. Two important examples of how to use backpropagation for training recurrent networks are the *backpropagation through time* and the *real-time-recurrent learning* algorithms.

392

In the backpropagation-through-time algorithm, described in [Williams90], an arbitrary recurrent network is expanded into a multi-layer feedforward network, where a new layer is added at each time step. The computation of the derivatives is then performed as usual in a feedforward backpropagation network. The main drawback of this algorithm is, obviously, that it requires the storage of data of the dynamic behaviour of the plant which may, in principle grow without bound when the network is assumed to run indefinitely.

In the real-time-recurrent-learning algorithm, the set of current derivatives of the outputs with respect to all the weights in the recurrent network is available at each time step. Whenever an error is observed in any of the output units, the values of the related derivatives are updated and so are the values of the weights in the network affected by these particular derivatives. The storage requirements of this method is equal to the product of the number of neurons times the number of weights. While this may still be an important requirement, it is comparatively low, when referred to that of the backpropagation through time, for the case of an indefinitely running network. When the number of time steps is reduced, though, it may be more memory-consuming to use the real-time-recurrent algorithm than the backpropagation through time.

4 Neural Controllers

The use of neural networks for implementing control systems has received a good deal of attention in the scientific community lately. These control systems, usually referred to as neural controllers, may consist of one or more neural networks and they may also contain linear elements such as integrators or differenciators

Several studies on neural-network controllers have recently been reported. [Anderson 89], [Chen 90], [Kuperstein 90], [Nguyen 90]. Most of these developments involve the use of feedforward neural networks with backpropagation or extensions thereof, but other important applications have been developed using recurrent neural networks as in [Nerrand 92].

4.1 Structural Design of Controllers

The design of a neural network organization, in terms of their nodes, layers and connections remains a non-trivial task. With the exception of very recent attempts (see [Sanner 91]), methods for determining suitable neural network architectures for particular applications have traditionally not been available. Most frequently, the choice of the number of input, hidden and output nodes, the number of hidden layers, and initial connection configurations and weights are determined by empirical considerations of the problem variables, structure and behaviour, followed by trial-and-error attempts with variations of the initially chosen network architecture. In general, several considerations concerning the plants and control structures must be borne in mind for design purposes.

The definition of the control schemes in terms of input and output of the controller and the plant greatly affects the design of the internal structure of the neural networks involved. For example, the choice of the number of tapped delays to be considered depends not only on the model of the controlled plant, but also on the type of input to the controller. A similar consideration is valid for the number of nodes and hidden layers.

393

Some empirical guidelines for selecting initial network configurations include the fact that a network with a single hidden layer can produce the same results as one with multiple hidden layers, although more training time may be required; and that, in general, hidden layers contain at least as many neurons as there are inputs, but not more than double the number of inputs, according to some literature e.g. [Maren 90]

4.2 Training Neural Controllers

Once the configuration of the neural networks has been defined, two major methods are available for the implementation of the neural controllers, in terms of their adaptation mechanism or training procedure ([Jordan 91] , [Barto 89]).

In both cases, the controller is closely related to the *inverse model* of the plant to be controlled. This is because the controller has the role of a model which must produce control signals as a function of the desired outputs and current states of the plant, whereas a *forward model* produces outputs and new states as a function of the current plant states and input control signals. Figure 3 illustrates this.

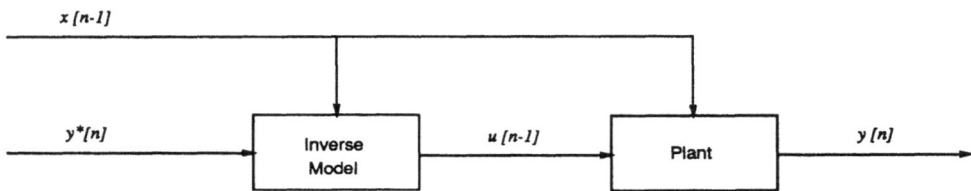

Figure 3: Inverse model of a plant as a controller. Adapted from [Jordan 91]

4.2.1 Direct-Inverse Modelling

The first method, usually referred to as *direct-inverse modelling* treats the training problem as a classical supervised-learning one. The main idea involved in this method is to use the observations of inputs and outputs of the plant and to train an inverse model simply by reversing their roles. Training is then carried out by presenting the neural network with the ouput and input observations to derive a mapping which relates them, as shown in Fig. 4.

Two main drawbacks which limit the usefulness of this approach have been reported. On the one hand, this technique may prove unable to find an inverse in the case of multiple-input/single-output plants, or, in general when many-to-one mappings exist between the inputs and outputs of the plant. On the other, this method is only applicable to the learning of inverse models and it does not allow for more general formulations of learning involving evaluations of the output other than the error between the desired and the actual values.

394

x [n-1]

u [n-1]

Plant

y [n]

+

-

Inverse
Model

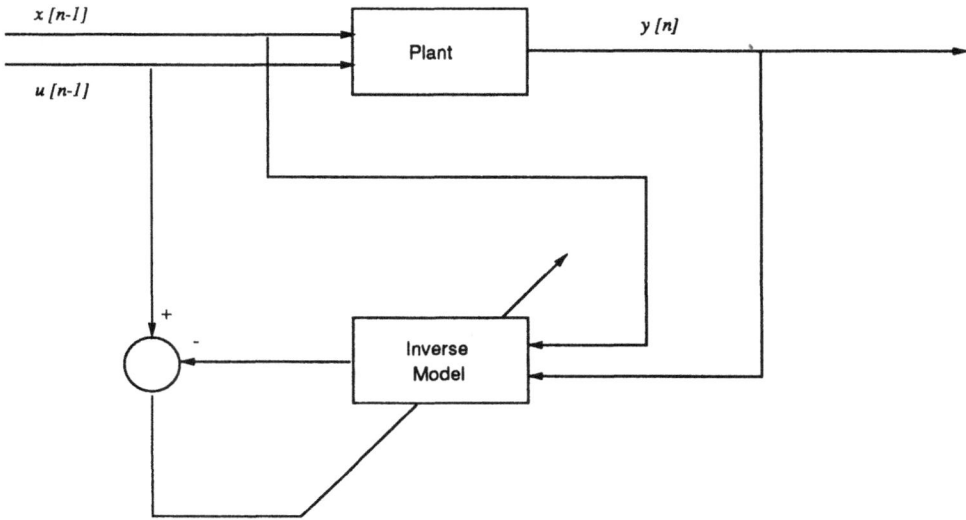

Figure 4: Scheme for direct-inverse modelling of neural controllers. Adapted from [Jordan 91]

4.2.2 Forward Modelling

The second method, sometimes known as *forward modelling* seeks to overcome the drawbacks of the direct-inverse modelling by proceeding in two phases. In the first phase a forward model of the plant is obtained using a supervised learning algorithm. In the second, the controller model and the plant model are combined, so that a desired behaviour of the system is learned across the composed network(s), as shown in Fig. 5. The backpropagation of the performance index in the second phase causes only the weights of the controller to change. When the performance index is the usual quadratic error between the desired and the actual outputs, then this method has been reported to make the controller model converge to an inverse even with many-to-one mappings in the plant model.

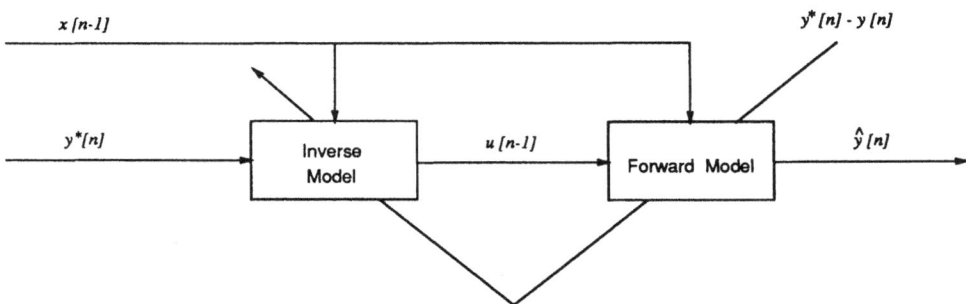

x [n-1]

$y^*[n] - y[n]$

$y^*[n]$

Inverse
Model

u [n-1]

Forward Model

$\hat{y}[n]$

Figure 5: Scheme for forward modelling of neural controllers. Adapted from [Jordan 91]

395

5 Adaptive Control using NN

Adaptive Control techniques seek to derive controllers for time-varying or partially unknown plants, which may be stably adapted during the plant operation to achieve some performance objectives. This is generally achieved using controllers with fixed structures, but variable parameters. These parameters are updated continually during the operation of the plant, by optimizing some performance index (usually an error measure) related to the actual output of the plant and the desired behaviour.

The inherent adaptability of neural networks makes them ideal candidates for the implementation of adaptive controllers and, indeed many of the studies in neural network learning have their counterparts in adaptive control theory. An excellent approach to the unification of the results in both fields may be found in [Nerrand 92].

In many neural control applications, neural networks have been used in a "supervised" mode, i.e. they are first trained with a finite number of cases and then they are used with fixed coefficients. However, neural controllers may also be designed to operate adaptively, i.e., to be trained continually, on-line with the plant. Just like conventional adaptive controllers, neural control structures will incorporate neural networks with fixed architectures in terms of node configuration and connections, but which must perform on-line weight adjustment.

5.1 Direct vs. Indirect Adaptive Control using NN

Two different approaches to the adaptive control of dynamic systems have been used extensively, namely, direct and indirect control. In direct adaptive control structures, the controller parameters are adjusted by directly seeking to reduce some evaluation function of the output error (the error between the actual and the desired behaviour of the plant). In indirect adaptive control structures, a plant identifier model is included. Then, the adjustment of the controller parameters is performed by first adjusting the parameters of the identifier and then computing the values of the controller parameters to achieve some control objective, considering the current values of the estimated parameters of the identifier as the true values of the plant. Both direct and indirect adaptive control schemes produce an overall nonlinear behaviour, even when linear time invariant plants are considered. Figure 6 and Fig. 7 illustrate direct and indirect adaptive control structures respectively. In these diagrams, y_p represents the output of the plant, e_o is the output error, e_i is the identification error and ϕ is the set of parameters of the identifier.

When using neural networks for adaptive controllers, indirect control structures containing neural identifiers are preferred. This is because the parameter or weight adjustment in the neural controller requires a mapping between these controller weights and the output error. For example, if feedforward backpropagation networks are used, this means that some error function of the output must be *"backpropagated" on-line* through the controller in order to adapt the weights of the neural networks.

Therefore, a model of the plant must be used, in order to be able to compute how a change in the parameters of the controller affects the output error. One usual approach in backpropagation neural controllers is to have a neural-network model of the plant.

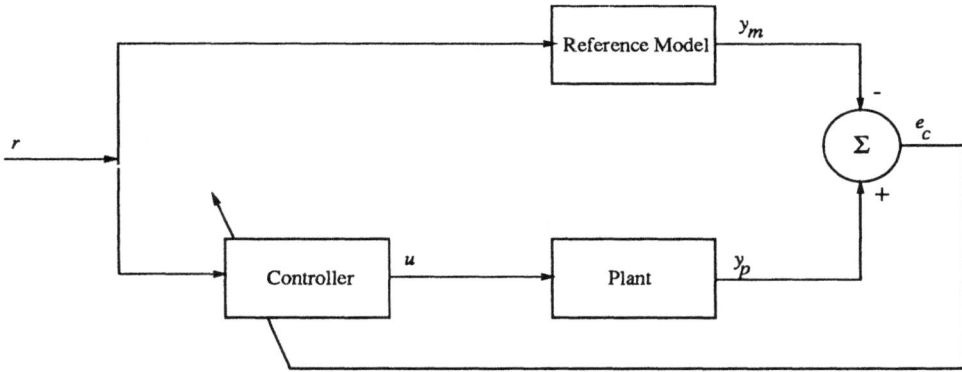

Figure 6: Direct adaptive control structure

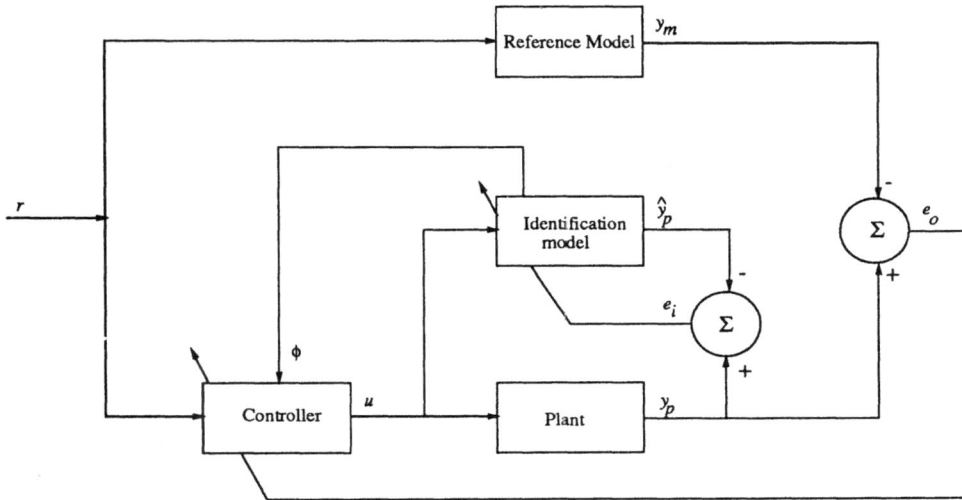

Figure 7: Indirect adaptive control structure

5.2 Some Possible Adaptive Control Structures with NN

One of the most useful approaches in Adaptive Control is to establish a desired behaviour of the plant by means of a *reference model*. The controller parameters are continually updated so as to reduce some error function which compares the actual and the desired output. This approach is generally known as Model reference Adaptive Control (MRAC) (For a more detailed exposition see [Astrom 89]).

A form of Model Reference Adaptive Control with a neural controller is proposed in Fig. 8. The desired behaviour of the plant must be provided explicitly as the reference model. The choice of this reference model must be such that it represents a reasonable (feasible) response of the controlled system.

Two neural-network blocks are involved in this structure, namely, a neural identifier (or emulator) and a neural controller. The neural network(s) in the identifier update its/(their) weights by backpropagating the error between the actual and the predicted values of the output, e_1, while the neural network(s) in the controller adjust its/(their) weights through

397

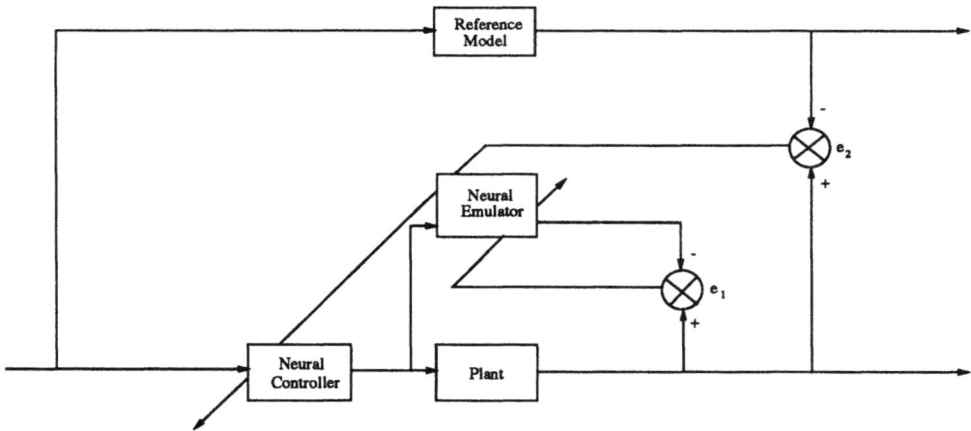

Figure 8: Indirect model reference adaptive control with NN

backpropagation of the error between the actual and the reference-model output, e_2.

The line across the neural emulator to the neural controller represents the backpropagation of the error function e_2 *across* the neural identifier. This means that the latest values of the parameters of the identifier are used for the backward computation of the derivatives of the error function e_2. However, the identifier parameters are not changed in this process.

Another possible structure for neural adaptive control includes a feedback loop from the output of the plant to the controller, which employs the error function between the set-points and the output e_3 as the input to the controller (see Fig. 9). A neural emulator of the dynamic system is also included in this structure.

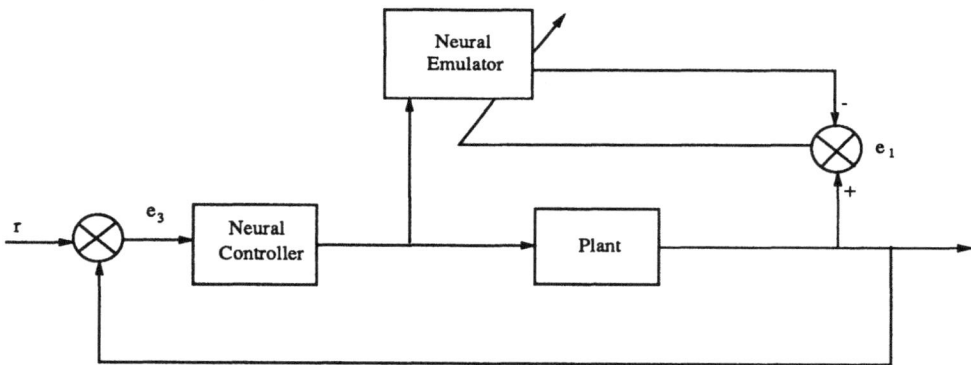

Figure 9: Alternative indirect structure for adaptive control with NN

This structure involves the adequate choice of a performance index to be backpropagated, which need not necessarily be the quadratic error function used in standard backpropagation. It is important to remark that if the standard quadratic error function between the set-points and the actual output were strictly used, this would, in practice, amount to requiring the controlled system to follow an implicit reference model with a transfer function equal to identity. This choice would demand the maximum of the controller, i.e., zero error over the whole bandwidth. Hence, more reasonable performance indexes must be chosen which evaluate the output of the controlled system against feasible desired behaviours.

5.3   Persistence of Excitation

The importance of the data used for training a neural network, generally known as training sets, has been widely described in the context of supervised learning. Generally speaking, it has been shown that the training sets must be representative of all the possible inputs which may be presented to the network after training, for the network to reach a good approximation of the desired behaviour.

The concept of a training set is no longer valid when dealing with continually adapting systems, but the degree to which the input signals to the neural networks include representative information of the desired dynamic behaviour is still a crucial issue. Indeed, when an adaptive identification (and/or control) system is implemented, the goal is to obtain an adaptive change in the parameters which leads them to converge to to the set of values which best approximate the real system. In the case of time invariant systems, these are sometimes called nominal values.

The ability of an adaptive system to converge to a set of nominal values depends not only on the procedure for adaptively changing weights to minimize some error function of the real and desired outputs of the system, but also, very importantly, on the particular choice of the inputs presented to the system and the model in order to produce measurements of the real and desired outputs. This is very clear if one thinks of a constant input being presented to a linear time-invariant SISO (single-input-single-output) plant and its model. One observation of the input and the outputs of the model and the plant, provides one equation to compute the values of the parameters, but subsequent observations will not provide any more relevant information. Whenever the model requires the estimation of more than one parameter, a constant input signal will not provide enough information about the dynamic behaviour of the real plant.

When input signals do provide sufficiently rich information on the dynamic behaviour of a system, so that the parameters of its model may be estimated, these signals are said to be *persistently exciting* for the said system. The conditions for persistence of excitation in linear systems are very clearly established in the frequencial and time domains (see for example [Sastry 89]). While no similar results have yet been achieved for nonlinear plants, it may be reasonable to extend that at least these necessary conditions to the input signals of nonlinear neural identifiers, in order that the models may be used as predictors in adaptive control schemes.

6   Some Relevant Applications

A number of interesting applications of neural identification and control of dynamic systems have been reported lately, especially in the fields of motion and navigation control. These fields are also the object of important on-going research projects.

Robot-arm control

One of the challenging areas of motion control, which has produced so far a number of very interesting results is the control of robot-arms. The main goal in this area has been to train robot arms to perform certain tasks in partially unknown environments, where the robot must take motion decisions on-line, based on sensory information. Special emphasis is currently being placed in the design of controllers which may direct the end effector of a robot to grasp objects, avoiding obstacles, based on vision and/or proximity sensors. Important achievements in training robots to relate goals (e.g. positions of the end effector) with control actions (e.g. movements of the joints), usually described in robotics as the direct and inverse kinematics and dynamics, have been described in [Kawato 90], [Jordan 90], [Martinetz 90]. Other relevant results have been obtained in path planning and trajectory tracking, based on sensory information, as shown, for example, in [Millán 92] and [Mel 90].

Vehicle navigation control Other appealing applications in motion control are satellite orbit control, currently under research in the European Space Agency and vehicle navigation in general. An illustrative example is the truck backer-upper developed by Nguyen and Widrow and coworkers [Nguyen 90], who trained an unmanned trailer-truck to back-up to a dock.

Other control problems

Several other control problems are currently the object of research and, since neural control is still in its infancy, no general methodologies or appraches exist. In an attempt to challenge experimenters in the field to compare the performance of different approaches with a common test-bed, a collection of control problems presented in [Anderson and Miller 90], with a complete problem statement and the results available so far. The problems include balancing a pole, steering a ship, landing an aircraft and controlling bacterial growth in a bioreactor.

Acknowledgements

Research in neural networks and their applications to process control at the Instituto de Cibernética is partially supported by the research grant CICYT-TIC91-0423 of the Spanish Science and Technology Council.

The contribution of Mr. Joan Alex Ferré in the preparation of this text and the helpful advice and guidance of Prof. Carme Torras are gratefully acknowledged.

7 References

[**Anderson 89**] Anderson, C. W., "Learning to Control an Inverted Pendulum Using Neural Networks", *IEEE Control Systems Magazine*, April 1989.

[**Anderson and Miller 90**] Anderson C. W., Miller W. T., III, "Challenging Control Problems" *Neural Networks for Control, T. Miller, R.S. Sutton, P. J. Werbos, editors*, MIT Press, 1990

[**Astrom 89**] Astrom, K. J., and Wittenmark, B., *Adaptive Control*, Addison-Wesley, 1989.

[**Barto 89**] Barto, A. G., "Connectionist Learning for Control: An Overview", COINS Technical Report 89-89, Dept. of Computer Science. Univ. of Massachussetts, 1989.

[**Chen 90**] Chen, F., "Backpropagation Neural Networks for Nonlinear Self-Tuning Adaptive Control", *IEEE Control Systems Magazine*, April 1990.

[Funahashi 89] Funahashi, K., "On the approximate realization of continuous mappings by neural networks", *Neural networks*, Vol. 2, pp. 183-192, 1989.

[Girosi 89] Girosi, F., and Poggio, T., "Networks and the best approximation property", *Artificial Intelligence Lab. Memo*, No. 1164, MIT, Cambridge, MA, October 1989.

[Guez 88] Guez, A., Eilbert, J. L. and Kam, M., "Neural Network Architecture for Control", *IEEE Control Systems Magazine*, April 1988.

[Jordan 91] Jordan, M. L., and Rumelhart, D. E., "Forward Models: Supervised Learning with a Distal Teacher", Occasional Paper 40, MIT Center for Cognitive Science, 1991.

[Kawato 90] Kawato M., "Computational Schemes and Neural Network Models for Formation and Control of a Multijoint Arm Trajectory" , *Neural Networks for Control, T. Miller, R.S. Sutton, P. J. Werbos, editors*, MIT Press, 1990

[Kuperstein 90] Kuperstein, M. and Rubinstein, J., "Implementation of an Adaptive Neural Controller for Sensory-Motor Coordination", *IEEE Control Systems Magazine*, April 1989.

[Maren 90] Maren, A., Harston, C. and Pap, R., *Handbook of Neural Computing Applications*, Academic Press, 1990.

[Martinetz 90] Martinetz T. M., Ritter H. J. and Schulten K. J., "Three-Dimensional Neural Net for Learning Visuomotor Coordination of a Robot Arm" *IEEE Trans. on Neural Networks*, Vol 1, No. 1, March 1990.

[Mel 90] Mel, B., "Vision-Based Robot Motion Planning", *Neural Networks for Control, T. Miller, R.S. Sutton, P. J. Werbos, editors*, MIT Press, 1990

[Millán 92] Millán, J. del R., and Torras, C., "A Reinforcement Connectionist Approach to Robot Path Finding in Non-maze-like Environments", *Machine Learning* vol. 8 No. 3 and 4, May 1992

[Narendra and Anwaswamy 89] Narendra, K. S. and Anwaswamy, A. M., *Stable Adaptive Systems*, Prentice Hall International, 1989.

[Narendra and Parthasarathy 90] Narendra, K. S. and Parthasarathy, K., "Identification and Control of Dynamical Systems Using Neural Networks", *IEEE Transactions on Neural Networks*, Vol. 1, No. 1, 1990.

[Nerrand 92] Nerrand O., Roussel-Ragot P., Personnaz L., Dreyfus G., "Neural Networks and Nonlinear Adaptive Filtering: Unifying Concepts and New Algorithms", *Neural Computation*, to be published.

[Nguyen 90] Nguyen, D. H. and Widrow, B., "Neural Networks for Self-Learning Control Systems", *IEEE Control Systems Magazine*, April 1990.

[Polycarpou 91] Polycarpou, M. M., and Ioannou, P. A., "Identification and Control of Nonlinear systems Using Neural Network Models: Design and stability Analysis", Technical Report 91-90-01, Department of Electrical engineering-Systems, University of Southern California, September 1991.

[Sanner 91] Sanner, R. M., and Slotine, J.-J. E., "Direct Adaptive Control with Gaussian Networks," *NSL Report*, No. 910303, March 1991.

[Sastry 89] Sastry S., Bodson M., K., *"Adaptive Control. Stability, Convergence and Robustness"* Prentice-Hall Intrernational 1989

401

[**Tzirkel-Hancock 91**] Tzirkel-Hancock, E. and Fallside, F., "Stable Control of Nonlinear Systems Using Neural Networks", Technical Report CUED/F-INFENG/TR.81, Cambridge University Engineering Department, Cambridge, England, July 1991.

[**Williams 90**] Williams, R. J., "Adaptive Representation and Estimation using Recurrent Connectionist Networks", *Neural Networks for Control, T. Miller, R.S. Sutton, P. J. Werbos, editors*, MIT Press, 1990

Associative Memories: The CMAC approach

A.J. Krijgsman

Delft University of Technology,
Department of Electrical Engineering, Control Laboratory,
P.O. Box 5031, 2600 GA Delft, the Netherlands.
Email: a.j.krijgsman@et.tudelft.nl

Abstract: The use of associative memories is an attractive alternative for normal, more frequently used type of networks. It has a major advantage in its convergence speed. The CMAC (Cerebellar Model Articulation Controller) approach is introduced. CMAC has proved to be very useful for both modelling and control.

Keywords: neural networks; assosiative memories; learning; intelligent control.

1 Introduction

In these lecture notes we will focus on the use of a specific kind of neural networks for control purposes: associative memories. Many aspects of neural networks indicate that they are very valuable for control purposes. Especially the ability to represent arbitrary nonlinear mapping is an interesting feature for complex nonlinear control problems.
The most relevant features for control concerning neural networks are:

(i) arbitrary nonlinear relations can be represented

(i) learning and adaptation in unknown and uncertain systems, provided by weight adaptation

(iii) information is transformed into an internal (weight) representation allowing for data fusion of both quantitative and qualitative signals

(iv) because of the parallel architecture for data processing these methods can be used in large scale dynamical systems.

(v) the system is fault tolerant so it provides a degree of robustness.

Many control design methods, based on linear control theory, meet some of the above mentioned properties but neural networks meet them all! The ability of neural networks to represent arbitrary nonlinear relations has been prooved by many researchers ([Cybenko 1989], [Cotter 1990]).

2 CMAC

Many of the ideas of AI techniques are based on knowledge about the functioning of biological systems, man in particular. With AI techniques attempts are made to implement "intelligent behaviour" on computer systems. Intelligent behaviour roughly consists of two components: the ability to reason with knowledge and to store past experiences as knowledge. Expert systems are part of the AI techniques which deal with reasoning, while artificial neural networks deal with aspects of learning.

The CMAC algorithm is a less known technique which, just as artificial neural networks, can be used to learn efficiently. In 1975 J.S.Albus published his ideas of CMAC (Cerebellar Model Articulation Controller) [Albus 1975a, Albus75b]. The CMAC algorithm is based on the functioning of the small brains, which is responsible for the coordination of our limbs. Albus originally designed the algorithm for controlling robot systems. It is essentially based on lookup table techniques and has the ability to learn vector functions by storing data from the past in a clever way. The basic form of CMAC is directly useful to model static functions. However in control engineering one is mainly interested in controlling and modelling dynamic functions. To give CMAC dynamic characteristics the basic form has to be extended.

CMAC is a perceptron-like lookup table technique. Normal lookup table techniques transform an input vector into a pointer, which is used to index a table. With this pointer the input vector is uniquely coupled to a weight (table location). Using normal lookup table techniques a number of problems occur. The main problems are:

- The input signals must be quantized. Quantization adds quantizing noise to the input signals. The quantization interval can not be chosen too small because of the increasing table size.

- The size of the table increases exponentially as the number of inputs increases.

- Output values corresponding to nearby input vectors are not correlated because every weight is uniquely coupled to a point in the quantized input space. Many real functions are continuous and can be approximated by rather smooth functions. Normal lookup table techniques do not use this property to there advantage.

- The pointer mechanism is not flexible enough for practical purposes mainly because of two reasons. Firstly, the range of the input is not always known in advance. Secondly, the shape of the reachable input space can be very complex. It is very hard to find an appropriate mapping algorithm which is both fast and efficient. In this context efficiency concerns the ratio of reachable points in the quantized input space and the number of reserved weights in memory.

Despite these disadvantages of lookup table techniques, two major advantages exist. Firstly, a function of any shape can be learned because the function is stored numerically, so linearity is no constraint. Secondly, lookup table techniques can be very fast when simple indexing algorithms are used.

2.1   Construction of CMAC algorithm

The CMAC algorithm can be classified as an advanced lookup table technique, strongly related to the field of neural networks. The strongest correlation of CMAC is with the perceptron type of network, introduce by Rosenblatt. The algorithm can be characterized by three important features, which are extensions to the normal table lookup methods. These important extensions are: distributed storage, the generalization algorithm and a random mapping. These three extensions are described in more detail in the following sections.

Distributed storage and the generalization algorithm

In control engineering many functions are continuous and rather smooth. In this case nearby input vectors will produce similar outputs. Distributed storage combined with the generalization algorithm will give CMAC the property called generalization, which implies that output values of nearby input vectors are correlated. Distributed storage

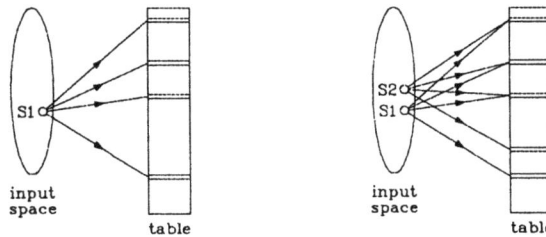

Figure 1: Illustration of (a) distributed storage and (b) generalization in CMAC

means that the output value of CMAC is stored as the sum of a number of weights. In this case the indexing mechanism has to generate the same number of pointers. Figure 1 shows the effect of distributed storage. The input vector S_1 results in the selection of four weights.

The generalization algorithm uses the distributed storage feature in a clever way: nearby input vectors address many weights commonly. Figure 1 reflects the effect of generalization by using distributed storage. An arbitrary input vector S_2 in the neighbourhood of S_1 shares three out of four weights with S_1. Thus the output values generated by S_1 and S_2 are strongly correlated. A nice feature of the generalization algorithm is that it will work for arbitrary dimensions of the input vector.

Random mapping

Because of an unknown range of a component of the input vector, a complex shape of the input space or a sparse quantized input space, it is very difficult to construct an appropriate indexing mechanism. These problems often result in very sparsely filled tables. The CMAC algorithm deals with this problem in a very elegant way. The generalization algorithm generates for every quantized input vector a unique set of pointers. These pointers are n-dimensional vectors pointing in a virtual table space. CMAC uses a random mapping to map these vectors into the real linear table space. The essence of random mapping is similar to a hash function in efficient symbol table addressing techniques.

Using a random mapping implies the occurrence of mapping collisions, which gives rise to undesired generalization. However, because distributed storage is used, a collision results only in a minor error in the output value. The major advantage of using random mapping is its flexibility, because irrespective of signal ranges and the shape and sparseness of the input space, the size of the table can be chosen solely on the number of reachable input vectors.

2.2 Review of the complete algorithm

The CMAC algorithm can be split in two parts. These two parts are similar to Rosenblatt's Perceptron architecture, meaning that the stimulus/response mapping is split in two functions:

$$f \; : \; S \Rightarrow A \qquad\qquad (1)$$

$$g \; : \; A \Rightarrow R \qquad\qquad (2)$$

where S is the stimulus space, A is the association cell space and R is the response space. In this mapping $f()$ is a fixed algorithm but $g()$ depends on one layer of weights, which can be adapted (see figure 2).

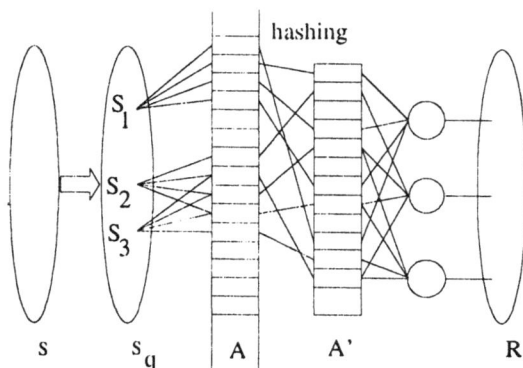

Figure 2: The basic CMAC algorithm

In the stimulus to association cell mapping two properties are fullfilled: only a small number ρ of the cells will have a non-zero value for a specific input and secondly similar vectors in the stimulus space will map to similar association cell vectors. This last feature is already described as generalization and ρ is call the generalization number. In normal applications $\rho > 1$, for $\rho = 1$ the algorithm reduces to a simple lookup table.

An important part of the algorithm is the part which determines which association cells are addressed. When a generalization parameter ρ has been determined, quantization much be chosen. This results in R_i intervals on each component s_i of the input vector. So the original stimulus space is mapped to a quantized space S_q:

$$f_1 : S \Rightarrow S_q \tag{3}$$

in which S_q is the quantized stimulus space. In the original CMAC algorithm of Albus this space $S_q \in I^n$. This result in a many to one mapping. S_q can be generalized as $S_q \in \Re^n$ to a bounded convex space which results in a unique one to one mapping. The algorithm then forms ρ layers each consisting of n-dimensional hypercubes of volume ρ^n, such that the quantized stimulus space is just covered. Every vector s_q maps to ρ n-dimensional hypercubes. This translates the problem of the mapping into the positioning of ρ layers on S_q such that the following condition holds:

- for points sq_k which lie within a certain neigbourhood of a point sq_j, the Hamming distance between the corresponding association vectors a_k and a_j should be linearly proportional to the Euclidean distance between these two points sq_k and sq_j.

It can be proved that this problem is the same as choosing $\rho - 1$ integer coordinate vectors $(1 \times n)$ inside a basic hypercube (size ρ^n), where one point is defined to be the origin $(s = 0)$ in such a way that"

a) only one point lies on each hyperplane $s_i = 1, 2, \cdots, \rho - 1$ for $i = 1, 2, \cdots, n$

b) the points and their corresponding images modulo ρ in the neighbouring n dimensional hypercubes (size ρ^n) are uniformly distributed as far as possible

This last condition can be interpreted as distributing the $\rho - 1$ points such that they are lying on a lattice with the neigbouring points as far apart as possible.
These conditions are satisfied by searching for an n-dimensional vector d, where d_i is co-prime to ρ, which maximises the minimum distance (min l_{pq}) between pairs of neighbouring points p, q, to find d which satisfies the condition

$$D(d) = \max\{\min l_{pq}\} \tag{4}$$

Hence the kth point $(0 \le k \le \rho - 1)$ is generated using modulo arithmetic by:

$$p_k{}^k = \mathrm{mod}(k * d_i, \rho) \qquad i = 1, 2, \cdots, n \tag{5}$$

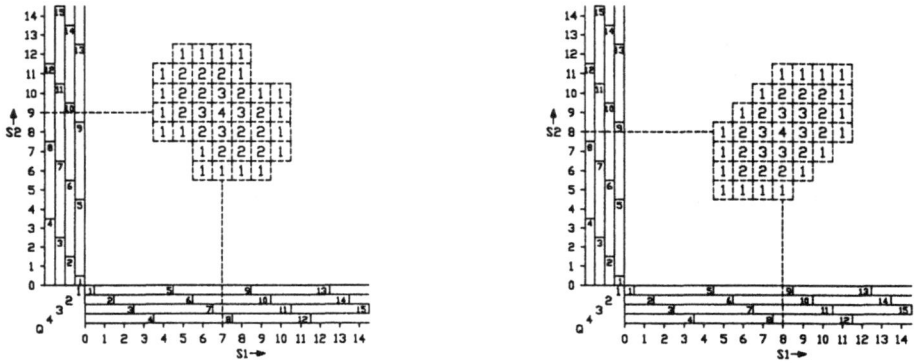

Figure 3: Two dimensional distribution of weights

For the original CMAC algorithm as described by Albus $d_i = 1, \forall i$ is chosen.

To illustrate the effect of this part of the CMAC algorithm a two-dimensional example has been worked out. In this example the original CMAC mapping has been used. In figure 3 this example has been depicted by displaying the number of shared weights (table locations) with the centre $(S_1, S_2) = (7, 9)$. Along the axes the quantization functions are depicted, corresponding to the original CMAC. From the figure we can see that the point $(8, 8)$ has two weights in common with $(7, 9)$. The resulting generalization is more or less optimal because a kind of circular distribution around the point $(7, 9)$ is generated. But when we take a look at another base coordinate $(8, 8)$ and construct the same picture for the shared weights we conclude that the algorithm with these settings is sensitive for changes in the input vector (e.g. the input space). It is obvious (see figure 3) that for this point the distribution is far from circular, but rather nonlinear. Other distributions for example can be obtained by choosing $d_i = 2$.

Kernel functions in CMAC

The next step in the CMAC is the mapping from the association cell space to the response space. This mapping is determined by so-called Kernel functions used to cover the association cells and a single layer network. The input to this single layer network consists of the ρ Kernel functions whose output is non-zero. The response r_i is formed by the normalized sum. So for a specific stimulus the response of the CMAC is given by:

$$r_i = \frac{\sum_{j=1}^{\rho} K(\|S_q - C_{q_j}\|_\infty) * w_{ij}}{\sum_{j=1}^{\rho} K(\|S_q - C_{q_j}\|_\infty)} \qquad (6)$$

in which r_i is the CMAC response for the ith coordinate. The Kernel function is denoted as $K(\|S_q - C_{q_j}\|_\infty)$ in which C_{q_j} is its centre.

The output of the Kernel function lies in the interval $[0, 1]$. The weight associated with association cell j for the ith output is denoted as w_{ij}.

408

The choice of the Kernel function $K()$ strongly determines the type of interpolation between the training centers. In the original CMAC Albus chose a kind of binary step function given by:

$$K(\|S_q - C_{q_j}\|_\infty) = \begin{cases} 1 & \text{if } \|S_q - C_{q_j}\|_\infty < \frac{\rho}{2} \\ 0 & \text{otherwise} \end{cases} \tag{7}$$

The result of this function is a piecewise constant output. Of course it is possible to choose any kind of Kernel function, for example using a triangular hat function:

$$K(\|S_q - C_{q_j}\|_\infty) = \begin{cases} 1 - \|S_q - C_{q_j}\|^2 \div \rho & \text{if } \|S_q - C_{q_j}\|_\infty < \frac{\rho}{2} \\ 0 & \text{otherwise} \end{cases} \tag{8}$$

It is possible to choose the Kernel function in relation to the kind of problem dealt with.

Learning in CMAC

The error between the actual CMAC output and the desired output is given as $(\hat{r}_i - r_i)$. The weight update is performed by a so-called Widrow-Hoff type of updating rule:

$$\Delta w_{ij} = \beta_j(\hat{r}_i - r_i)K(\|S_q - C_{q_j}\|_\infty) \tag{9}$$

in which $\beta \in [0, 1]$ is the learning rate for the jth association cell. The convergence of the weights to a set w^*, such that the output error is minimized, is guaranteed as long as the the following conditions for $\beta_j(t)$ hold:

$$\sum_{t=1}^{\infty} \beta_j(t) = \infty \tag{10}$$

$$\sum_{t=1}^{\infty} \beta_j^2(t) < \infty \tag{11}$$

The first condition ensures that new inputs can be stored, while the second constraint ensures that learing proceeds in a stable manner.
A typical learning rate, such that these constraints hold is given by:

$$\beta_j(t) = \frac{c}{c+t} \tag{12}$$

in which c is an integer constant and t is the number of times the association cell j has been updated. In general this indicates that the learning rate decreases (or has to decrease) during the learning phase.

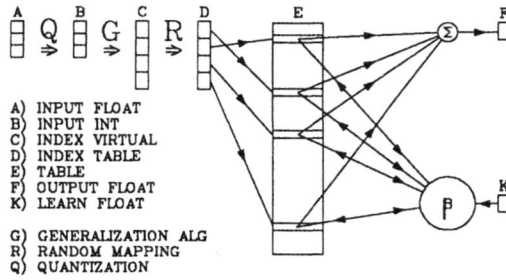

A) INPUT FLOAT
B) INPUT INT
C) INDEX VIRTUAL
D) INDEX TABLE
E) TABLE
F) OUTPUT FLOAT
K) LEARN FLOAT

G) GENERALIZATION ALG
R) RANDOM MAPPING
Q) QUANTIZATION

Figure 4: CMAC calculation steps

2.3 Properties and parameters of CMAC

In the previous paragraphs a full description of the CMAC algorithm is given. In this section we will give some attention to the implementation details of the algorithm. In figure 4 all calculation steps have been depicted. Firstly, the weights must be addressed by pointers. In figure 4, vector A is the 3-dimensional floating point input vector, which is supplied by the user. This vector is quantized into an integer vector B by operation Q. Vector B is then processed by the generalization algorithm G which generates a pointer C in the virtual table space. Next, pointer vector C is scrambled by the random mapping R into vector D, pointing to the real linear table space of table E. Secondly, the weights can be accessed in two ways:

- a retrieve operation calculates an output value F by simply summing the selected weights.

- a store operation is possible which adapts the weights according to the desired output value in vector K. This adaptation (learning) is done in a cautious way using a Widrow-Hoff type of adaptation rule is used (the delta learning rule).

The main properties of CMAC are:

- Input quantization which adds quantizing noise to the inputs. To reduce this noise, the quantization interval must be chosen as small as possible. Because the limited size of the table, the quantization interval cannot be chosen too small.

- Generalization which correlates outputs of nearby input vectors. Generalization is strongly related to interpolation and smoothing. It reduces the required table size and, in contrast with the lookup table it avoids that all possible input vectors in the input space have to be learned.

- The random mapping maps an arbitrary vector in the input space into the linear table space without problems.

410

- Using the storing concept, CMAC is able to learn arbitrary relatively smooth functions. However, because function values are simply stored, CMAC can only generate a reasonable output in an area where learning has been performed.

To use CMAC a (small) number of parameters have to be set. Some important parameters are:

- Input resolution. The input resolution (= quantization interval) affects the amount of quantization noise, the table size and the generalization distance in the input space. The generalization distance is the distance between two input vectors where correlation vanishes.

- Generalization number. The generalization number is the number of weights which are addressed simultaneously. It affects the generalization distance and the speed of the CMAC algorithm.

- The size of the CMAC table. The number of weights in the CMAC table can not be chosen arbitrary. The size of the table is directly related to the number of reachable vectors in the input space. When we consider a n input, m output CMAC and a quantization which results in R_i intervals for the i_{th} input ($i = 1, 2, \cdots, n$), and we take a generalization number ρ the number of assocociation cells is given by:

$$p = \sum_{j=1}^{\rho} \prod_{i=1}^{n} (\frac{R_i - j}{\rho} + 1) \tag{13}$$

The physical size of the table is determined by the hashing scheme and the number of collisions that is still acceptable. Therefore a more practical lower bound of the table size can be obtained.

- The learning factor β. This parameter affects the speed at which the weights are changed to meet the desired output. The choice for this parameter is a trade off between distortion of the surrounding output values and speed of learning (adaptation, initialisation).

2.4   Example - Learning using CMAC

In this example the learning behaviour of the CMAC is investigated, by learning a static function $f(x_1, x_2)$. The function to be learned is described by:

$$f(x_1, x_2) = \sin^2(x_1 * \pi).\sin^2(x_2 * \pi), x_{1,2} \in [0,1] \tag{14}$$

In figure 5 this function is depicted. To illustrate the learning behaviour the results after a number of training runs are depicted. The input resolution has been chosen such that there are 10^6 possible input points. Each training run consists of 50 training points. Figure 6 depicts the result of the CMAC after a number of training runs. For the CMAC we have chosen some default values: $\rho = 4$, table size $= 2^{14}$, input resolution $= 0.001$, output resolution $= 0.001$, $\beta = 0.8$ and binary kernel functions. To illustrate the effect of one of the parameters of CMAC an experiment has been performed with a table size which is too small compared to the function to be learned. This leads to collisions in the table space. This give rise to rather big distortions in the table space (see figure 7).

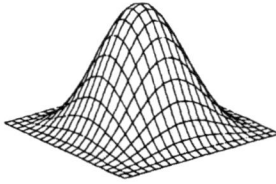

Figure 5: Function to be learned

AMS - 1 training run

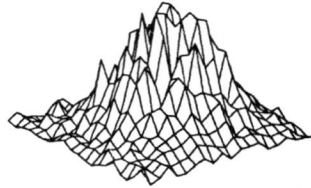

AMS - 2 training runs

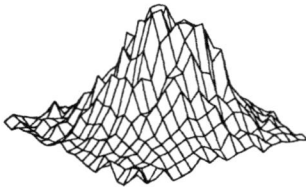

AMS - 3 training runs

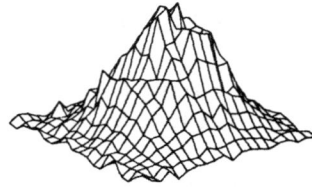

AMS - 5 training runs

Figure 6: Illustration of learning

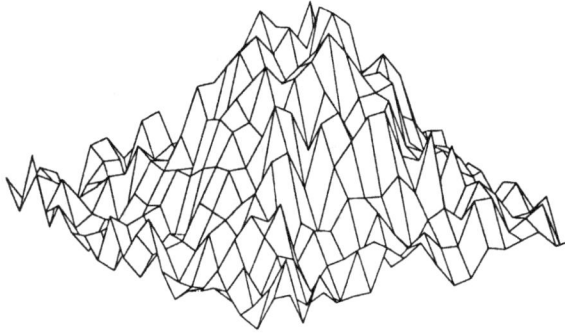

AMS - 5 training run(s)

Figure 7: Illustration of collisions

3 Comparison with classic neural networks

To give the reader some idea about the properties of this method, compared to other more conventional method like backpropagation networks, a comparison is made between two neurally inspired models: CMAC and backpropagation networks. CMAC belongs to the catagory of Associative Memories (AMS), which uses local interpolations for modeling. Backpropagation networks use global approximations to model a function. These two neural models can be seen as a contradiction with more classical models, which use polynomial approximations. One of the problems with these classical methods is that they have (big) problems with highly nonlinear systems. The number of coefficients in the polynomial increases with:

$$n_c(n,q) = \frac{(n+q)!}{n!q!} \tag{15}$$

in which q is the order of the polynomial and n the dimension of the input. As an example: if $n = 5$ and $q = 4$ then the result is $n_c = 126$. In literature some methods are known which use a local interpolation between a certain number of known data points near an asked point. This is performed by calculating the least square plane following a procedure suggested in [McLain 1974]. This method is known as the McClain type of interpolation. An associative memory based on this principle is called a MIAS (McClain Type Interpolating Associative Memory). The main problem is that all data points have to be stored and found. Therefore the memory capacity and the access time are increasing with the number of training points t. In table 1 the four methods mentioned have been summarized, while in table 2 their main characteristics have been shown. From this point of view most advantages are found in the field of AMS, with respect to the control applications. In the previous sections however we have seen advantages and disadvantages of both methods. The choice for a specific method depends on the type of application.

413

	local interpolation	global approximation
Classic	MIAS	polynomial approximation
Neural	AMS	backpropagation networks

Table 1: *Classical and neural models.*

	AMS	backpropagation networks	MIAS	multidimensional polynomials
training indication	yes	no	yes	no
calculation time per iteration	nearly constant	$O(n)$	$O(t)$	$O(\frac{(n+q)!}{n!q!})$
convergence speed	high	very low	high	low
calculation effort	low	medium	medium	high
memory consumption	high	low	$O(t)$	medium
noise filtering	possible	possible	difficult	possible
adaptations on scattered data	no	no	yes	no

Table 2: *Classical and neural models.*

4 Modelling and control using CMAC

In this section CMAC is introduced as a proper device for modelling and control of dynamical systems. As in the previous sections the characteristics of CMAC are described, the reader is assumed to be more or less familiar with the most important CMAC parameters.

4.1 CMAC as a modelling device

Modelling dynamic systems using neural networks needs a special configuration in which information about the history of the input and output of the plant is used. In figure 8 the general identification structure using neural networks is depicted. Of course many types of neural networks are suitable for use this configuration. The general objective of this configuration is to describe a general analytic system Σ with a nonlinear difference equation:

$$\Sigma : y(k+1) = f(y(k), \cdots, y(k-n); u(k), \cdots, u(k-m)) \tag{16}$$

in which $n+1$ is the order of the system, with $m \leq n$. If insufficient time-lagged values of $u(t)$ and $y(t)$ are assigned as inputs to the network, it cannot generate the necessary dynamic terms and this results in a poor model fit, especially in those cases where the system to be modelled is completely unknown. Therefore we have to use model validation techniques to be sure that the model developed is acceptable and accurate enough to be used in a control situation in which the control action depend on the model. Generally it is 'easy' (although time consuming) to train a neural network which predicts efficiently over the estimation set, but this does not mean that the network provides us with a correct and adequate description of the system. It is difficult to get analytical results but

414

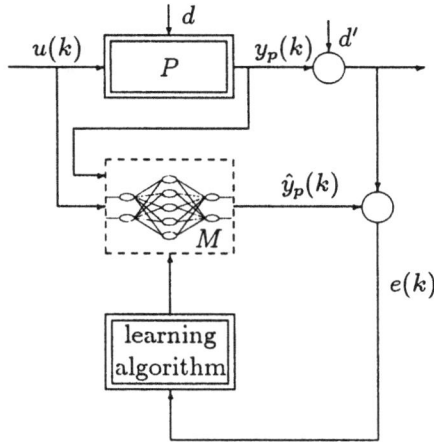

Figure 8: Plant identification

in [Billings 1992] some model validity tests are introduced which were proved to be useful experimentally.

To illustrate the (fast) learning of CMAC for output prediction an experiment has been carried out in which CMAC is used to predict 4 samples ahead. The process to be estimated is a second order process described by:

$$H(s) = \frac{K_{dc}\omega_n{}^2}{s^2 + 2\zeta\omega_n + \omega_n{}^2} \qquad (17)$$

The result of the prediction is compared to the actual output of the process after the prediction horizon, under the assumption that the control signal is kept constant. In the upper figure the output of the process and the reference signal are depicted, while in the other figure the predicted output is shown. Note the very quick learning procedure of this associative memory element

4.2 CMAC control

Using a neural network as a general 'black box' control component is extremely interesting. There are several possibilities for neural control, which can all be thought of as the implementation of internal model control (figure 10), in which a model M is used to control plant P using a controller C. The structure is completed with a filter F to map the control error into the space learned by CMAC.

- **inverse control:** the model of the plant to be controlled is used to train the controller in a so-called "specialized learning architecture". The model, obtained by observation, is used to generate error corrections for the controller. Under special conditions this kind of control can be guaranteed to be stable ([Billings 1992]): the

415

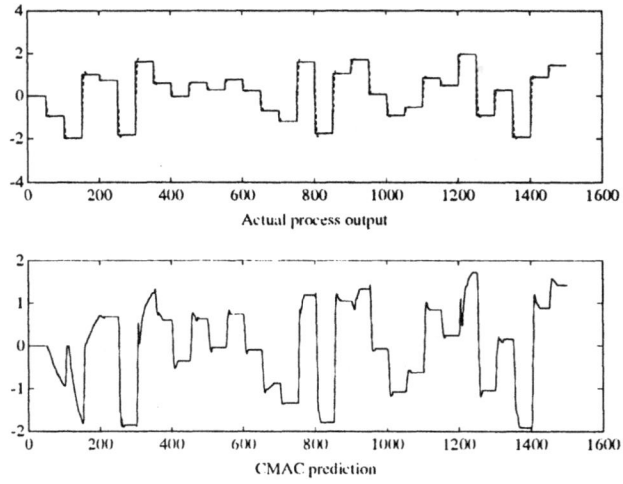

Figure 9: Prediction using CMAC

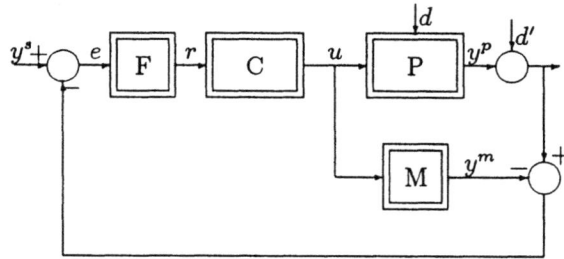

Figure 10: Internal Model Control

416

system has to be invertible and monotonic. In the next section an example of this kind of control using the inverse of a system is shown.

- **model-based control:** the model of the plant to be controlled is used in a conventional (e.g. predictive) control scheme to calculate the proper control actions. One of the possibilities is to implement an iterative method to calculate the inverse. Other possibilities are the use of a neural model as the basis of an on-line optimization routine to calculate the optimal control action.
  A special situation exists when the controller is also a neural network, while the neural model (emulator) is used to generate the Jacobian of the process. This gradient information is then used to adapt the controller. The method is known as the "error backpropagation" method [Narendra 1990], [Narendra 1991].

- **reinforcement control:** no neural model is derived and used but the controller is adapted according to a criterion function which evaluates the desired behaviour.

The effect of this kind of control is that limitations concerning the linearity of the system to be controlled are released. The choice of the best approach, and determining when the training procedure is completed are the main issues of concern.

A CMAC based controller

Once the inverse model of the process is learned, it can be used to control the process. Using CMAC two phases can be distinguished: the controller phase and the learning phase. During the controller phase a control signal U is retrieved from CMAC at the

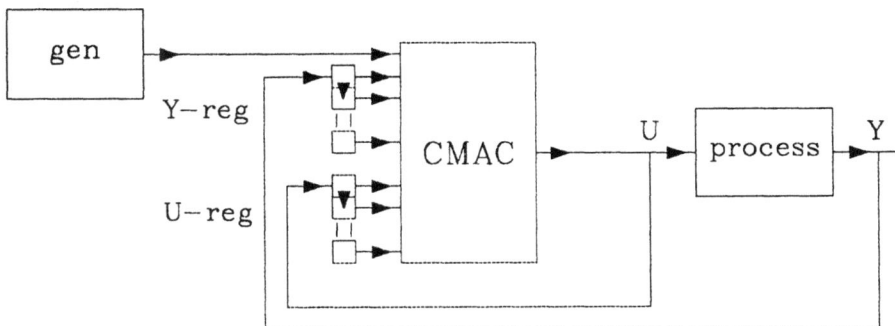

Figure 11: CMAC control configuration

current input state. This phase is depicted in figure 11 where 'gen' is the reference generator. In the learning phase CMAC stores the reaction of the process to the former control signal(s). The learning phase is for building an initial model and to adapt and tune the existing model in real time. This learning feature provides the controller with adaptive properties.

417

Problems using the CMAC controller

Using the former described CMAC controller depicted in figure 11, we are faced with a number of problems. Certain problems are CMAC specific, while other problems are related to learning in general.

Start up problem. Initially the CMAC table is empty so every retrieve operation yields an output value equal to zero. When CMAC is used as a controller, the proces is not controlled at all and the control signal is not very suitable for learning proces behaviour.

Learning in a subset of the input space. The part of the input space which is learned depends strongly on the characteristics of the process input signal. When CMAC is trained using an unsuitable input signal, it only learns control values in a small subset of the input space. Therefore, CMAC cannot retrieve a control signal at input vectors that are reachable (and occur during the control period) but lay outside this known subset.

Table size. The size of the table is limited because of the amount of memory available. The size strongly depends on the generalization number, input resolution and input dimension. An indication about the number of memory locations that are needed is given by equation 4: In this equation N is the input dimension and G is the generalization number. From this equation it can be derived that the size of the table increases exponentially with the input dimension. The order of the ARMA filter used to model the process has a strong influens on the table size. The modelling order is reflected in the length of the shift registers.

Input resolution. For a good control performance the input resolution must be as small as possible. A relative large resolution gives rise to a noisy control signal, while a relative small resolution causes impractical table sizes. A trade off has to be made between the accuracy and a practical size of the table.

Time delay. In order to model a process with a delay, the amount of delay must be known. When the time delay is known the controller configuration can be adjusted so the correct causal relations are modelled. Whenever the time delay varies or is unknown, the use of a CMAC controller is difficult.

A practical CMAC controller configuration

To extend the controller configuration for practical use, a number of problems must be solved. In figure 12 the complete configuration is shown.

The number of possible input combinations is limited by the introduction of a dynamic range. The dynamic range is defined as an upperlimit for the difference between successive samples of the process output, so it influences only the contents of the Y-reg in figure 12. Fast setpoint changes can easily result in input vectors outside this dynamic range. To avoid problems the setpointsignal must be preprocessed by the DRF (dynamic range filter) of figure 12.

418

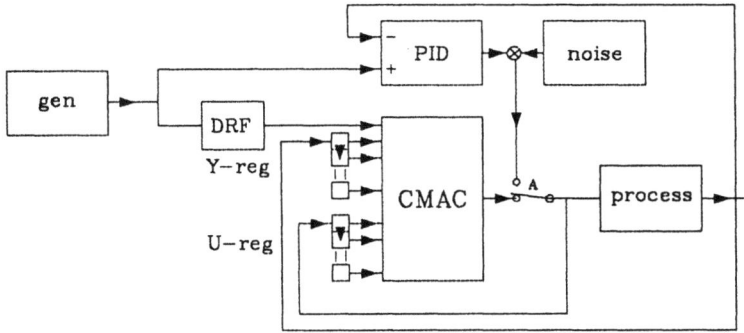

Figure 12: A practical CMAC controller configuration

The shift register of the process output is used in a different way. Only the first element is stored as the actual value, while the others are stored as the differences between successive signal values. The advantages are that a better quantization is obtained and that the differences relate directly to the dynamic range.

Different resolutions can be selected. first, a static resolution can be chosen for the first element of the Y-reg. Second, a dynamic resolution can be chosen for the difference components of the Y-reg. Different resolutions combined with the dynamic range, it is now possible to obtain a smooth control signal with a relatively small table.

In the control configuration a PID controller is used. This PID controller is meant to teach the CMAC table. Note that the CMAC is not modelling the PID controller! After the learning period control is switched over to the CMAC controller. In order to be independent of the dynamic characteristics of the control signal of the PID controller, probing noise is added.

Results

In this example CMAC is shown as a neural controller device for a nonlinear model. The configuration in which the controller is trained is based on the internal model control structure presented above. The system to be controlled is a nonlinear model of the pH value in a fluid. The model of this process is simplified and described by:

$$\frac{Y(s)}{U(s)} = k(y).\frac{1}{s} \tag{18}$$

$$k(y) = e^{-\alpha\|y\|} \tag{19}$$

First of all a model is derived via an identification procedure. After that an inverse control configuration is set up. In figure 13 the result of this nonlinear inverse controller is depicted. The behaviour of the controller depends on the settings of the CMAC module, especially the quantization interval.

419

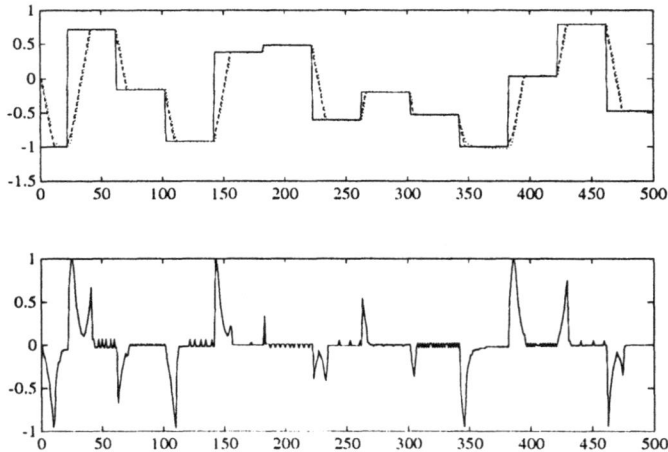

Figure 13: CMAC pH level control

5 Summary and concluding remarks

CMAC has proved to be a useful tool for modelling and control purposes. The parameters of the CMAC are of crucial importance for the correct functioning of this method. Associative memories are usefull modelling tools, including identification of nonlinear, unknown processes. Special attention must be paid to the choice of the CMAC parameters and the type of input-output mapping which is learned.

In control configuration it is very well possible to use associative memory elements (and other types of neural networks) to set up a model based control strategy, in order to overcome limitations in conventional approaches.

References

[Albus 1975a]   Albus J.S. (1975). *A New Approach to Manipulator Control: The Cerebellar Model Articulation Controller (CMAC)*. Transactions of the ASME, pp. 220-227, September.

[Albus75b]   Albus J.S. (1975). *Data Storage in the Cerebellar Model Articulation Controller (CMAC)*. Transactions of the ASME, pp. 228-233, September.

[Billings 1992]   Billings S.A., H.B. Jamaluddin and S. Chen (1992). *Properties of neural networks with applications to modelling nonlinear dynamical systems*. International Journal of Control, Vol. 55, No. 1, pp. 193-224.

[Cotter 1990]   Cotter N.E. (1990). *The Stone-Weierstrass Theorem and Its Application to Neural Networks*. IEEE Transactions on Neural Networks, Vol. 1, No. 4, pp.290-295, December.

420

[Cybenko 1989]   Cybenko G. (1989). *Approximations by superpositions of a sigmoidal function.* Mathematical Control Signal Systems, Vol. 2, pp. 303-314.

[Harris 1989]   C.J. Harris and C.G. Moore. *Intelligent identification and control for autonomous guided vehicles using adaptive fuzzy-based algorithms.* Engineering Applications of Artificial Intelligence, vol. 2, pp. 267-285, December 1989.

[McLain 1974]   McLain D.H. (1974). *Drawing contours from arbitrary data points.* Computing Journal, Vol. 17, 1974.

[Miller 1987]   Miller W.T. (1987). *Sensor-Based Control of Robotic Manipulators Using a General Learning Algorithm.* IEEE Journal of Robotics and Automation, Vol. RA-3, No. 2, April.

[Miller 1990]   Miller W.T., E. An, F.H. Glanz and M.J. Carter (1990). *The Design of CMAC Neural Networks for Control.* Yale Workshop on Adaptive and Learning Systems.

[Narendra 1990]   Narendra K.S. and K. Parthasarathy (1990). *Identification and Control of Dynamical Systems Using Neural Networks,* IEEE Transactions on Neural Networks, Vol.1, No. 1, March, pp.4-27.

[Narendra 1991]   Narendra K.S. and K. Parthasaraty (1991). *Gradient Methods for the Optimization of Dynmical Systems Containing Neural Networks.* IEEE Transactions on Neural Networks, Vol. 2, No. 2, pp. 252-262, March.

[Rosenblatt 1961]   Rosenblatt F. (1961). *Principles of Neurodynamics: Perceptrons and the Theory of Brain Mechanisms.* Spartan Books, Washington DC.

[Venema90]   Venema N. (1990). *Applications and practical aspects of the CMAC algorithm (in dutch).* Report A90.001, Control Laboratory, Fac. of El. Eng., Delft University of Technology.

[Vlothuizen88]   Vlothuizen W.J. (1988). *The CMAC Method.* Report A88.0 A88.071, Control Laboratory, Fac. of El. Eng., Delft University of Technology.

PART VI

SUPERVISORY CONTROL /

MONITORING /

OPTIMIZATION

Road Map.

The sixth part of the course deals with *Supervisory control / Monitoring / Optimization*. Basic knowledge on control engineering is expected. The application of artificial intelligence techniques in general to supervisory control will be introduced here. For an introduction to artificial intelligence techniques, the reader is referred to part I.

There are three chapters:

Supervised Adaptive Control. This chapter gives an introduction to supervised and adaptive control. The different components of the overall control system are introduced and the use of supervision in the controller is shown. Finally some attention has been given to implementation of the supervision function and the role that expert system techniques can play here.

Monitoring and Fault Diagnosis in Control Engineering. In this chapter another example is given were expert system techniques can be very helpful; monitoring and fault diagnosis. The layout and the architectures for knowledge bases for expert control systems are discussed. Then the requirements for a real-time expert monitoring system are given. The chapter concludes with a generic example of an expert system for fault diagnosis and applications in control engineering.

Expert Governed Numerical Optimization. The last chapter introduces numerical optimization techniques, that can be used at several stages of design activities. Here expert system techniques are used in combination with procedural programming languages to make numerical optimization more robust. From the user no longer a deep knowledge is required on the optimization techniques themselves, so that they can concentrate on their application.

This road map should enable the reader to make up his own choice in this basic course, depending on his skills and future use.

SUPERVISED ADAPTIVE CONTROL

F. Morant, M. Martínez and J. Picó
Dpto de Ingeniería de Sistemas, Computadores y Automática
Universidad Politécnica de Valencia.
Spain

1 Introduction.

In the last few years, there has been a great interest in order to apply the Artificial Intelligent (AI) techniques to process control. In the paper "Expert Control" (Åström, Anton & Arzen, 1986) is shown that when we are designing a control system, the code corresponding to the control system is a minimum part of the overall system. Handling of abnormal situations, communications, information treatment, starting-up and shutting-down the system,etc., are tasks that we have to consider in order to assure the good functioning of the overall system.

On the other hand, if we analyze the integral control of an industrial process, we can divide it in several levels:

1. **Controllers level.-** In this level we group all local controllers which are implemented using either simple control algorithms (PID) or more sofisticated algorithms (adaptive, GPC, etc).

2. **Controllers supervisor level.-** We have to deal with the supervision of local controllers. This task will be more or less complicated depending on the controller to be supervised.

3. **Plant supervision.-** This level requires a special treatment of the information coming from the lower levels, and carries out functions such as monitoring, displaying diagrams, fault detection and correction, operation aids, production optimization, which can include automatic generation of lower levels references.

4. **Management level.-** Where the production planning is carried out considering financial, safety and market aspects.

Analyzing each one of these levels and the tasks the levels have to realize, we come to the conclusion that the higher the level the less the importance of the algorithmic aspects in comparison to the heuristic ones. So, techniques characteristic of the AI seem more appropriate for the implementation of these levels. Even in the first level, where the local control of each process is undertaken, when the system to control is complex, with different functioning modes, with difficult access to the instrumentation and with the necessity of an expert in order to guarantee the good functioning of the overall system, the use of this kind of techniques is justified.

Let us consider the first and second level in an industrial complex system. Usually, when we are working with this kind of processes we have to deal with different sub-processes. Some of them are exactly known and its control can be done using traditional control techniques. Conversely, there are more complicated sub-processes such that its control have to be done using more sofisticated techniques (adaptive control, non-linear control, GPC, etc). Finally we can find sub-processes not exactly known, without an easy mathematical model, with a qualitative model, or even without any at all, such that the only way of doing its control is by using AI techniques.

So, these tools can be used both for the direct control of the processes aforementioned (first level) or for the monitoring and supervision of the classic control loops (second level).

In the supervisory level, the operator usually does some of the following tasks:

- Selecting the most appropriated controller and control structure.

- Tunin g the controller parameters.

- Changing the controller set points.

- Evaluating the performance of the controlled process.

- Starting-up and shutting-down the process.

- Performing some extra activities, such as addition of external excitation, computation of indicators, exchange of information between control loops, and so on.

- Logging of incidences and alarms.

Most of these activities are based on approximated knowledge, being suitable to be implemented using AI techniques.

As conclusion we can say that the use of AI techniques is justified in each one of the aforementioned tasks. In this chapter we will concentrate our attention in the use of AI techniques for the supervision of controllers and process model estimators in an adaptive loop (second level). The adaptive control is a well studied topic, and a number of available products can be found in the market. But, as it is well known, they present some problems for critical processes, i.e., processes with unknown delay, non minimum phase, variable noisy environments and so on. Some of these problems can be overcome through the appropiate selection of algorithms and tuning parameters, selections mostly based on approximated knowledge and where AI techniques are well suited for.

In an adaptive scheme the following basic blocks may be considered:

- **Parameter estimation module:** It looks for the most appropriated parameters, within a family of models, to fit the process dynamic behavior.

- **Controller design module:** According to the stated objectives and the process model previously obtained, offers at least one control law. Usually, the control solution is not unique and additional constraints, goals or requirements, will decide for the most suitable one.

- **Closed loop evaluation module:** It verifies the fulfillment of control specifications, looking for abnormal situations, such as bumping, hidden oscillations, long term drift, etc.

At the lower levels, the problems can be handled algorithmically. A number of indicators, such as prediction error, trace of estimated covariance matrix, poles location, cost indices, among many others, may be evaluated. These indicators, together with some experimental knowledge, should be used at the top level. The use of heuristics will lead to the best option at the right time.

In the following we are going to study how the second level (supervisor level) can be designed using AI techniques and we will use these conclusions in order to develop an adaptive control supervisor.

2 Adaptive Control.

Adaptive control is a kind of non linear feedback control in which the process states may be divided into two categories as a function of their change velocity. The states which change slowly are considered as the system parameters modeling the changing properties of the controlled process and its signals. The control system adapts its behavior to such changes.

The three basic schemes devised for the adaptive control of processes are (Åström & Wittenmark, 1989):

- Gain scheduling or feedforward adaptation

- Model reference control

- Self-tuning regulators

In all them we start up with an ordinary feedback control loop with a process and a regulator with adjustable parameters. The goal is to find out a suitable way to modify the regulator parameters in response to changes in the process and in the disturbance dynamics. The three schemes differ in the way this adaptation is carried out.

2.1 Gain Scheduling.

If changes in the process behavior are well correlated with auxiliary measurable variables, and if we know in advance how we can adjust the regulator parameters as a function of these auxiliary variables, a feedforward adaptation may be used (Åström & Wittenmark, 1989). The scheme, depicted in figure 1, has the advantage of adapting very quickly, but it also presents some important drawbacks:

1. Normally, a good theoretic knowledge of the process is required to find the auxiliary variables.

2. Feedforward adaptive systems are difficult to design, since the regulator parameters must be determined for a great number of operating conditions.

3. The main drawback lies on the open loop compensation nature of the scheme. No feedback compensating for an incorrect following.

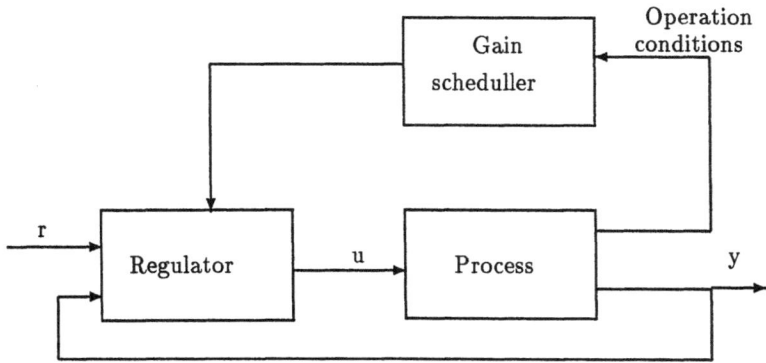

Figure 1: Gain Scheduling Adaptive loop.

2.2 Model Reference Adaptive Systems.

In this adaptive scheme, depicted in figure 2, the required information about the process is obtained through the determination of the closed loop behavior in comparison with that of the model reference desired. This last giving us how the process output should ideally behave in response to the input signal.

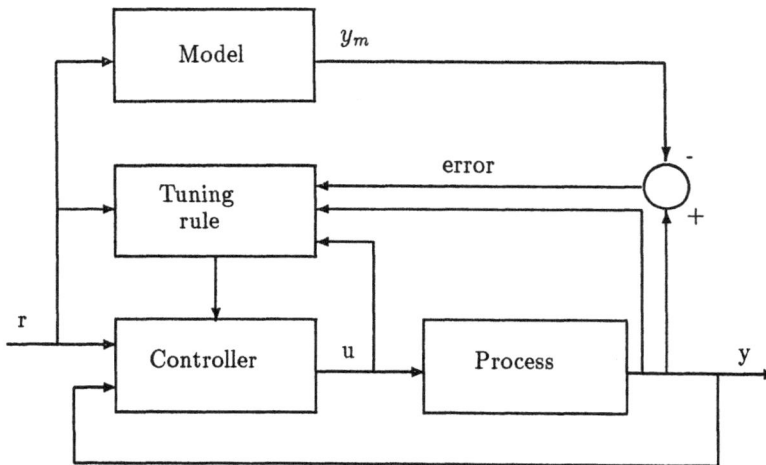

Figure 2: Model Reference Adaptive System.

The regulator may be viewed as a two loop one. The inner loop is an ordinary control loop, with the process and the regulator. The outer one adjusts the regulator parameters in such a way that the error between the model output (y_m) and the process output (y) is minimized.

2.3 Self-Tuning Regulators.

The basic scheme may also be viewed as an inner and an outer loop (figure 3).

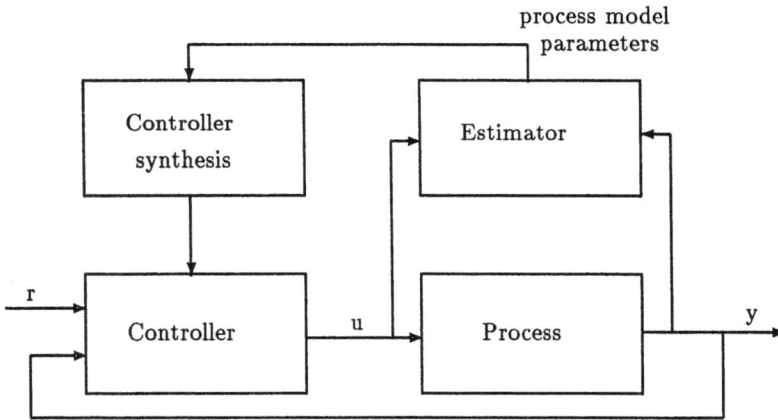

Figure 3: Self-Tuning Regulator.

The outer loop, which adjusts the regulator parameters, is made up of a process parameters estimator and a regulator parameters calculus block. Two different approaches can be followed for this calculus, giving rise to two self-tuning regulator schemes (Isermann, 1981):

Implicit: process parameters are not estimated, but directly those regulator.

Explicit: First, a parameterized model of the process is estimated, and then the regulator parameters are explicitly calculated as function a of the process ones.

This latest scheme has been widely used because of its flexibility, easy design and good behavior under the conditions that will be explained later.

The self-tuned adaptive controllers may be also classified (Isermann, 1982) according to:

1. *Process model:* Implementation on digital computers leads to discrete process models, mainly parametric models because the availability of appropriate powerful controller design methods.

2. *Process identification method:* Recursive parameter estimation methods, which can be framed under a generic estimation scheme, are preferably used for on-line identification.

3. *Information about the process:* If the estimated parameters are assumed to equal the real ones, i.e. no uncertainty is considered, the so called certain equivalence controllers are obtained. Conversely, cautious controllers are obtained.

4. *Controller design criteria:* The behavior of adaptive control systems based upon process identification leans heavily on the identification quality, and hence, on the process input signal. Dual controllers determine the control signal in such a way that a good present and future identification can be obtained. Non dual controllers only use present and past information about the process in the design criteria.

5. *Control algorithms:* Adaptive control algorithms must satisfy:

431

(a) closed loop identifiability conditions.

(b) applicability to different process and signal classes.

(c) low computation cost.

2.4   Self-tuning system example.

In the following paragraphs, some well known basic expressions for a recursive estimator
are given as reminder.

Let us consider a SISO stable process defined by the following difference equation:

$$A(z^{-1})y(k) = B(z^{-1})z^{-d}u(k) + v(k) \tag{1}$$

where $y(k)$, $u(k)$ y $v(k)$ are the input, output and disturbances at instant k and A
and B are polynomials defined by:

$$\begin{aligned} A(z^{-1}) &= 1 + a_1 z^{-1} + a_2 z^{-2} + ... + a_n z^{-n} \tag{2} \\ B(z^{-1}) &= b_1 z^{-1} + b_2 z^{-2} + ... + b_m z^{-m} \tag{3} \end{aligned}$$

The disturbances are described by the stochastic process:

$$v(k) = H(z^{-1})e(k) \tag{4}$$

where $H(z^{-1})$ is a rational function, and $\{e(k)\}$ is a white noise with:

$$\begin{aligned} E[e(k)] &= 0 \tag{5} \\ E[e(k)e(k+T)] &= \sigma^2 \delta(T) \tag{6} \end{aligned}$$

Many methods can be used for the recursive estimation of the model parameters,
among them (Ljung & Soderström, 1983):

1. Recursive Least Squares: RLS.

2. Extended Least Squares: RELS.

3. Instrumental Variable: RIV.

4. Maximum Likelihood: RML.

5. Stochastic approximation: STA.

6. etc.

The choice one or another depends on the nature of H. For instance, if $H(z) = 1$, the
Least Squares method gives good results for the estimation of A and B. In this case, the
process can be written as:

$$y(k) = \alpha^T(k)\theta(k) + e(k) \tag{7}$$

where:

432

$$\alpha^T(k) \quad = \quad [-y(k-1), .., -y(k-n), u(k-d-1), .., u(k-d-m)] \tag{8}$$

$$\theta \quad = \quad [a_1, .., a_n, b_1, .., b_m] \tag{9}$$

Is well known that the least squares algorithm provides the estimated parameters θ at instant k so that the criterion:

$$L(\theta, k) = \sum_{i=1}^{k} [y(i) - \alpha^T(i)\theta(i)]^2 \tag{10}$$

is minimized. The expression for the estimated parameters is (for the non recursive version):

$$\theta(k) = [\Psi^T(k)\Psi(k)]^{-1}\Psi^T(k)Y(k) \tag{11}$$

where:

$$\Psi(k) \quad = \quad (\alpha(1), \alpha(2), ..., \alpha(k))^T \tag{12}$$

$$Y(k) \quad = \quad (y(1), y(2), ..., y(k))^T \tag{13}$$

and its recursive form with forgetting factor λ is:

$$\theta(t+1) \quad = \quad \theta(t) + K(t+1)[y(t+1) - \alpha(t+1)^T\theta(t)] \tag{14}$$

$$K(t+1) \quad = \quad \frac{P(t+1)\alpha(t+1)}{\lambda + \alpha(t+1)^T P(t)\alpha(t+1)} \tag{15}$$

$$P(t+1) \quad = \quad \frac{1}{\lambda}\left[P(t) - \frac{P(t)\alpha(t+1)\alpha(t+1)^T P(t)}{\lambda + \alpha(t+1)^T P(t)\alpha(t+1)}\right] \tag{16}$$

It is important to remember that all the aforementioned estimators can be described in terms of a set of equations which basically follow the same general structure of (14), (15) and (16).

Once the process model parameters have been estimated at some time instant k, they can be used for the synthesis of the controller.

3   Adaptive Control and Supervision.

Practical experience with adaptive controllers has shown limitations in their applicability to real processes, due to problems arising in the system (Isermann & Lachmann, 1985) (Albertos, Martínez & Morant, 1989). Undesirable phenomena, like large variations in the control actions during the adaptation phase or parameter estimates bursting, are not unusual when the adaptive systems freely evolve. These problems arise specially because of violations in the pre-conditions for the process parameters estimation, and because the need of a priori setting some system design factors, such as sampling rate, process model structure, etc.

These and other problems show the need of a third level in the adaptive scheme, the so called supervision level, which monitors the evolution of the whole system, taking the appropriate actions when so needed.

In the next paragraphs we will briefly introduce the main problems we may find in an adaptive system and the tasks to be made by the supervisor level. We will focus our attention on the explicit self-tuning regulators, since they are widely used, and in them we can see most of the problems arising in any adaptive scheme. The basic tasks performed by the supervisor will be shown, and the concept of supervisor indicator introduced. In later sections we will come through an example and we will review the implementation of the supervisor using AI technics.

The supervisor tasks can be grouped into four main sets (Isermann & Lachmann ,1985) attending to the system part they affect to:

1. Starting-up or pre-identification.

2. Parameters estimation.

3. Controller synthesis.

4. Loop supervision.

3.1   Starting-up or pre-identification.

Strictly speaking, this is not a supervisor task, but it will give some design factors which will be supervised on-line (for instance, the model order and delay estimates), or will be used by the supervisor.

If the adaptive controller was applied from the very beginning, we would face a lack of process parameters estimates, which in turn would produce undetermined and possibly unacceptable control actions. So, a previous estimation of the process model order, delay and parameters is made in open loop.

Model order estimation may be done by several methods, including:

1. Cost function test.

2. Poles-zeroes cancellation test.

3. Residues independence test.

4. Determinants ratio test.

Determination of the model delay is usually carried out either by the cost function test or by using a model with expanded numerator, the delay being the number of terms in the numerator which can be neglected (Isermann & Lachmann, 1985).

Once the process model structure is determined, the disturbances model structure must be tested. The noise auto-correlation vector is calculated, the noise being estimated as the prediction error of the identification procedure. If the disturbance present in the process can be modeled as a white noise, then:

$$cor[e(k), e(k)] = \sigma_e^2 \neq 0$$
$$cor[e(k), e(k + \tau)] \approx 0 \quad ; \tau > 0 \tag{17}$$

If the noise is coloured, there will be ν terms in the second expression far from being zero, ν being the order of the coloured noise which modelizes the process disturbance.

3.2 Parameters estimation.

The estimation methods constitute the center of the self-tuning regulator. The reliability of the controller leans to a great extent on a correct estimation (Schumann & others, 1989). Aspects to be looked after include:

3.2.1 Signal and estimates filtering.

Input and output signals must be filtered so that the estimation does not become noise corrupted.

For instance, to avoid that abrupt sudden disturbances corrupt the estimation, the expected process output may be used. The process output and the current model output for the current input are compared. If the difference appears to be unacceptable, the expected output instead of the real one can be used. Obviously, a sudden change at the process output may have been origined by a real sudden change in the process parameters, with the process parameter estimates having not yet converged to the real ones. The origin of the difference should be detected or, as typically we will have to do with most supervision tasks, a trade off must be established between the rejection to sudden abrupt output signal changes and the estimation reaction rapidity facing sudden changes in the process parameters.

3.2.2 Estimator suitability to signals excitability.

Persistent loop excitability is needed for a correct estimation. This persistence is guaranteed by a control signal which sufficiently excites all the dynamic process modes. If this condition fails to be accomplished, for instance when the controller is well tuned, linearly dependent rows appear in the estimator variances-covariances matrix. So, this matrix will become singular or nearly singular in practice, and the estimator equations will become ill-conditioned.

When forgetting factors are used, this problem can be seen from another point of view. From the equation of the inverse variance-covariance matrix (16) , if no new significative information enters to the system the regression vector will contain negligible terms, so:

$$P(k + 1) \approx \frac{P(k)}{\lambda} \tag{18}$$

P will grow exponentially if $\lambda < 1$, and from equations 14 and 15, the parameter estimates will have large excursions, they will *burst*. So, if no new significative information is present in the data we could stop the estimation.

If new significative information is available, this must be detected, usually by using functions of the prediction error variance and/or average.

435

Estimation switching on and off can not face the non uniformity of the excitability in the parameters space. This fact produces different behaviors of the different parameter estimates. So, for instance, bursting can be localized only in some parameters. To tackle with this problem several methods have been devised, including variable and/or vectorial forgetting factor (Fortescue & others, 1981), direct actuation over the elements of the matrix P or weighting of additional terms (usually regarding known statistical properties of the process parameters) in the cost indice to be minimized by the estimator (Simó et. al. , 1981).

3.2.3 Algorithm exhaustion.

From equation 16 and assuming $\lambda \approx 1$, since the second term of the right side is quadratic, if the system is properly excited, the values of the P elements will be decreasing . This means that the estimation algorithm will extinguish, making impossible further time varying parameter estimations. For the correct tracking of variable process parameters a minimum estimation energy must be kept in the algorithm.

This effect may be monitorized through the P matrix trace. When its value is under certain threshold two basic actions can be taken:

1. **Matrix P reinitialization.** This solution has the disadvantage of modifying the algorithm search direction, with corresponding transients.

2. **Forgetting factor modification.** This modification can be progressive, thus not producing rough transients.If the trace of P is too high, with the corresponding parameters bursting danger, the forgetting factor λ should be increased. Conversely, decreasing λ increases the sensitivity of the algorithm to the new data. One possibility related is the use of constant trace estimation algorithms (Fortescue et. al. , 1981).

3.2.4 Forgetting factor and parameters bursting.

The use of forgetting factor to *"give energy"* to the estimation algorithm may produce parameters bursting if not carefully used. As said before, a small continuous modification of λ forces the algorithm into a search direction which, in general, does not coincide with the natural one that the algorithm would take. This makes worse the estimates, but without the large transients produced by P reinitializations or by large discrete variations of λ.

Anyway, we will have to trade off between the prevention of large estimated parameters variance (P small, λ large), and the maintenance of enough estimation energy (P large, λ small).

3.2.5 Sudden changes in process parameters.

This is not an unusual situation in real processes. We may think for instance in process configuration changes (opening or closing valves,...). Besides, real processes are almost never absolutely linear. The non-linearities produce sudden changes in the linearized process model parameters when faced to step reference changes.

The prediction error can be used to inform about sudden process parameter changes, the traditional action being taken by the supervisor, by decreasing the forgetting factor.

436

3.2.6 Identifiability conditions.

When working in closed loop, the orders of the estimator and controller polynomials must satisfy certain relationships for the estimated parameters to converge to the real ones (Isermann, 1981).

3.3 Controller synthesis.

The goal here is to avoid the synthesis of a new controller (either by its parameters change or by its structural change including the change of controller type), when this may degrade the control.

Among the functions associated with this supervision task we can find:

3.3.1 Poles and zeroes test.

Both process and disturbance model poles and zeroes are calculated. Depending on them different control strategies will be applied. For instance, minimum variance controllers with control action weighting must be used if the process is a minimum phase one, etc.

3.3.2 Stability test.

Before updating the controller parameters, the closed loop characteristic equation should be solved in order to determine the stability of the system. Parameters such as the control action weighting in minimum variance controllers can be modified accordingly by the supervisor.

3.3.3 New controller parameters usefulness verification.

The controller synthesis should be avoided if the new estimated process parameters have not varied appreciably. It should be taken into account that if the controller is updated too often, the control signal will be too affected by the estimation imperfection in spite of using filtered estimates. If the variance of these last overcomes certain threshold the controller parameters are not changed. Other criteria are based on the comparison between the prediction error with the current process estimated parameters and the new ones.

3.3.4 Model quality analysis.

The decision upon the controller updating may be based on the improvement of the loop behavior with the new parameters (Isermann & Lachmann, 1985). If so done, the behavior for the two models:

- The identified

- The one used to synthesis the current controller

are compared.

The behavior indicator can be the loop error, i.e. the difference between the desired reference and the process output.

3.4 Loop supervision.

The aim of this task is the detection of unstabilities in the process input and output signals. If required, the loop control is transferred to the back-up controller.

4 Indicators: a Tool for Supervision.

4.1 Introduction.

As it has been seen in the previous section, the third level, the so-called supervisory level, has two different goals 1) to check on-line the aforementioned conditions and 2) to take the appropiated actions in order to assure a good functioning of the overall system when some of these conditions fail. In this section we analyze the information set needed at this level in order to take the appropriate actions.

The concept of index or indicator, as the element capable of detecting these violations, is introduced (Isermann & Lachmann, 1985) and justified because of the impossibility of getting general solutions to the problems of stability and robustness of the adaptive algorithms.

Furthermore, the special characteristics of industrial environments, with the presence of uncertainty and imprecise knowledge, make especially useful the so-called heuristic indicators, due to the easeness for interpreting high level qualitative aspects within the supervision methodology.

4.2 Study and Selection of Indicators for Supervision

A supervisory kernel, taking the signals coming from a control loop (reference, output, control actions...), calculates the indicators associated to the adaptive control.

These indices or indicators may be associated in groups related to:

1. Start-up procedure: pre-identification and model verification.

   - Sampling time test.
   - On-line estimation of order.
   - On-line estimation of delay.
   - Unmodelled dynamics test.

2. Parameter estimation.

   - Prediction error indicators.
   - V-C matrix indicators.
   - Estimated parameters indicators.

3. Controller design.

   - Indicators for the poles and zeroes of the model.
   - Loop error indicators.

4. Closed loop.

- Output signal indicators.

- Characteristic equation indicators.

Each one of these groups incorporates statistic indicators (means, variances, covariances, norms, trends, accelerations, etc...), being possible to establish which ones give more information about the violations outcoming during the control. In Appendix A, a complete description of numeric indicators is presented.

4.3   Evaluation of Indicators

The validation of the indicators was studied through a simulation package allowing for different estimation and control combinations in closed loop, in addition to processes and signal modeling simulation.

In this section we show some results that allow us to connect the numeric information with heuristic information. A complete study of these relations is in (Martínez, 1991).

Order test:

A quadratic function of the prediction error is used in order to test the more suitable process order estimation.

In figure 4 a stable overdamped second order process is tested. Three estimators assuming different process orders run simultaneously. As shown, the index for an assumed first order process blows up, and only the corresponding to second and higher orders keep bounded. This implies that all least a second order should be considered for the process. A subsequent test for the delay will let us to decide upon the correct order.

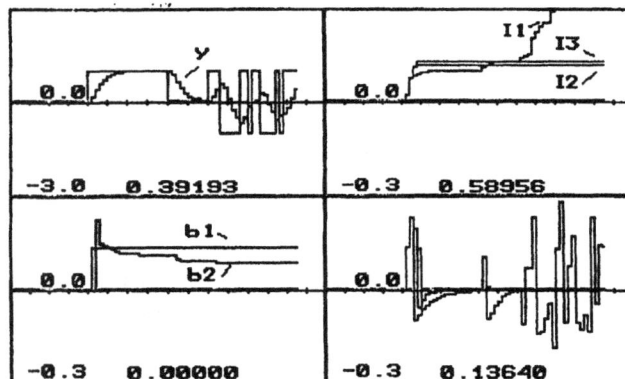

Figure 4: Process order test.

Delay test:

Several tests for the time delay estimation are currently used in the literature. Among them, the easiest to implement consists of an estimation algorithm in which a model with an expanded numerator is estimated.

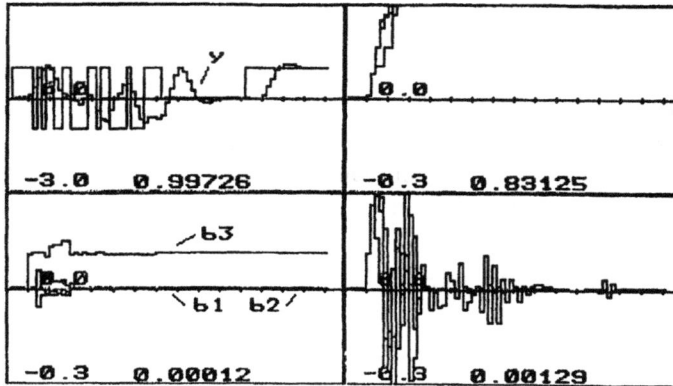

Figure 5: Process delay test.

In figure 5, a third order process with a time delay equivalent to three sampling periods has been used. The three first values of the coefficients b_i estimated with an increased order estimator are almost negligible, which indicates the presence of the afore mentioned delay. In this case, the parameters estimation does not work properly due to the use of an expanded estimator without delay. The delay detection will show us which estimator has to be used afterwards.

Unmodelled dynamics:

The prediction error autocorrelation is used as an indicator for detecting the presence of unmodelled dynamics. As known, if the prediction error can be modelled as a white noise, the autocorrelation factor will be zero for those signals shifted by one or more sampling periods. If it can be modeled as a coloured noise, as many factors as the order of the corresponding filter will be different from zero (figures 6 and 7).

Process parameters change:

In order to detect changes in the process parameters, the cross-correlated factors between the prediction error and the control action and process output signals respectively can be used.

These indices inform us quickly, not only about a generic parameters change, but also (under some assumptions) about which specific parameter has changed. This is very important, since allows taking specific actions for the parameter or group of parameters under change, while untouching the others.

When the estimation is undertaken in closed loop, the information given by these indicators is not so clear as in the open loop. This is specially true for the process denominator coefficients, due to the feedback (the output an implicit and explicit function of the parameters), which makes difficult to specify which one is the changing parameter. Yet, under a qualitative point of view, where same imprecision may be admitted, it is possible to handle the information given by these indicators (figure 8).

440

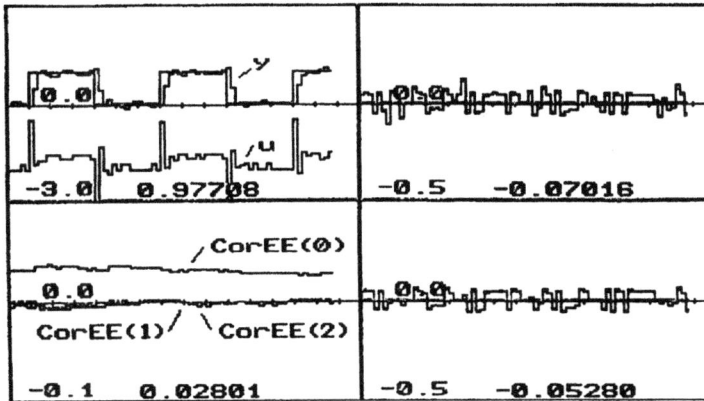

Figure 6: Unmodelled dynamics structure. Second order overdamped process. RLS. 1) Ref.,input,output. 2) A priori prediction error. 3) Autocorrelation $\Phi_{ee}(\tau)$ coefficients. 4) A posteriori prediction error. The only autocorrelation coefficient different from zero is $\Phi_{ee}(0)$.

4.4 Minimum set.

The selection of a minimum set of indicators that represent the process state depends on the individual viewpoint and on the process to be controlled.

In table 1 we describe the Numeric Indicators that have been studied in the evaluation step. In the same way, table 2 shows all the Heuristic indicators analyzed. Of course there is a relation between both kind of indicators, this relation being expressed in table 3. In the last table, we can see that certain numerical indicators supply similar information about the same heuristic aspects whereas others supply complementary information. We propose a minimum set of indicators (table 4) which is more extensive that the set propose by (Isermann & Lachmann, 1985).

4.5 Functions and Tasks of the Supervisory Level.

Once the missfunctioning has been detected by the indicators (either individually or together), we need to correlate their response with some function which takes actions in order to correct the problem.

In practice these supervision functions can be classified in diferenciated groups according to their actions (Martínez, 91):

1. Updating of filtering factors:

   o of signals (inputs and outputs)

   o of parameters

2. Fitting process structure changes:

   o increment/decrement of model delay

   o increment/decrement of model order

441

o change of estimation algorithm

3. Handling of process parameter changes:

   o variable forgetting factor
   • estimation algorithm re-initialization

4. Ensuring identificability conditions:

   o fixed forgetting factor
   o estimator switch on/switch off
   o extra signal addition

5. Controller choice/tuning:

   o choice of compatible controllers
   o scheduling of the design factors
   o change to back-up controller

6. Ensuring loop stability:

   o robust design
   • back-up controller change

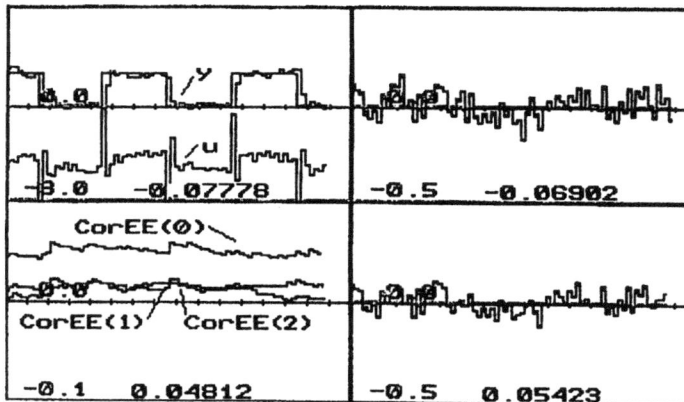

Figure 7: Same process as before, but with a coloured perturbation.

Each supervisory function is the response of the supervisory system to the anomaly which may appear during the operation conditions and are detected by the indices. One single index or some together can detected the anomaly. The information of the indices set determines the state of the overall system and, accordingly with this state, the more appropiated functions are activated.

442

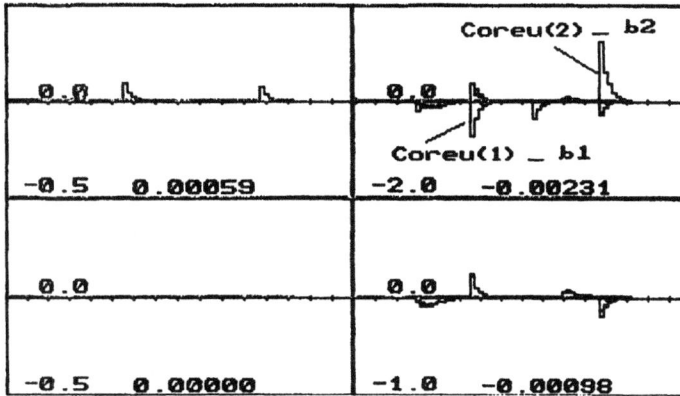

Figure 8: Cross-correlation factors evolution when b_1 and b_2 change. 1) $\Phi_{ee}(\tau)$ 2) $\Phi_{eu}(\tau)$ 4) $\Phi_{ey}(\tau)$.

Supervisory functions together with their suitable indicators have been studied, each supervisory function being associated to a state. For instance:

1. **State:** process structure change.
   Indices:
   - poles and zeroes cancellation.
   - prediction errors.
   - test for time delay detection.
   Function:
   - model structure change.
   - controller structure change.

2. **State:** parameter process change.
   Indices:
   - prediction error correlation.
   - estimated parameters norm.
   - estimated parameters variance.
   Function:
   - forgetting factor modification.
   - estimation algorithm starting.
   - parameters filtering.

3. **State:** Signal excitation.
   Indices:
   - V-C matrix trace.
   - V-C matrix trace trend and acceleration.
   Function:
   - forgetting factor modification.

443

- extra control signal.

The indicator which is more suitably associated to a supervisory function depends on the process to control as well as the environment where the overall system must work.

5   Implementation of Supervision Functions.

5.1   A Classical Approach.

Modern programming languages are very powerful and flexible, embodying useful aids for programming complex functions due to the powerful data structures, real time programming, easy hardware level programming (peripheries), and storing and graphic capabilities that they incorporate.

Therefore, it is possible to implement a supervisor using one of such languages, fitted with a set of additional capabilities, such as the aforementioned, making possible the experimentation with the developed algorithms in a real time environment and on real processes. Furthermore, it is possible to compare the efficiency of these algorithms when distributed on different computers with a suitable communication.

Since these tools allow for different approaches, we have studied various supervisor schemes. There exist a quite generalized tendency to implement supervisors under the form of an unique procedure, in which all supervision functions are grouped. This, usually lets an easier programming, yet practice demonstrates that some supervision functions should be run at specific points within the control loop and immediately (in time). The mentioned scheme hardly held this kind of imperatives.

Then, it seems more logic to design the algorithm that embody the supervision functions in such a way that it can stand alone. Its grouping (generally under the form of concurrent or parallel processes) could then obey to the temporal limit in which each function should respond.

5.2   An Artificial Intelligence Approach.

Numeric information can be converted to heuristic one using some technics (fuzzy theory). If besides, we consider the natural presence of heuristics in the control, moreover at industrial environments, it is reasonable to conclude that Artificial Intelligence tools, and particularly the Expert Systems, will be very suitable to handle this information and to solve the supervision problems. Within this new point of view still it is needed to solve many problems. Expert Systems have been up to now mainly developed for off-line applications, where time is not a critical magnitude, conversely to real time systems situation. An additional problem comes out with the data consistency when examined by an ES (Gerhardt, 1986).

As for our application concerns we find that programming languages dealing with E.S. are oriented to conventional problems, being little versatile and with very limited mathematical capability. That is why additional programming efforts are necessary to get E.S. environments by general purpose languages.

In spite of the difficulties we think that this is the correct research line, because an E.S. with its constitutive elements (knowledge base, inference engine, data base) appears to be the best way to model the big quantity of imprecise and vague knowledge present in the industrial processes to be controlled (Morant et. al. , 1989), (Morant et. al. , 1992).

In figure 9 we can see the control structure we have used.

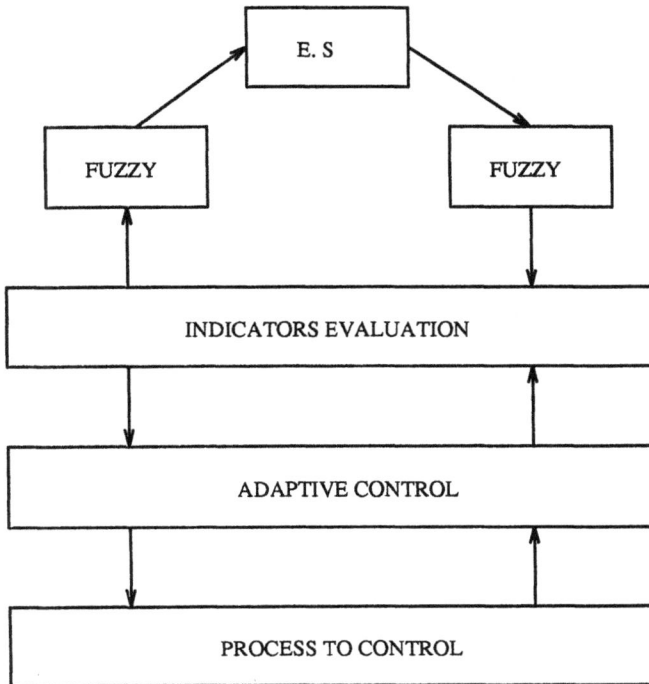

Figure 4: Control Structure.

5.2.1 Numeric-heuristic conversions.

ES deals with qualitative heuristic information, therefore we must regard the different supervision indicators in an heuristic fashion. But, data generated in the control loop are basically of numeric nature. The technics of the fuzzy logic can be used to make the translations between the two contexts (figure 10).

5.3 Expert System Evaluation.

The experience obtained working with Adaptive Control in industrial environments permits to conclude some practical aspects dealing with the implementation of this kind of control. We have analyzed the most appropriate index set to carry out the supervisory functions.

Based on this idea, an expert system prototype for real time control, which supervises the correct behavior of the adaptive loop has been developed. The use of a processes simulator allows for the validation and adjustment of the expert system. Some of the most outstanding results are showed in this section.

Forgetting factor scheduling:

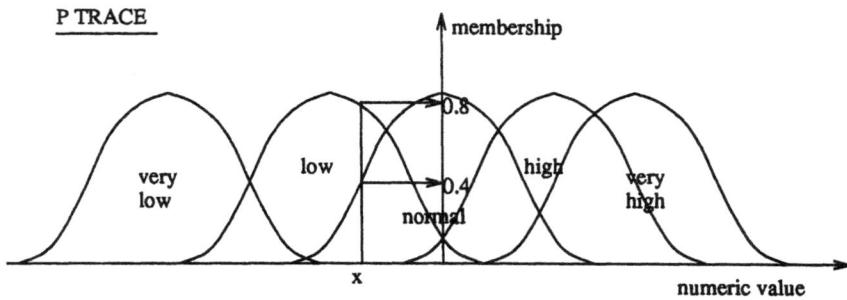

P TRACE

Numeric value: x

Heuristic values: Trace P = low, m=0.8
                   Trace P = normal, m=0.4

Figure 5: Numeric to heuristic conversion.

The E.S. changes the value of the estimation forgetting factor to cope with different specified events. These events are detected by the heuristic indicators:

1. parameters blowing-up.

2. algorithm extinction.

3. lack of excitation.

4. parameters change.

5. parameters convergence.

which are then used by a set of rules in the E.S. The state of the heuristic indicators is determined by the corresponding numeric indicators in a monitoring table, and translated through an events table using fuzzy logic (figure 11).

Estimator scheduling:

The E.S. tries to detect an insuitable choice of the estimator type. Two estimators with wide applicability, a RLS and a RELS, run in parallel, and rules based on the corresponding autocorrelation of the prediction errors are used to change of estimator (figure 12).

446

NUMERIC INDICATORS	
RELATED WITH THE ESTIMATION	
Related with the prediction error	A Priori error
	A Posteriori error
	Mean/variance error
	Trend and acceleration of the error
	Correlation function
	Crosscorrelation u/e, y/e
	Error cost function
Related with the v-c matrix	v-c matrix
	Inverse of the v-c matrix
	Trace of the In. v-c matrix
	Trace/acceleration of the v-c trace
Related with the estimated parameters	mean of the parameters
	Variance of the parameters
	Norm of the parameters
	Trend of the norm
	Acceleration of the norm
RELATED WITH THE REGULATORS	
Related with poles and zeroes of the model	Poles and zeroes : Modula, arguments, real, imaginary part
	Trend of the poles and zeroes modula
	Acceleration of the poles and zeroes modula
Related with the error	Error
	Mean of the error
	Variance of the error
	Trend of the error variance
	Acceleration of the error variance
RELATED WITH THE CLOSED LOOP	
Related with the characteristic equation roots	Ch. eq. roots :
	Modula, argument, real and imag. parts
	Trend of the modula of the ch. eq. roots.
	Acceleration of the ch. eq. roots
Related with the output signal	Mean, variance, trend of the output
	Acceleration of the output signal
	Saturation test

Table 1

NUMERIC	HEURISTIC INDICATORS
ESTIMATION	
Numeric indicator	Heuristic indicator
Prediction error index Cancelation test of poles and zeroes Delay test	Model structure change
Cross-correlation function of the prediction error/input	Process zeroes change
Cross-correlation function of the prediction error/output	Process poles change
Mean and variance of the prediction error (trend and aceleration)	Convergence speed of the parameters
Correlation function of the prediction error	Unmodelled structure
Matrix P trace (trend and acceleration)	Signal excitability bursting and energy of the algorithm
Prediction error Norm (trend and acceleration) and variance of the parameters estimated	Change of the parameters
REGULATOR CALCULATION	
Numerical indicator	Heuristic indicator
Equation roots of the process model	Incompatible designs Unstable roots
Root speed of the charact. equation Loop error (acceleration and speed)	Sliding towards incompatible designs A posteriori fitting design
CLOSED LOOP	
Numerical indicator	Heuristic indicator
Ch. equ. roots Saturation test of the control action Output value (mean and variance)	Closed loop stability

Table 2

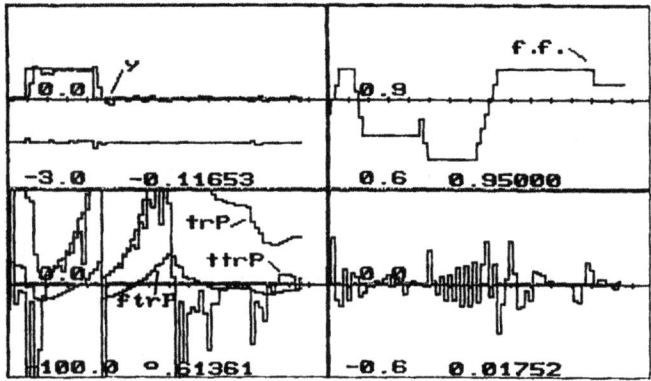

Figure 11: Forgetting factor scheduling.

HEURISTIC INDICATORS	
Related with the pre-identification	
	Inadequate sampling rate Wrong election of unmodelled dynamics Presence of unmodelled dynamics
Related with the estimation	
	Change of the model structure Change in the parameters of the model Convergence of the parameters of the model Unmodelled dynamic structure Estimation bias Energy of the estimation algorithm Bursting of the parameters estimates Input signal excitability Closed loop stability Presence of disturbances
Related with the regulator	
	Position of the poles of the model Position of the zeroes of the model Trend of the poles change Trend of the zeroes change Set of admissible regulators
Related with the closed loop	
	Ch. equation roots position Trend of the roots Control signal saturation

Table 3

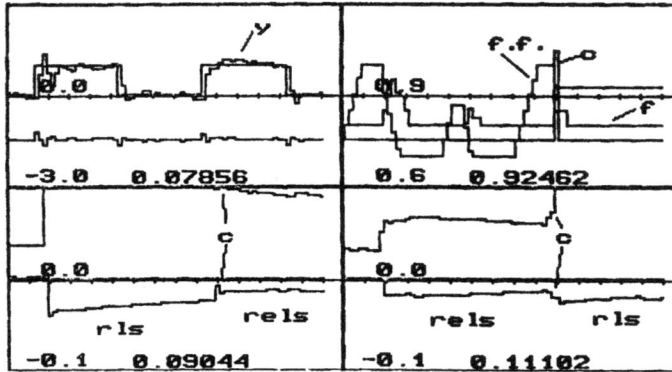

Figure 12: Estimator scheduling.

HEURISTIC INDICATOR	NUMERICAL INDICATOR
Validity of the process order	Prediction error index
Validity of the process delay	Delay test
Parameters process change	Norma/trend of the parameters
Zeroes process change	Cross-correlation function p. error/input
Poles process change	Cross-correlation func. p. error/output
Convergence of the parameters	Variance of the prediction error
Estimation bias	Correlation function of the prediction error
Bursting of parameters	Matrix P trace
Energy of the algorithm	Trace of the matrix P
Signal excitability	Trend of the matrix P
Unmodelled dynamics	Variance of the parameters
Admissible set of regulators	Ch. equation roots
Closed loop stability	Variance of the output

Table 4

A Indicators in the adaptive control supervision.

A.1 Numeric indicators.

A.1.1 Associated to the pre-identification.

The pre-identification process is the responsible for an adecuated adaptive control starting-up. Hence, it must provide enough information in order to avoid the violation of the system pre-conditions. During the pre-identification, which is carried out in open loop, the initial design factors for the on-line control starting-up will be proposed by the Numeric Indicators associated.

Indicators:

1. **Sampling rate test**: allows for the most adecuated sampling rate to be chosen from the determination of the 95% of the settling time (Isermann & Lachmann, 1985).

$$\frac{1}{10}T_{95} \leq T_0 \leq \frac{1}{4}T_{95} \tag{19}$$

2. **Process order test**: the prediction error quadratic function may be used to propose the process model order m:

$$V(m) = e(m)^T e(m) \tag{20}$$

Alternative tests, such as the poles-zeroes test or the residues test exit, but they are more time and memory consuming.

3. **Process delay test**: based on the B polynomial coefficients which can be neglected:

$$| \hat{b}_1 |, | \hat{b}_2 |, ..., | \hat{b}_\beta | \ll \sum_{i=1}^{m} \hat{b}_i$$
$$| \hat{b}_{\beta+1} | \gg | \hat{b}_\beta | \tag{21}$$

4. **Unmodelled dynamics test**: with this test the disturbances model can be determined, being either white or colored noise, so the most adecuated estimator can be chosen. The interpretation of the prediction error autocorrelation $\Phi_{ee}(\tau)$ is used for this purpose.

A.1.2 Associated to the estimation.

In this case, the identification is carried out in real time on-line conditions. Just the other way round of the general situation for the pre-identification stage.

Indicators:

- **a) Associated to the prediction error:**

    1. A priori error:
    $$e(k) = y(k) - \varphi(k)^T \hat{\theta}(k-1) \tag{22}$$

    2. A posteriori error:
    $$e(k) = y(k) - \varphi(k)^T \hat{\theta}(k) \tag{23}$$

    3. Error mean value (heretofore may be referred both to the a priori and to the a posteriori error):
    $$\bar{e}(k) = E\{e(k)\} \tag{24}$$

    4. Error variance:
    $$\sigma_e^2(k) = E\{[e(k) - \bar{e}(k)]^2\} \tag{25}$$

451

5. Error variance trend:

$$t_e(k) = \alpha \frac{\sigma_e^2(k) - \sigma_e^2(k-1)}{T} + (1-\alpha)t_e(k-1) \qquad (26)$$

6. Error variance acceleration:

$$a_e(k) = \alpha \frac{t_e(k) - t_e(k-1)}{T} + (1-\alpha)a_e(k-1) \qquad (27)$$

7. Autocorrelation function:

$$\Phi_{ee}(\tau) = E\{[e(k)e(k+\tau)]\} \qquad (28)$$

8. Crossed correlation function u/e:

$$\Phi_{ue}(\tau) = E\{[u(k)e(k+\tau)]\} \qquad (29)$$

9. Crossed correlation function y/e

$$\Phi_{ye}(\tau) = E\{[y(k)e(k+\tau)]\} \qquad (30)$$

10. Cost function:

$$J(\tau) = e(\tau)^T e(\tau) \qquad (31)$$

- b) **Associated to the v-c matrix:**

  1. V-C matrix:
  $$H(k) = [\Psi(k)^T \Psi(k)] \qquad (32)$$

  2. V-C matrix inverse:
  $$P(k) = [\Psi(k)^T \Psi(k)]^{-1} \qquad (33)$$

  3. Trace of the V-C matrix inverse:
  $$tr P(k) = \sum_{i=1}^{2n} p_{ii} \qquad (34)$$

  4. Trace trend:
  $$t_p(k) = \alpha \frac{tr P(k) - tr P(k-1)}{T} + (1-\alpha)t_p(k-1) \qquad (35)$$

  5. Acceleration of the trace.
  $$a_p(k) = \alpha \frac{t_p(k) - t_p(k-1)}{T} + (1-\alpha)a_p(k-1) \qquad (36)$$

- c) **Associated to the estimated parameters:**

  1. Estimated parameters average:
  $$\bar{\hat{\theta}}_i(k) = E\{\hat{\theta}_i(k)\} \qquad (37)$$

452

2. Estimated parameters variance:

$$\sigma_{\hat{\theta}_i} = E\{[\hat{\theta}_i(k) - \overline{\hat{\theta}}_i(k)]^2\} \tag{38}$$

3. Estimated parameters norm:

$$N_\theta(k) = \sqrt{\sum_{i=1}^{m+n} (\hat{\theta}_i(k) - \overline{\hat{\theta}}_i(k))^2} \tag{39}$$

4. Norm trend:

$$t_N(k) = \alpha \frac{N_\theta(k) - N_\theta(k-1)}{T} + (1 - \alpha)t_N(k-1) \tag{40}$$

5. Norm acceleration:

$$a_N(k) = \alpha \frac{t_N(k) - t_N(k-1)}{T} + (1 - \alpha)a_N(k-1) \tag{41}$$

A.1.3 Associated to the controller calculus.

The indicators associated to the controller calculus in an adaptive loop, are mainly orien-ted to the detection of incompatibilities between the estimated process poles and zeros and the controller to be designed (Isermann & Lachmann, 1985).

Indicators:

- **a) Associated to the model polynomials poles and zeros.**

    1. Poles and zeroes: module, arguments, real and imaginary parts.
    2. Poles and zeroes modulus trend.
    3. Poles and zeroes modulus acceleration.

- **b) Associated to the loop error:**

    1. Loop error:
    $$e_b(k) = r(k) - y(k) \tag{42}$$

    2. Loop error average:
    $$\overline{e}_b(k) = E\{e_b(k)\} \tag{43}$$

    3. Loop error variance:
    $$\sigma_{e_b}^2(k) = E\{[e_b(k) - \overline{e}_b(k)]^2\} \tag{44}$$

    4. Loop error variance trend:
    $$t_{e_b}(k) = \alpha \frac{\sigma_{e_b}^2(k) - \sigma_{e_b}^2(k-1)}{T} + (1 - \alpha)t_{e_b}(k-1) \tag{45}$$

    5. Loop error variance acceleration:
    $$a_{e_b}(k) = \alpha \frac{t_{e_b}(k) - t_{e_b}(k-1)}{T} + (1 - \alpha)a_{e_b}(k-1) \tag{46}$$

A.1.4 Associated to the closed loop.

They mainly try to detect the loop stability evolution.

Indicators:

- **a) Associated to the characteristic equation roots.**

    1. Characteristic equation roots: module, arguments, real and imaginary parts.
    2. Characteristic equation roots module trend.
    3. Characteristic equation roots acceleration.

- **b) Associated to the output signal:**

    1. Mean:
    $$\bar{y}(k) = E\{y(k)\} \tag{47}$$

    2. Variance:
    $$\sigma_y^2(k) = E\{[y(k) - \bar{y}(k)]^2\} \tag{48}$$

    3. Output trend:
    $$t_y(k) = \alpha \frac{y(k) - y(k-1)}{T} + (1 - \alpha)t_y(k-1) \tag{49}$$

    4. Output acceleration:
    $$a_y(k) = \alpha \frac{t_y(k) - t_y(k-1)}{T} + (1 - \alpha)a_y(k-1) \tag{50}$$

    5. Control signal saturation test: its goal is to determine whether the control signal is saturated during a pre-determined number of sampling periods and simultaneously changing in sign (Isermann & Lachman, 1985).

References

[1] Albertos,P.; Martinez,M.; Morant,F. (1989). Supervision level in Adaptive Control. *LCA'89. IFAC Sym. Milan. Nov 89.*

[2] Åström, K.J. and others (1986). Expert Control. *Automatica Vol 22 No 3.*

[3] Åström,K.J.; Wittenmark,B. (1989). Adaptive Control. *Addison-Wesley*

[4] Fortescue, T.L. and others (1981). Implementation of Self Tuning Regulator with Variable Forgetting Factors. *Automatica, Vol. 17.*

[5] Gerhardt, R.W. (1986). *An Expert System for Real Time Control.* Thesis. Rensselaer Polytechnic Institute. Troy. N.Y.

[6] Isermann, R. (1982). Parameter Adaptive Control Algorithms-A tutorial. *Automatica, Vol. 18 pp. 513-528.*

[7] Isermann, R.; Lachmann, R. (1985). Parameter-Adaptive Control with Configuration Aids and Supervision Function. *Automatica. Vol. 21 No 6.*

[8] Isermann, R. (1981). *Digital Control Systems.* Springer-Verlag NY.

[9] Ljung, L.; Soderström, T. (1983). *Theory and Practice of Recursive Identification.* MIT Press.

[10] Martinez, M. (1991). Indicators for Adaptive Control Supervision. Implementation by means of Expert System Methodology. *PhD. Thesis (in spanish)*

[11] Morant, F. and others (1989). Hierarchical Expert System as Supervisory Level in an Adaptive Control. *4hr IEEE I. Sym. on Intelligent Control. Albany, September 89.*

[12] Morant, F. and others (19992). Expert System for an Adaptive Real-Time Control with Supervisory Functions. *Procc. SICICI,92. Singapore, February 92.*

[13] Simó,J.; Picó,J.; Vivó,R.; Navarro,J.L. A Software Package for Integral Slurry Milling Control in Cement Production Plants. *Procc. ADCHEM,91. Toulouse, October 91.*

[14] Schumann, R. and others (1981). Towards Applicability of Parameter Adaptive Control Algorithms. *Procc. IFAC Congress Kyoto. Pergamon Press.*

MONITORING AND FAULT DIAGNOSIS IN CONTROL ENGINEERING

Dr. Ranjan Vepa

Department of Aeronautical Engineering, Queen Mary and Westfield College,
University of London, LONDON, E1 4NS, U.K.

1 Introduction

He that will not apply new remedies must expect new evils, for time is the greatest innovator.

Francis Bacon

This course is about the application of AI techniques to process control. Before actually discussing this aspect it is important to understand how AI in general and Expert systems in particular can be used to a help a group of operators run a process plant efficiently.

A very important technique in AI is based on the idea that one can learn from examples. And there is no better example of a process plant than an aircraft with the pilot and his crew being roughly equivalent to the operators running the process plant.

"For Example" is not proof.

Anon

What then are the advantages of using real time expert systems, i.e. expert systems that perform well within the constraints imposed by the need for time critical solutions, in the cockpit of an aircraft. The principal reason for using real time expert systems would be to reduce the cognitive load on the crew so as to enable them to be free to take decisions in time critical emergency situations without the cognitive load on them increasing. This in fact applies to any process plant and its group of operators. The principal indicators that a real time expert system may be appropriate especially when approaches using conventional techniques have

456

failed or are impractical, include problem solving situations where humans
1) suffer from cognitive overload,
2) are unable to monitor effectively all available information and to resolve conflicting constraints,
3) capable of performing the tasks are scarce and training new experts is expensive,
4) make high cost mistakes,
5) miss high revenue opportunities,
6) cannot simultaneously manipulate all the relevant information to obtain acceptable or low cost solutions
7) or cannot provide a solution fast enough.
A particular implementation of a real time expert system may be used in one or several of these situations.

1.1. Expert Systems for monitoring and fault diagnosis:
There are no facts, only interpretations.

Friedrich Nietzsche

The main objectives for applying expert system techniques in aircraft are:
1) To prevent pilots from cognitive overload situations that arise when the rate of alarm messages in critical situations goes beyond the pilot's capability of interpretation
2) To quickly locate faults when they occur and bypass them with the least interruptions and generally to assess the internal status of all aircraft sub-systems
3) To optimally use the resources in the cockpit and in fact in the entire aircraft to deal with critical and emergency situations
4) Planning the flight route or re-planning to respond to changes in the flight plan or to accommodate equipment faults.

These objectives would in turn require various sub-tasks to be addressed such diagnose observable malfunctions as and when they occur, interpret the data to assess the prevailing situation and predict the short and long term response of the aircraft and the demands it therefore places on the pilot and his crew. These problems have been addressed over the years using classical control and system theory techniques but with limited success. The major difficulty is in understanding the complete model of the aircraft, particularly in critical situations and in the presence of failures not only in the aircraft but also on the part of the pilots. A numerical approach to such situations is totally inadequate as one is interested in the causal sequences by means of which one understands the device's functions.

Expert system techniques however are by there very nature designed to deal with incomplete, vague, behavioural, functional, causal or qualitatively describable information.

A considerable number of powerful and sophisticated information processing systems may be introduced to aid a pilot. Since the pilot is normally not interested in the very low level sensor data, the main difficulties of these information processing systems is to collect the low level information and combine and interpret it. This high level information then allows the pilot to quickly understand the current situation and to respond to demanding conditions.

Conventional control and systems analysis techniques, are capable of very substantial digital processing and analysis are limited in their ability to provide an understanding of the causal relations of the measured data. Although conventional analysis or design systems can ensure the completeness of the information, expert interpretive type systems have to have a thorough understanding of the situations and may not need all the information all the time as long as they can have access to it when necessary.

One could envisage a typical configuration where the control device would do the monitoring and provide the observables to the display system as well as performing the control functions. The coupled expert system on the other hand would be responsible for the entire problem solving process. It's first task is the assessment of the situation and the second is the abstraction of the data and the diagnosis in case of faults. The expert system would not be "problem triggered", i.e. only activated when some sort of irregular behaviour or fault is detected; rather it will be continuously active, monitoring the essential observables continuously. This monitoring task offers permanent help to the pilot.

In contrast to application areas that most expert systems have so far focused upon, the class of expert systems designed to work as an outer control loop are characterized by:
1) high organizational complexity compounded by a large number of

constituent parts

2) a large number of input/output interactions

3) a dynamic situation to reason upon, requiring time dependent reasoning mechanisms to account for delays between the occurrence of events and their notification, commands to the aircraft, their response, situation notification and the final diagnosis

4) the need to incorporate learning in order to deal with changes in the aircraft sub-systems, initially incomplete knowledge and the possibility of early transient knowledge being false.

One further feature of aircraft control applications and the most difficult to solve within the framework of AI is that the entire reasoning process must be completed very fast, i.e. in real time. The control system must not only be able to take decisions towards the correct solutions but also take these decisions fast enough so that the performance of the controlled aircraft is not adversely affected by the delays in the reasoning process.

This feature imposes a very different structure on expert systems used in control or in real time monitoring. The common concept of expert systems is based on interaction with the user in that the user has to provide all the information the system requires for its reasoning process. By incorporating the user in this way the system usually deals with fairly high level information which includes the users own judgment and opinions based on his extensive database. The user often combines and interprets the information presented to him before making it available to the expert system.

In real time aircraft control applications such a concept is not sensible since it is not really possible to supply the expert system with all the relevant data interactively. Not only is the amount of data too large but also because it is changing with great rapidity, manual input would not really be feasible, apart from the fact that it would cause unacceptable delays. Thus the expert system has to communicate directly with the plant i.e. the aircraft and access low level data for doing time dependent and time critical situation assessment or fault diagnosis. This requirement establishes the need for real time monitoring. In summary we may define an expert controller as the combination of a real time expert system and a feedback system. Some form of situation assessment or fault diagnosis is usually present and this requires an extensive monitoring of all data available.

459

1.2 Components of a Real Time Expert System: The real time expert obtains its knowledge from a knowledge source directly connected to the plant as well as a global knowledge base and routes this information to a local knowledge data base. The information in the local knowledge data base is made available to a reasoning system. The feedback system tries to modify either the plant inputs or the plant structure on the basis of a pre-defined set of rules, depending on the conclusions arrived at by the

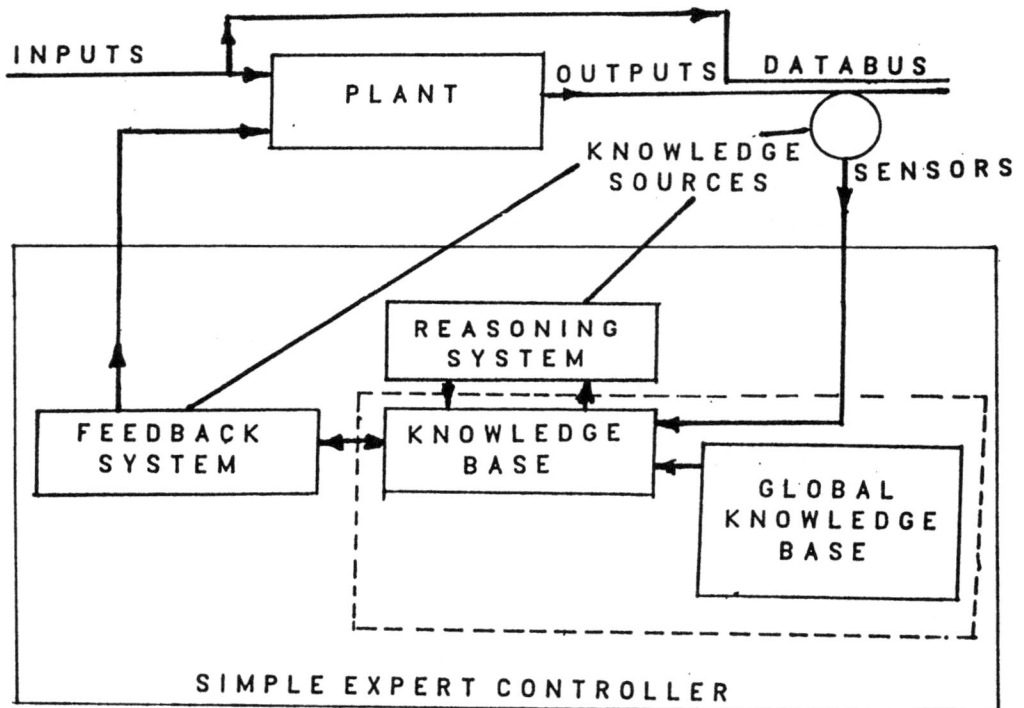

Fig. 1

real time reasoning system. This in its simplest form is an expert controller and is illustrated in Fig. 1.

The reasoning system is based on formal classical logic as well as the associated calculus of reasoning and practical implementations and interpretations. Deductive and inductive reasoning based on non-monotonic and monotonic logic, practical reasoning based on heuristic methods, the generate and test paradigm, forward or backward chaining, the recognize and

460

act paradigm etc and the associated methods of generating hypotheses form the basis of all reasoning systems. The reasoning process is generally problem independent.

1.3. General structure of the Knowledge Base: On the other hand the knowledge base which not only includes the low level data but also the various problem dependent techniques associated with the processing of the data is naturally unique to every different application. Thus the successful application of the expert systems approach depends on a proper understanding of this knowledge base and its use in relation to a expert system.

It is important to understand the various classifications of the knowledge base, the representation of the knowledge using data structures and the features of such data structures. The knowledge base can be divided into two major categories: i) Knowledge about the particular aspects of the problem (Domain Knowledge) ii) Knowledge about the problem solving approaches (Meta Knowledge) [Pau(1986)].

Domain Knowledge can be seem to consist of two main blocks: Experiential and Fundamental. Experiential Knowledge is the shallow emphirical knowledge that an expert acquires over a period of time based on experience. Generally in this block there are no extensive cause and effect links between the conditions with events or conclusions and observations or actions except through empiricism and experience. Although there may be a cause and effect relationship the deeper intermediate reasons for this are generally unknown. For example a sensor may put out erroneous data. First it is important to recognise this fact. Then one may ask the question as to what system fault may have caused this to happen and experience may well give us the answer. On the other hand a more complete approach would be to ask as to what changes in the system description have caused the data to put out erroneous data. Then we may subsequently ask the question as to what system fault was responsible for the system description to change in the manner it did.

The above example illustrates the importance of the deeper fundamental knowledge which is model oriented and a form of the system description. On the other hand fundamental knowledge by itself is not adequate to rationalize the experiential knowledge. There is also a need for the system

461

relational knowledge which relates the experiential knowledge to the system description. In our example system relational knowledge would explain what changes in the system description led to erroneous sensor outputs and what system faults are responsible for the system description to change as it did.

Thus a broad picture of the knowledge base now begins to emerge. This is illustrated in the figure 2 below.

If one examines the experiential knowledge block on can easily identify a multi level structure. This multi level structure is illustrated below (figure 3) for the experiential and causal knowledge blocks and is generally true for the third block as well although exact nature of the various levels is very much problem oriented. The structure presented is fairly general. However there may be situations in practice when a system relational knowledge may not be available and has to be constructed by the tools provided for by the meta knowledge block.

It is in fact quite a difficult task to present a detailed picture of the meta-knowledge block for it does depend to a very large extent on the class or category of expert system we are dealing with.

Knowledge can be represented in a number of ways including algorithms, symbols, fact lists, logic predicates, data capture routines, rules or production systems, frames, knowledge objects etc. The most popular representations, frames, production systems and semantic nets have formed the basis of many expert systems. Knowledge acquisition and representation are closely linked as the representation chosen strongly influences the type of knowledge that can be acquired. In real time expert systems one has to represent knowledge dealing with uncertain reasoning as well as temporal knowledge. Thus knowledge representation structures must be adequate to describe a particular system or information to the appropriate depth that may be essential.

In summary one can divide the knowledge bases into several sub-bases. A knowledge base (KB) alone does not constitute an expert system. A reasoning system must be available to provide the problem solving capability. A reasoning system has three basic components:
i) A communication sub-system that allows communication between KB nodes

462

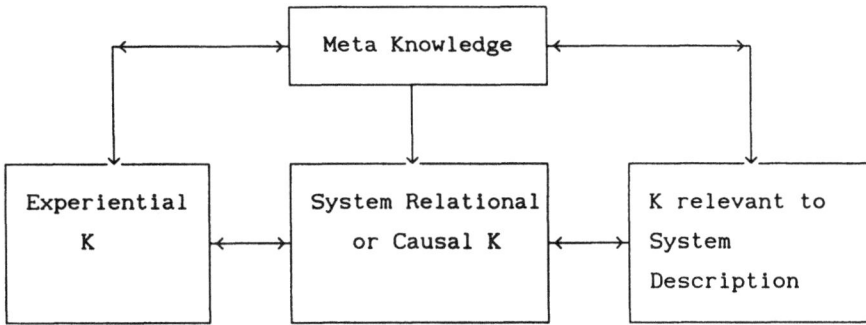

Fig. 2. General Structure of the Knowledge (K) base

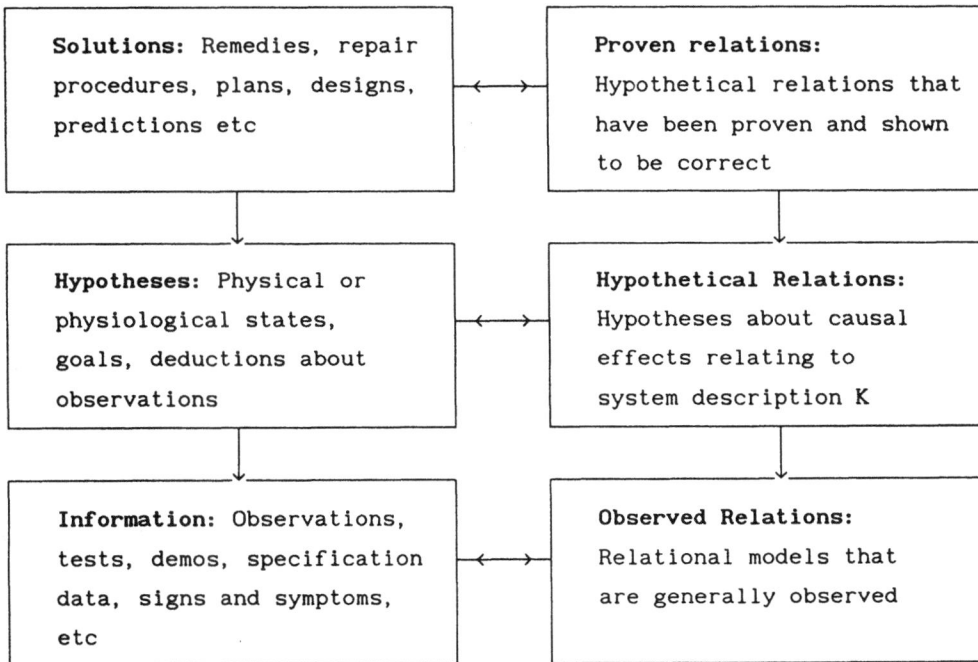

Fig. 3 Layered Nature of Experiential and Causal Knowledge

ii) A control structure that allow the scheduling of the various problem solving techniques that need to be applied in the right sequence in order that an acceptable solution be obtained and

iii) finally an inference engine or inference mechanism that can use the knowledge in the KB to solve a specific problem by emulating the reasoning process of an human expert. Such an inference engine is usually built around practical reasoning techniques.

Communication is extraordinarily broad and complex. It basically involves transactions between people. As it involves a flow of information, communication constitutes an essential component of system coordination and control. It involves three sectors. One sector, semantics, concerns itself with the meaning of symbols and words. Another sector, syntax, is concerned with the systematic properties of communication, that is the relations between the symbols used. A third part of communication concerns itself with pragmatics and focuses on the relationship between communication and purposeful action. The central question from the point of view of an expert system is 'How does information move through the various nodes of the knowledge base?'. While there are no simple solutions to this question the generally accepted trend is to use a hierarchical communication structure. It is particularly important to reduce errors in communication such as cross-talk and information loss. One approach to do this is the use of redundancy. Another important concept related to cross communication is the concept of 'Blackboard' or the use of a blackboard mechanism which is discussed in a latter section.

Control structure in an expert system needs to be very efficient. Although the control structure may be apparent in most problems in a broad sense the scheduling of methods in the right order may not be known a priori especially when several alternatives are available. Thus the problem reduces to one of 'Conflict resolution' and the control structure may have to generated in real time using a feedback process. Generally the issues of communication and control are interrelated, because of the need to achieve 'Coordination'. Blackboard mechanisms are particularly useful for conflict resolution problems involving coordination.

1.4 Reasoning Techniques: Reasoning involves a search for principles which will solve problems. This search for principles, which is a primary aspect of reasoning proceeds properly when logic is used, that is when certain rules are followed. A student of control engineering who has mastered all conventional control system design techniques as well as optimal control and H_{∞} control and has acquired only partially the skills of problem identification, reasoning and logic may feel sufficiently well equipped to enter the arena of the real world for battling with real problems. Yet on entering that formidable battleground he is certain encounter many problems that cannot be solved by any of his repertoire of tool-kit techniques. He will often resort at that stage heuristic problem solving techniques particularly if he is faced with time constraints and deadlines. Intelligent compromises are often made in the form of heuristic solutions and these are often the only alternatives given the constraints and the situation. An example is important to stress the point. Conventional optimal control techniques are more than adequate for solving standard flight control design problems such as stability augmentation, automatic piloting, navigation and guidance. However what poses a real problem in the design of flight controllers that must perform in extremely emergency situations or in spite of the presence of some faults. The study of heuristic problem solving strategies is important as one can then identify a number of specialized and practical reasoning techniques that can be applied to a whole class of problems.

Heuristic problem solving involves inventing a set of rules that will be of assistance in or that will guide the search for one or more satisfactory solutions to a specific problem. The emphasis is not on an optimal solution but on a satisfactory solution. This may arise from a need to meet a multitude of objectives so as to maximize payoff and minimize risk. The rules are inferred to as heuristics (from the Greek word 'Heuriskien' meaning to discover). Most often heuristics are implicitly used to reduce the problem space sufficiently so that an analytical procedure or simulation may be applied while in other instances heuristics may be themselves sufficient to produce an acceptable feasible solution.

Whereas analytical procedures are based on deductive reasoning supported by mathematical proofs and known properties (eg. Adaptive control) heuristic problem solving techniques are based on inductive reasoning techniques involving generalization from examples and related to human characteristics of problem solving including learning, insight, intuition and creativity. A particular heuristic is followed simply because it appears promising on the basis of past experience or intuition. The heuristic may not only aid the search for the solution but also for a better rule and if in the process of problem solving a better rule is found the older rule is discarded. So not only does Heuristic problem solving involve the use of currently accepted rules but also involves a search for even better rules to replace

465

them. In fact one may identify six basic classes of practical reasoning techniques:

1) Simple Heuristic Search: This is the classical approach to reasoning in expert systems and is based on the premise that in applications such as fault diagnosis simple heuristic search can diagnose the problem more easily than others. This ability is due to the special domain specific knowledge acquired as a result of experience. The knowledge always manifests itself as rules of thumb or heuristics. This type of knowledge, if it is actually available or if it can be compiled is as basic to reasoning systems as a control law or control filter is to control systems. The objective of practically all other reasoning techniques is essentially to compile such a knowledge base. For example, if the battery is not charging during the running of a car it is either the battery itself or the alternator that is faulty. If one the examines the battery and establishes that it is not faulty, the alternator can be identified as the source of the problem.

2) Generate and Test: In this approach heuristics are generated as they are needed on the basis of a set of hypotheses and the validity of these hypotheses are evaluated from solutions that can be generated as the system proceeds.

3) Forward Chaining/Backward Chaining: there are situations where it may become necessary if not desirable to identify a set of hypotheses from a larger set or global set, that can be supported by a given database. In such cases the reasoning can proceed from the database or evidence to establish the truth of the hypotheses or alternatively from the hypotheses to establish the validity of the data generated by comparison with the evidence on the premise that the hypotheses is true. Thus in forward chaining given the data or evidence conditions for establishing the truth of the hypotheses are tested. This process is akin to pruning a decision tree. On the other hand in backward chaining the truth of certain hypotheses is assumed and then data is generated to test the hypotheses. If the generated data is in conflict with the initially available evidence then the hypotheses is assumed to be false and all the conclusions that resulted from the false hypotheses are deleted.

4) Recognize/Act : This approach is applicable to certain classes of systems where the occurrence of certain patterns or features in the data or its characteristics necessitates that certain conclusions be drawn and certain actions be initiated. Such an approach is commonplace in statistical detection . This approach is particularly useful in integrating

466

statistical approaches to detection with knowledge based reasoning techniques.

5) Constraint Directed: This approach in the basis of many design and planning techniques. It begins with an initial predefined plan that is actually generically defined which may not meet all the specifications and the plan is recursively updated so as to satisfy certain constraints till all the specifications are met. Typical examples are the certain geometric approaches to the Travelling Salesman problem.

6) Metarules : These can most easily be explained as a collection of rules not for the actual problem at hand, but for generating and testing a number of hypotheses in order to build up a compiled knowledge base. typically a meta rule begins with the generate and test paradigm and then provides actions when the test proves to be false and true. Methods of generating hypotheses are based on analogy, experience, abduction or by default.

2. Architecture of knowledge bases required for implementing Expert control systems

> Knowledge is of two kinds; we know a subject ourselves, or we know where we can find information upon it.

> *Samuel Johnson*

2.1 Introduction

The general structure of expert systems should be well understood by now. They usually consist of a set of Knowledge sources, a knowledge base including a meta-knowledge base and a reasoning system or an inference mechanism. The latter is sometimes referred to as an "inference engine". Expert system shells are today available that actually provide all the inference mechanisms built in, so one can easily implement an expert controller provided one can generate the required knowledge base (KB). The generation of such a knowledge base which may take the form of a rule base is not an easy problem and in fact it is the central issue in the development of an expert system. For, in the AI style of thinking, what is required is not an algorithm but a paradigm i.e. the specification of a goal and the definition and identification of the methods to achieve that goal.

The architecture of a knowledge base is also important when one is considering a large system with a number of inferencing blocks within it. Generally two distinct approaches [Grimshaw and Anken(1991)] may be followed:

467

a) a centralized knowledge base

b) a decentralized knowledge base

The centralized architecture is based on the concept that all information transactions and data exchanges take place through a common agent or agents. There are three broad classes of centralized architectures namely:

1) The Blackboard architecture

2) The information broker architecture and

3) The meta level controller architecture.

The Blackboard architecture was developed for a speech understanding system in the 70's known as the HEARSAY-II system [Hayes-Roth(1985)]. It was based on the idea of a common workplace or blackboard where a problem can be solved progressively. This a particularly important feature for reasoning with temporal data within real time constraints. The problem solving system is partitioned into a number of 'knowledge sources' which are the decision aids having a two-way access to all the information in the Blackboard. This gives the partial solving tasks the ability to access partial solutions during the solution process. It also means that some form of control must be exercised on the databases in the Blackboard to prevent the occurrence of collisions, deadlocks, etc and to determine which of the problem solving tasks or knowledge sources must access the data in the Blackboard first or execute first. The Blackboard concept is playing a fundamental role in the design of expert systems for monitoring and fault diagnosis and is discussed further in the next section.

The information broker concept seeks to partition the knowledge bases into several 'brokers' while simultaneously constraining each broker to deal only with similar information or knowledge. This each broker is only allowed to deal with knowledge or information belonging to the same broad class. Two types of architectures have evolved based on this concept and are illustrated in Fig. 4.

In the meta-level controller architecture the knowledge base is also partitioned as in the information broker architecture. In addition the control mechanism is also partitioned so that each of the problem solving tasks or decision aids have access to the knowledge base through a dedicated controller. The approach is illustrated in Fig. 5.

Decentralized techniques do not use such common agents for information transactions are generally one of two types.

468

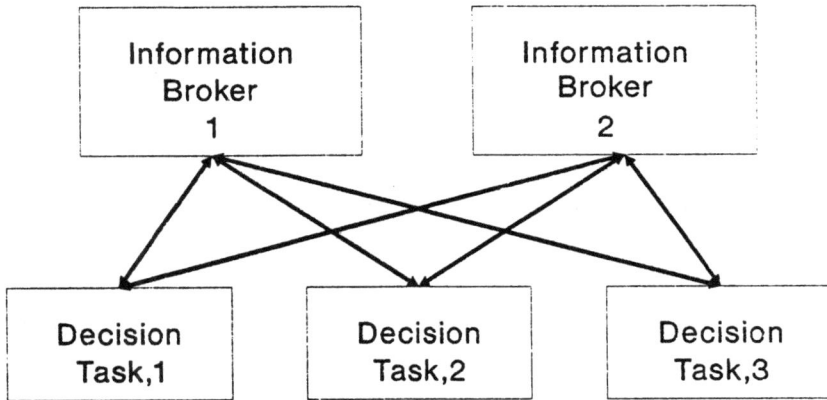

Fig. 4a. Information Broker Architectures: Networked architecture

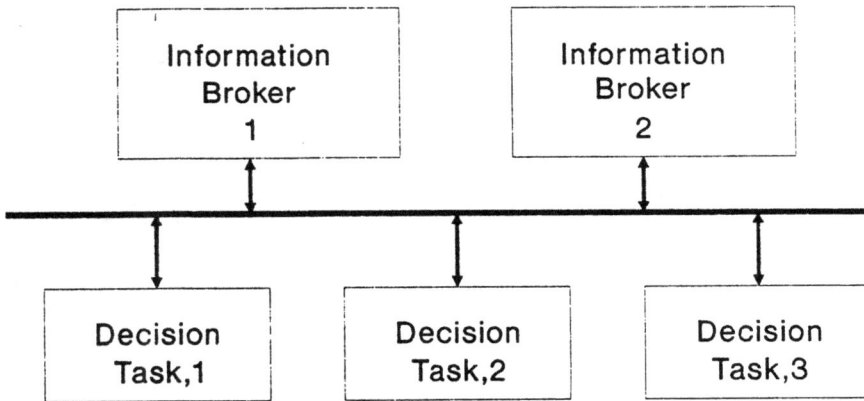

Fig. 4b. Information Broker Architectures: Broadcast architecture

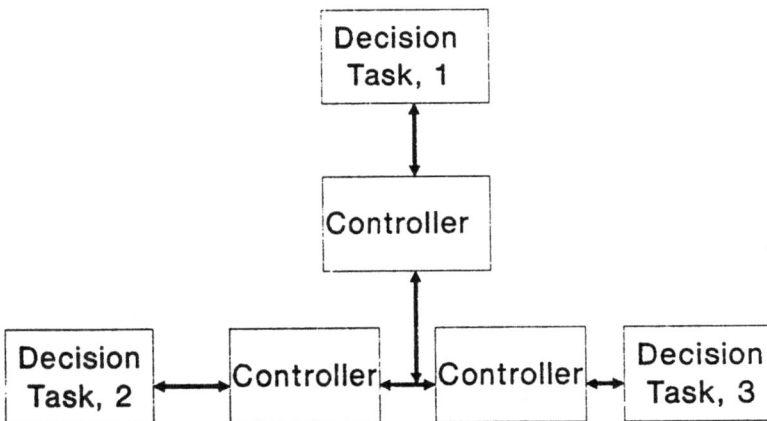

Fig. 5 The Meta-level Controller architecture

469

1) Hardwired architecture: The hardwired architecture consists of a number of tasks cooperating through direct calls to one another. Because of this each of the problem solving tasks in the network is expected to be aware of the specific nature of all other tasks in the network it will interact with. The actual interaction is based on function calls which are embedded at the time of development. Thus the architecture is 'hardwired' and is difficult to modify at run time. It is illustrated in Fig. 6.

The broadcast architecture is based on the idea that all tasks broadcast the data on a bus as and when required by specific user tasks. The Broadcast architecture is illustrated in Fig. 7.

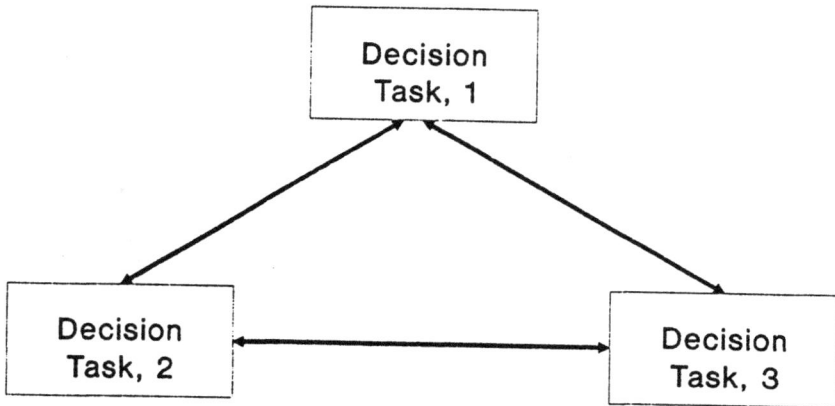

Fig. 6 Hardwired Decentralized Architecture

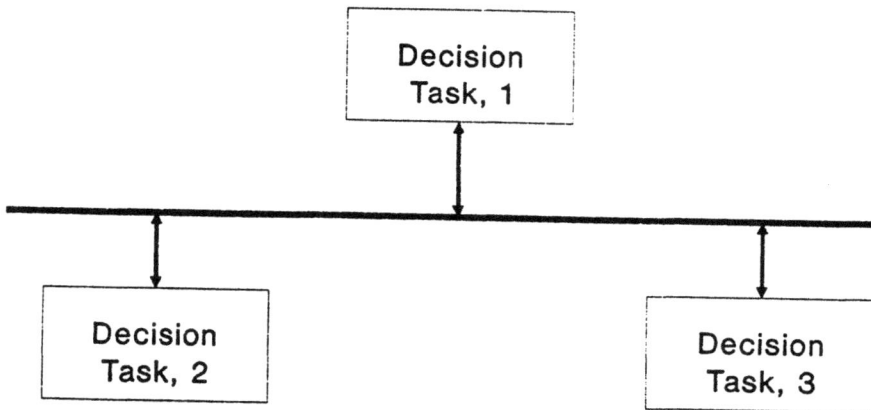

Fig. 7 Broadcast Type Decentralized architecture

470

The process of control system design is a well understood, structured and iterative process. One form of real time control involves real time design and in this approach to expert control the controller is designed or updated on a real time basis. It is this approach to expert control that is considered here as an example to explain the Blackboard concept in greater detail. However many of the concepts apply to other approaches as well.

2.2 The Blackboard Concept: The process of design involves the application of a sequence of problem solving methods. This sequencing of tasks or software control is performed on the basis of a scheduler and each task selects an appropriate method based on a rule base. The generation and explicit specification of such a rule base is considerably simplified if the meta-knowledge base has a structure to it. The requirements can be stated as follows:

1) It must be possible to record all solutions and intermediate solutions generated during the problem solving process into the global data base.

2) Independent processes called knowledge sources must have access to this global database to record solution elements or data generated by them.

3) A scheduling mechanism must be provided that can choose a particular knowledge source to execute its action and access the global data base.

Such a global database is called a "blackboard" and is illustrated in Figure 8. The implementation of such a blackboard using an object oriented approach is considered. The concept is particularly useful for software control [Hayes-Roth(1985), Pomeroy, Spang and Dausch(1990), Pomeroy and Irving(1990)].

2.3 Classes in the Blackboard

Considering the first of the requirements and from an understanding of the real time redesign process we may define five basic classes in KB (apart from the scheduler and the knowledge sources). Each of these classes is characterized by a typical object. Thus corresponding to the five classes we have five basic objects [Favier and Senges (1989)] which are defined as: plant object, model object, controller-model object, controller-design object and the specifications object. The model object is fairly atomic in that it is fundamental. Methods are available within each object and may be "public" or "private" in that the former are available for operation on other objects while the latter can only be used to operate on the particular object to which they are available. These methods are then

471

referred to respectively as into model synthesis methods, model analysis methods, controller synthesis methods, controller implementation methods and design validation methods. The model synthesis methods are responsible for generating the model object from the plant object. The model analysis tools are private to the model object. It is important to realize that these objects are essentially super classes and there are other objects within them representing sub-classes. On the other hand these objects

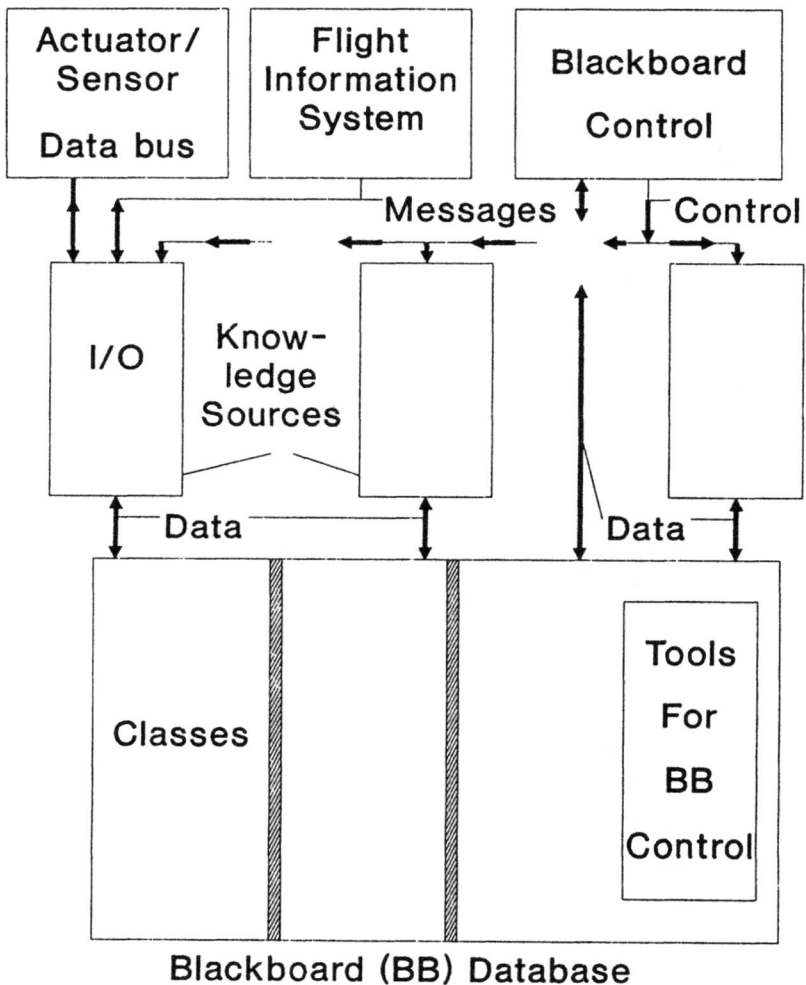

Fig. 8 The Blackboard Architecture

represent the principal sub-classes of the blackboard object. Methods for software control of the objects in the blackboard are available within the blackboard. Such a view can be translated into computer code quite easily and provides for efficient and structured software integration.

To actually establish a rule base for controller design a further sub-classification of the methods is essential. The methods available to each object are then classified as: 1) Methods for parameter identification 2) methods for providing actions 3) Diagnostic methods 4) Methods for making decisions 5) Methods for planning and 6) Methods for classification (including benchmark tests and tools for comparing methods in each class)

2.4 Specification of the objects and methods

A detailed specification of the five objects and the methods available in each class are now listed. The rules must operate on the objects and if certain premises are satisfied, certain conclusions are reached and the associated actions are initiated. When there is an ambiguity the rule must provide an alternative action to resolve the ambiguity.

Examining the objects and given a particular design problem we then establish a rule base for choosing the appropriate methods in the design process. The rule base is grouped into tasks so that the design process can be efficiently realized. Based on these tasks and the design process the scheduler is written. The knowledge sources are assumed to be disc files which are used to initialize the objects and the tasks which are built around the rule base. These tasks include, object initialization, selection of design approach, generation of secondary data for design, synthesis of controller, controller implementation (optional), and controller evaluation. The tasks are processed by an inference engine to reach the appropriate conclusions on the selection of methods. Appropriate methods of message passing are used to communicate with the scheduler. At present each task has its own inference engine to process the condition- action rules.

A general KB structure for expert control based on the "blackboard" concept and object oriented design are emerging as fundamental concepts in the implementation of practical expert control systems. It is being used to develop several expert control system packages that can be used in process control as well as in the avionics industry. It is clearly indicating the following:

1) The emergence of the Blackboard concept as a definite method of providing for efficient software control and software integration.

2) It can be easily integrated with the C language which has proven advantages in real time control giving the programmer complete flexibility.
3) Coupled with the advantages of object orientation, objectives of software engineering and the emulation of some special features of AI languages (LISP, PROLOG etc) can be realized by C++ .

3 Requirements for a Real-Time Expert Monitoring System: The overall function of monitoring system is to relieve the operator of as much low level data processing as possible and to ensure the presentation of data at a sufficiently high level in a form that it can be easily comprehended without the need for further interpretation. The data itself may be information from a variety of sensors or transducers such as Piezoelectric devices, mechanical strain gauges, variable reluctance pickups, variable capacitance pickups, magnetic pickups, photo voltaic devices, photo conductive devices, thermocouples, charge coupled devices, thermistors, signal condition circuits and a variety of other of other digital or analog sensing devices. Further the sensing devices may be spread out at various different locations in the plant. The monitoring system must permit the operator to focus his attention on specific sensors without distraction from the overall task at hand and without having to be concerned about the tedium of low level data processing or the interpretation of the data.

To achieve this a monitoring system must
i) understand the objectives of monitoring
ii) must have access to data from within the system it is monitoring as well as the environment it is operating in
iii) anticipate decisions that the operator must make and on that basis prioritize and shift attention as well as
iv) be capable of 'what if' type reasoning.

These features it must possess in addition to its ability to manage all data, simultaneously manage all data sources and data users if any, provide solutions in real time so they can be valuable to the operator and finally respond to the operators commands. Thus we may identify a set of requirements for real time monitoring systems[Laffey *et. al.*(1988)]. These are:
a) symbolic-numeric integration
b) focus of attention
c) continuous operation and interrupt handling

474

d) optimal environment integration

e) predictability

f) temporal reasoning and truth maintenance in real time.

3.1 Symbolic-Numeric Integration: Most methods associated with monitoring in control engineering are based on based on sophisticated numerically intensive signal conditioning techniques such as Kalman Filtering, Luenberger Observers, Matched Filters, etc. To be able to continue using these techniques at a lower level it is essential to integrate symbolic and numeric processing. Fuzzy Logic and Neural Networks coupled with CMAC and associative memories permit such symbolic numeric integration. The performance requirement for a monitoring system is usually specified at a high level in a symbolic manner. These are then interpreted as Fuzzy sets and defuzzified to produce numerical values. The numerical values are then used to perform the appropriate signal processing on the monitored data. A similar approach is possible with Neural Nets.

3.2 Focus of Attention: This concept is fundamental when a large number of sensors or data channels are being monitored. The monitoring system must be capable of 'shifting its attention' from one group of data channels to another depending on the current situation. Situations are generally assumed to be triggered by 'enabling events'[Jakob and Suslenschi(1990)]. This is particularly true in 'Alarm Monitoring'. Enabling events in this context may not themselves be cause for alarm. However they may they be indicators of an impending alarm situation or may be the cause be the cause for a subsequent alarm for which an exception message must be posted. Thus by continuously trying to detect the occurrence of enabling events the monitoring system can be designed to shift its 'attention' from one set of sensors to another. When enabling events do happen the monitoring system may have resort to 'what if' type of reasoning to predict the onset of a potentially alarming situation.

3.3 Continuous Operation and Interrupt Handling: The monitoring system must be capable of continuous operation irrespective of the occurrence of a fault or alarm while being able to respond to prioritized interrupts. A typical example is the 'black box' carried on board a civil aircraft monitoring all the activities of the flight crew in the cockpit, cockpit instrumentation etc which is often used to perform a 'post mortem' examination in the event of an airliner crashing.

475

3.4 Optimal Environment Integration: It is often essential that a monitoring system fully utilize the features of the environment it is operating in. This must be achieved not only in an effective and opportunistic manner but also in an optimal way. For this reason the environment may also have to be monitored to a sufficient degree depending on the nature of the problem. As general rule the number of attributes of the environment that are monitored must be of the same order of magnitude as the number of attributes of the system being monitored. The data or knowledge thus generated must also be efficiently integrated with the data or knowledge generated by the sensors monitoring the system attributes.

3.5 Predictability: Predictability is also an important feature of a monitoring system. In fairly general terms this means that the entire behaviour of the monitoring system must be predictable. This of course implies several important features. First the response times must be completely deterministic and predictable apart from being fast enough (real time performance). Second all the errors during the monitoring process must be completely predictable so that the monitored quantities can be appropriated corrected. Predictability also implies that when there are decision loops and logic statements the behaviour related to these decision loops and logic statements must be entirely predictable. Thus predictability generally implies a strong requirement which not only implies that the behavioural response must be deterministic but also that this response must be verifiable. Performance and behavioural verification is a field of study in its own right and will not be discussed here any further.

3.6 Temporal Reasoning and Truth Maintenance in Real-Time: Temporal reasoning i.e. reasoning with time dependent data is an important feature of expert systems used to monitor control systems. It is important that temporal relationships be identified and effectively utilized during monitoring. There are a number of approaches in the artificial intelligence literature to implement temporal reasoning features. However the constraints imposed by the need to perform in real time are not met when most of the techniques of temporal reasoning are implemented. One approach now is practically a standard in control engineering applications. This based on time-stamping all data and applying the normal logic methods to the data as well as to the time stamps. However such an approach may at

476

time lead to a vast amount of data and in order to meet the real time constraints certain restrictions must be imposed. Although there are several solutions to resolve this problem, three techniques are particularly important and they form the basis of progressive reasoning. The three techniques are:

i) Progressive deepening

ii) Progressive widening

iii) Variable precision logic

Progressive deepening relies on the reorganisation of knowledge into the three basic levels as:

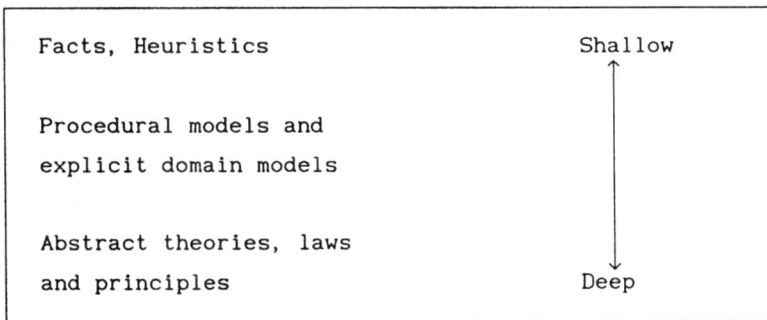

```
┌──────────────────────────────────────────────────────────┐
│                                                            │
│   Facts, Heuristics                     Shallow            │
│                                            ↑               │
│   Procedural models and                    │              │
│   explicit domain models                    │             │
│                                             │              │
│   Abstract theories, laws                   ↓              │
│   and principles                        Deep              │
│                                                            │
└──────────────────────────────────────────────────────────┘
```

Fig. 9 Three levels of Knowledge

1) heuristic or shallow knowledge

2) domain knowledge (includes procedural knowledge)

3) deep or theoretical knowledge

This is illustrated in Fig. 9. To illustrate this knowledge classification scheme we may consider how a user would react if a typical servo-motor control unit did not function. First he might check to see if it is properly connected to a power source and the check to see if all the other connections are proper. He may also perform certain standard verification tests to ensure that his conclusions are correct. If however he has the failed to identify what is wrong with the unit he may the take to a licenced repair shop. The user has only level 1 knowledge of this servo.

The repair technician in the shop has much more knowledge of the servo-motor although he may not have a degree in electronics. The form of the knowledge the technician has is procedural so as to achieve the goal of repairing the servo-motor. He usually also possess the level 1 knowledge and uses this knowledge effectively in a step by step manner to repair the servo-motor.

However if one is interested in designing a new servo-motor controller the user may have to find someone in an engineering department of an university who understands not only how servo-motors work but also how they can be designed step by step. Further he may not always be able to repair the users servo-motor controller for a variety of reasons but he does have a deeper knowledge of the physics and electronics of control systems.

From a point of view of reasoning we may associate methodologies with each level of knowledge. We may associate case based reasoning with facts and rule based reasoning with heuristics (and facts). Conventional reasoning uses procedural knowledge and models where as model based reasoning uses explicit domain models. (Much of classical time and frequency domain control theory may be classified as model based reasoning.) Analogical and automated reasoning systems (theorem provers etc) are associated with deep knowledge. This relationship between reasoning techniques and levels in knowledge is shown in Figure 10.

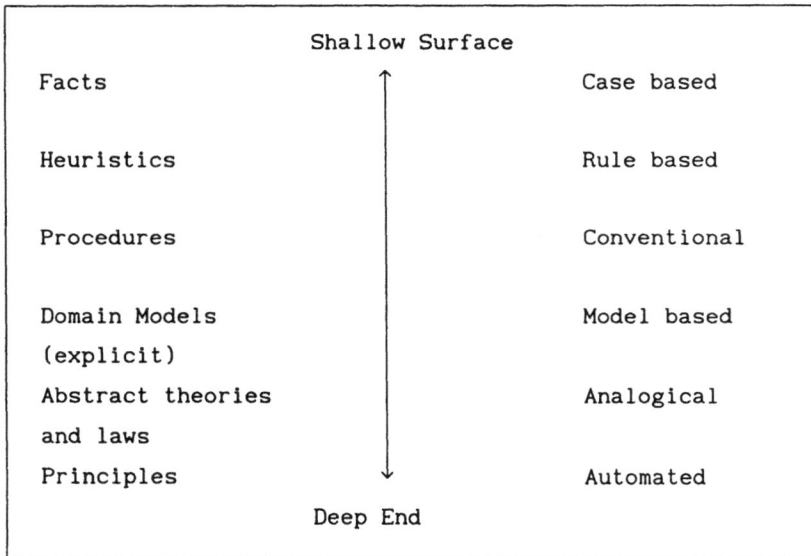

```
┌─────────────────────────────────────────────────────────────┐
│                    Shallow Surface                           │
│  Facts                      ↑            Case based          │
│                             │                                │
│  Heuristics                 │            Rule based          │
│                             │                                │
│  Procedures                 │            Conventional        │
│                             │                                │
│  Domain Models              │            Model based         │
│  (explicit)                 │                                │
│  Abstract theories          │            Analogical          │
│  and laws                   │                                │
│  Principles                 ↓            Automated           │
│                    Deep End                                  │
└─────────────────────────────────────────────────────────────┘
```

Fig. 10 Reasoning Techniques associated with levels of knowledge

The basic idea behind progressive deepening is to gradually apply reasoning techniques associated with deeper levels of knowledge [Skinner and Luger (1991)] depending on the time available. Of course this implicitly assumes

478

that the use of deeper knowledge models and the associated reasoning techniques is more time consuming than the use of the reasoning techniques associated with shallow knowledge models. This may not always be the case!

Progressive Widening is based on the idea that while restricting oneself to a certain level in the knowledge hierarchy one may widen the the scope of the knowledge base gradually within the real-time constraints. One approach used in the control theory literature is to start with aggregate models when the real time constraints are tight and use more sophisticated models if relatively more time is available for computation. This is a form of progressive widening.

Variable precision logic is based on the concept of using a coarse set of heuristics or rules and subsequently using a finer set of rules. The finer set of rules are initially sensored. Thus this gives the reasoning system the ability to reason with variable precision. While it may be argued that this is a form of progressive widening, in fact it is possible to combine both the progressive deepening and widening concepts and design variable precision systems that maintain an effective balance between depth and width. From a practical point of view this is highly desirable and variable precision logic is finding several practical applications.

Another related and important feature required of a real time monitoring system especially when dealing with uncertain data or knowledge is a truth maintenance facility. Such a facility is intimately related to temporal reasoning due to the real time constraints and cannot be discussed separately.

As an example if one considers the generate and test approach, reasoning and hypothesis generation are complicated by uncertainty in the symptom data or evidence, the availability of the data as well as the rules relating the symptoms to the hypotheses and the hypotheses themselves. There is also the question of belief or disbelief. Even if a certain piece of data is completely available without uncertainty, the 'reasoning system' may not believe in this piece of data simply because it does not fit with its various models of rational behaviour. For this reason a truth maintenance facility is essential. Truth maintenance systems can usually be designed on the basis of four basic theories of uncertainty management [Ng and Abramson(1990)] namely:

479

1) Bayesian theories based on probability and statistics
2) Fuzzy logic and possibility theory
3) The Dempster-Shafer theory of belief
4) Others such as the theory of endorsements

Of these four techniques the probability based and fuzzy logic based techniques have both found several applications. While fuzzy logic has proved to extremely useful in such applications as controller synthesis, probabilistic techniques have proved to particularly useful in diagnosis. While fuzzy logic techniques can be naturally associated with rule based reasoning, the probabilistic techniques have been naturally associated with model based reasoning although this need not be considered to be a restriction or a constraint. Probabilistic techniques of truth maintenance are based on Bayes rule which allows one to update one's belief in a hypothesis based on the probability of new evidence appearing in the knowledge base.

A practical application of truth maintenance is the use of 'Certainty factors' in expert systems. Certainty factors associate a degree of certainty with each fact and rule and as the expert system generates new facts or rules certainity factors are also associated with new data as well. Thus conclusions are also associated with certainty factors.

3.7. Examples of Real Time Expert Monitoring Systems in Aerospace
There a number of real time expert monitoring systems that have been developed for aerospace applications. Some of these are listed here and briefly described.
1) Flight Expert System (FLES): This is an expert cockpit monitoring system designed to assist the pilot in fault diagnosis. It uses a blackboard type architecture and the diagnosis is based on the generate and test paradigm.
2) SECURE: SECURE is a flight environment monitoring system designed to be used in the cockpits of commercial aircraft and has been successfully tested on a DC-10 simulator. It is designed for situation assessment and to identify any possible faults that could threaten the safety of the flight.
3) The Pilot's Associate: This is control advisory system designed to assist the pilot in the cockpit of military aircraft for a host of purposes including threat assessment, mission planning and cockpit management.
4) Predictive Monitoring System (PREMON): PREMON is a model-based monitoring system designed for a space simulator at the Jet Propulsion

Laboratory. It uses behavioural models to the sub-systems to predict their behaviour, manage the sensors and to generate alarms when necessary.

5) Smart Processing of Real-Time Telemetry (SPORT): This a classic example of a monitoring system designed to manage the data downlinked by a spacecraft telemetry system.

6) The Lockheed Satellite Telemetry Analysis in Real Time (L*STAR): L*STAR is again a telemetry data management system designed to be used with the HUBBLE space telescope. The inference engine is written in C and the system is interfaced to a simulator.

7) The Expert System for Inertial Measurement Unit (EXIMU) : This a expert monitoring system designed to manage the real time telemetry data from the space shuttle's Inertial Measuring Unit (IMU) and uses forward chaining to evaluate if there are any departures from nominal expected performance.

8) The Liquid Oxygen Expert System (LES): As the loading of liquid oxygen into the space shuttle is a complex sequence of operations with a high potential for accidents and disasters, the LES is used to monitor measurements from the launch processing system during this phase. The system is an alarm monitoring system used to monitor the process controller.

9) The Spacecraft Control Anomaly Resolution Expert System (SCARES): SCARES is designed for autonomous operation of spacecraft attitude control systems and for this purpose it monitors the downlinked telemetry data relating to the attitude control system.

4. Fault Diagnosis: A Generic Example of an Expert System

To understand the concept of fault diagnosis it is important to distinguish between failures, errors and faults. When a certain system, which is normally expected to provide a particular service to a user, is perceived by the user to have ceased to provide that service then the system can be said to have failed. In special cases systems may be expected to provide certain signals to other systems. When there is an unacceptable change in some of the characteristics of the signal from what is normally expected then the signal is said to be in error. Thus errors can be considered to be special cases of failures. In many systems failures can be related to a set of signal errors. A fault on the other hand is the cause for the failure. Thus failure detection involves the process of monitoring certain signal and comparing their relevant characteristics with what is normally expected to establish the occurrence of a set of errors. On this basis a decision can be made whether or not the system has failed. Fault diagnosis is the process of establishing the cause of the failure. There are usually several reasons for establishing the cause of the failure. Typically in flight or process control system diagnosis of a fault may be essential in establishing the changed dynamic structure of the system. Thus it is possible to identify the impact of the failure on the dynamic system easily if the cause of the failure is established. Knowing the new dynamic structure of the system after the occurrence of a failure it is normally possible to initiate corrective action to meet the performance objectives. Failure detection in practice may involve the monitoring of certain characteristics of several signals which may in turn throw some light on the cause of the fault. Thus in most practical situations a certain amount of fault diagnosis is normally possible during the failure detection phase.

There are several approaches to failure detection based on statistical and probabilistic techniques, pattern recognition analysis, rule based reasoning or connectionist methods based on neural networks. All these approaches are based on alternate approaches to decision theory. Numerical-Symbolic integration, where necessary, is achieved by the use sophisticated signal processing filters such as observers, Kalman Filters, Matched filters etc with specially designed sensitivity or robustness properties. They are generally used to compare the characteristics of the monitored signal with dynamically defined standards and generate residuals which are in a sense the indicators or errors. It is important in the failure detection stage to check if these residuals exceed certain thresholds. The thresholds themselves may be defined on the basis of statistical techniques, fuzzy sets or neural networks.

Fault diagnosis is a process of reasoning whether it is based on the sensitivity properties processed signals with and without the fault or any other features of the signals and the residuals. The reasoning systems are generally based on logic be it two-valued or multi-valued logic. Propositional calculus, which can be used to assess the truth or falsehood of propositions or Predicate calculus, which can be used to reason about the structure of the propositions are generally used in classical reasoning systems to draw conclusions. Thus classical fault diagnosis schemes can be viewed as inference engines.

4.1. Approaches to Fault Diagnosis
Generally speaking there are several approaches to fault diagnosis. (See

482

for example Patton et al(1989), Tsafestas (1989), Milne(1987), Frank(1990), Gertler(1988).] However there are a number of common features. Most diagnostic problems start with observations or monitoring of certain responses either continuously or intermittently. the next stage is the detection of a malfunction, error or failure i.e. a deviation from the expected or desirable. When a malfunction, error or failure has been successfully detected the diagnostician has to usually hypothesize about the cause of the failure. This stage therefore involves the generation of a hypothesis or a family of hypotheses. Typically this may in some form be to identify or to isolate a component or subsystem as the cause of the malfunction. It could be that the component or sub-system has failed or that it is no longer interfaced properly to the other component or sub-systems. Of course this presupposes that the system being diagnosed is an interconnected dynamic system. In fact one can view all dynamic systems as inter-connected dynamic systems and this is not a bad assumption especially if one is interested in the application of these concepts to automatically controlled systems. Once a set of hypotheses have been generated the actual solution of the diagnosis problem is relatively easily as it can progress along one of the known approaches to reasoning. More important is the actual generation of the hypotheses. A practical approach to hypotheses generation based on analogy is simulation. It can be considered as one form of fault testing. In fact some form of fault testing is unavoidable in the process of hypothesis generation. However the process can be time consuming in the context of a real time system and may have to be performed very fast. The end result is usually a set of hypotheses based entirely on the characterisation of the observations.

At the next stage in the fault diagnosis process, in most situations, one must identify a reduced set of hypotheses belonging to a restricted class relating strictly to the faults in the system and corresponding closely to the situation at hand. By repeatedly applying this principle of reducing the number of hypotheses in the set one may eventually identify the one particular hypothesis that is most appropriate to the failure under consideration. This process of repeatedly reducing the number of hypotheses may involve the generation of additional tests or real time simulations possibly at a different level of abstraction or at a sub-system level. The objective of these tests and simulations is to be able to discriminate between the effects of the different hypothesized faults and failure at hand. One approach is to precompile the effects of each of hypothesized

483

faults as symptoms corresponding to each of the faulty models of the
system. These are in the form of causal relationships between the existence
or occurrence of the fault in each of the fault models and the
characteristics of the observations. Such a precompiled list of
cause-effect relationships could be maintained in memory in the form of a
fault library. Such a database could in general be both quantitative or
qualitative or a hybrid combination of both. The problem then reduces to a
form of pattern classification.

The pattern classification approach may be applied both to faults within
the subsystem or component level or to faults at the interfacing level.
Naturally it is reasonable to ensure that each subsystem or component is
fault free and before considering the faults due to interfacing. Faults
within the subsystem or at the component level can be identified by
behavioural modelling. Interfacing faults on the other hand may require
teleological reasoning which is based on the belief that events and
developments are a result of the purpose and design that is served by them.
For example if one considers an analog computer, the behaviour of the
component op-amps may be characterised in terms of voltages, currents,
impedances and voltage gain, while a typical computing block containing an
op-amp may be described as a summer, integrator, sign-inverter, multiplier
etc. To go from a level of description in terms of voltages, currents,
impedances and voltage gain to one of summing, differencing, integration or
multiplication requires a process of abstraction which often requires an
ordered representation between the function of the computing block and the
structure of the block. Thus the teleology of the computing block must be
understood before any interfacing blocks can be identified.

Thus it is apparent that there are several approaches to fault diagnosis.
We may classify the various approaches to fault diagnosis in a multilevel
fashion based on the levels in the knowledge base and the levels in the
reasoning methodology. It is therefore important to examine the levels in
the knowledge bases available for fault diagnosis as well the levels in
diagnostic reasoning techniques to identify the approaches to fault
diagnosis.

4.2. Levels in the Knowledge Base
The fundamental layered nature of the knowledge base has already been
discussed in section 1.3 in relation to expert systems in general. When

considering fault diagnosis it is possible to identify a number of levels in the meta-knowledge base. Three basic levels may be identified within the meta-knowledge block. At the highest level we have the techniques of sequential control or scheduling of the various meta knowledge tasks. At the next level we have the various diagnostic tasks or activities. At the lowest level in the meta-knowledge base we have the tools for accessing the the knowledge in the knowledge base which may be in the form of experiential knowledge, causal knowledge and the knowledge relating to the description of the system. The model accessing tools are usually in the form of simulation analysis tools including techniques for simulating interconnected dynamic systems, connectivity analysis which facilitate not only the simulation of the structural model of the system but also incorporate the appropriate behavioural and functional models. At the diagnostic tasks level one has tools for fault detection, hypothesis generation and testing techniques relating to various abstraction levels. Tools are also available here to isolate the particular component or sub-system responsible for the fault that is identified by the process of diagnosis. At the highest level in the meta-knowledge base we have a control mechanism that is responsible for scheduling the tasks in the correct sequence so as to obtain meaningful results.

4.3. Levels in Diagnostic Reasoning

One can identify several levels in the approaches to diagnostic reasoning based. Each of the lower levels can be associated with the structural model, behavioural model and the functional model of the system. At the highest level is the pattern compilation process which allows one to develop an appropriate set of rules or a library of case histories based on the features of the knowledge base generated by the lower levels and a list of shallow assertions. This is illustrated in Fig. 11 and is based on Chandrasekhar and Milne(1985).

Thus for a interconnected dynamic system at the lowest level one has diagnostic reasoning based on connectivity. Based on the connectivity information as well as the behavioural models of each component one may construct a behavioural model of the system. One may use such behavioural models coupled with qualitative or quantitative reasoning to diagnose faults in the system behaviour. Given a specification of the functions of all relevant components and a teleological specification or an account of the intentions for which the device is used, one may also generate a

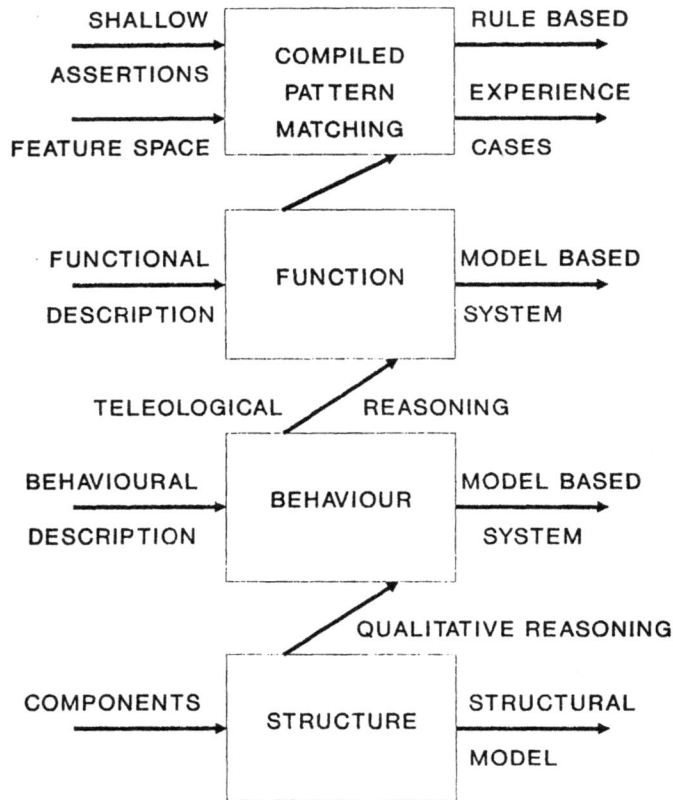

Fig. 11 Levels in Diagnostic Reasoning

functional model. The functional model may be used to simulate hypothesized functional faults and compare the resulting symptoms with the observations. The information or fault library the case histories may then be compiled to develop a rule based system for relating the observations to faults. Thus on has available both case based as well as rule based pattern classification methods of fault diagnosis.

4.4. Specialized Techniques of Diagnostic Reasoning

There are a number of specialized techniques for diagnostic reasoning which may be classed as:

1) Techniques which reason about the structure and connectivity

2) Qualitative reasoning about behaviour

3) Methods using 'deep' functional models

4) Methods using experience

5) Methods based on learning and

6) Diagnosis of the problem solving techniques rather than the device.

Structural Reasoning: There are techniques that verify conditions about the structure and connectivity within the system. A typical example of the in control systems is the use of the observability rank condition for failure detection. Such methods (parity space methods), although apparently quantitative, essentially verify the conditions of connectedness.

Qualitative Reasoning: The principal objective of qualitative reasoning is to be able to generalize and not mere data reduction. Real time diagnosis has always to be performed on data that is continuously appearing at certain information sources. These are normally translated to qualitative representations which can then be reasoned about more fully through the remainder of the expert system and also be able generalize to related situations. It is then only necessary to reason about a finite set of qualitative states as well as focus on certain states that are relevant to the behaviour or function of the device.

The use of deep models and the associated reasoning techniques has already been discussed earlier. Methods based on learning to perform a task better from the experience gained by performing it repeatedly as well as specialized techniques based on the diagnosis of the problem solving techniques are particularly relevant to control systems and closely related to intelligent control and identification techniques.

5. Applications to Control Systems

The control system designer is generally concerned with faults that have an impact on the closed loop performance and these could be further classified on the basis of control theory depending on whether they merely alter the coefficients of the mathematical model or manifest themselves as biases or spurious inputs which effectively alter the structure of the model. The first class of faults can be dealt with by *passive* techniques. Passive techniques of fault tolerance at the sensor or input level are based on the design of robust controllers which would guarantee performance even with the faults present. Algorithms that guarantee robust performance have been developed in the literature. On the other *active* techniques involve the concept of multi modular redundancy with voting or *analytical* redundancy which is essentially a sensor reconfiguration technique. Passive techniques at the output or actuator level are similar in principle to those at the sensor level. Active techniques at the output or actuator level are essentially based on dynamic *restructuring*, a dynamic reconfiguration technique.

From the point of view of digital hardware a flexible architecture for Integrated Control can be established. Then it is necessary to use fail safe parallel algorithms for reconfiguration, analytical redundancy and control filtering. These are essentially based on control theory.

The availability of relatively inexpensive digital multi-processors has led to the possibility of developing integrated control systems with a high level of fault tolerance. Monitoring, Fault detection Fault diagnosis and real time re-design algorithms have been developed and incorporated into self-repairing control systems.

A primary feature of any reconfiguration scheme is failure detection and fault diagnosis. From the point of view of Flight Control Systems, any abrupt change in the plant/actuator/sensor dynamics could be regarded as a fault. Failure detection and fault diagnosis is the first step before any corrective action can be taken. The design of a fault diagnosis scheme for a set of hypothesized faults involves several considerations. It must essentially be a fast system with minimum detection delay (and hence sensitive to the faults) while also ensuring that false alarms are minimized. On the other hand it must be robust to modelling errors and take into account existing system redundancy. Willsky (1976) and Basseville (1988) present a survey of failure detection systems applicable to Flight Control Systems.

Based on a number of proposed concepts and designs related to the design of restructurable and reconfigurable flight control systems a number of common underlying concepts related to monitoring and fault diagnosis in flight control (Montoya *et al.*, 1982; Stifel *et al.*, 1988.) may be identified. These ideas have been brought together and form the basis of the general concept of a self repairing flight controller. This concept of a self repairing aircraft flight controller is discussed in detail. The problem of a self-repairing aircraft controller can be stated as follows: Upon failure of a control element, the flight control system is to be reconfigured or repaired (restructured) in such a way that the aircraft recovers to a safe condition and can then be flown, either manually or automatically, to a

488

safe landing. It is assumed that this is possible in principle provided the flight control computer and the aircraft sensors are fully operational. The principle of a self-repairing flight control is illustrated in Fig. 12.

We may then classify the components of such a self-repairing system into

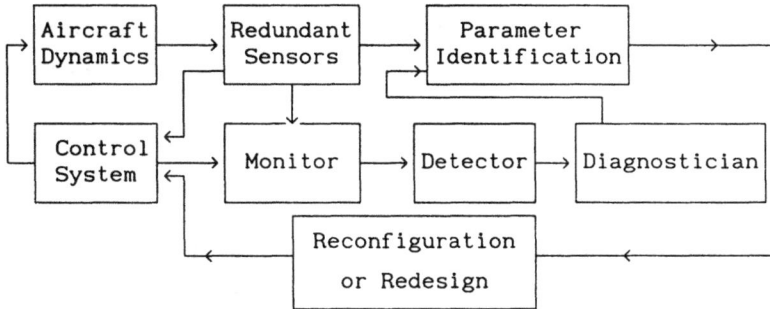

Fig. 12 Principle of a self-repairing flight control system

four basic sub-systems. These are:
a) The Monitoring system (Monitor)
b) The Failure detection system (Detector)
c) The Diagnostician
d) The repair/reconfiguration/redesign system (Redesign module)

Recovery from a failure as a result of a fault not only involves the assessment of the damage caused by it but is also followed up by an attempt to repair the fault or to reconfigure the system so as to mask its effects in some sense. In a flight control system this means that the "control laws" must be altered to account for the impact of the fault on the aircraft's motion. Damage treatment or fault treatment is affected by the provision of alternate control laws. Such a provision can be made by one of several different techniques such as, gain scheduling, controller scheduling, real time redesign, real time search techniques. While the first of these techniques is the simplest each of the latter techniques involves a degree of complexity not present in the earlier technique.

Clearly the system has the structure of an adaptive feedback loop. A detailed description of the first three subsystems is now presented.

489

Based on the concept of *observers* monitoring can be efficiently performed while the system is in its normal use, without introducing any significant perturbations into the behaviour of the system in question. One has to assume that both the input and and output are *accessible* for this purpose. However monitoring is possible even if the input is not accessible. The popular monitoring systems in use in avionics are basically observers that have been optimized so that the variances of the monitored variables are a minimum. Such optimized observers that reconstruct all the states of the system are also known as *Kalman filters*.

A fundamental requirement in most detection processes, the existence of a *threshold* is either known explicitly or is implied. For example in radar detection of an object is often described by stating that the signal flux at the receiving antenna has exceeded some pre-defined level.

Hypothesis testing is the very basis of statistical detection. In some respects the theory is formulated as one in hypothesis testing for analytical convenience. However experience has provided the logic and justification for such an approach. A complete theory is based on assigning penalties to incorrect decisions and payoffs for correct decisions and includes a measure for evaluating the performance of the detection process. The threshold concept is also naturally encountered in theory. The measure that is compared against a threshold is known as *likelihood ratio*.

A failure detection decision criterion will now be described which does take proper consideration of decision errors and penalties. For certain special situations the precise criterion can be related to the mean square error or the maximum signal to noise ratio. The detection criterion is motivated by the fact that the failure detection process is fundamentally a problem in *statistical* hypothesis testing. The act of determining the presence or absence of a signal in a noise environment is equivalent, as stated in the terminology of statistics, to testing the hypothesis H_0 that noise alone is present against the alternative hypothesis that signal plus noise is present.

To use such failure detection criteria in flight control one computes the residual or difference between an output from a sensor and a corresponding reconstructed signal from the monitor and the failure detection methodology is applied to this residual. The detector thus establishes the existence of

490

a failed condition.

In some cases, where the outcome detection process is known with certainty, the type of decision problem is deterministic and the reasoning process can be implemented using well known techniques. If, on the other hand, the outcome from a given detector could still be influenced by chance or random events then the decision could be probabilistic (statistical), possibilistic (fuzzy decision making) or believable (other techniques).

A number of simple methods based on hypothesis testing are commonly used in statistical detection. Some of the more advanced techniques include the Page-Hinkley stopping rule, the maximum aposteriori probability (MAP) detector, the Maximum likelihood detector (ML), the Sequential Probability ratio test, the Generalized likelihood ratio test and the application of multiple hypothesis testing to banks of Kalman filters. These and other related topics are reviewed by Willsky (1976), and Basseville (1988).

At this stage it is important to isolate the failed element responsible for the particular failure detected (e.g. an actuator, a sensor, a control surface etc..) From a quantification of the impact of various failures on the aircraft's motion, the impact of the detected failure on the aircraft's motion is established. These can be represented as failure constraints on the mathematical model of the aircraft. This is basically an exercise involving a knowledge base of the impact of various failures on the aircraft's motion and on the mathematical model representing the aircraft. Thus it involves an Expert systems approach in general and can be considered to be a form of *damage assessment*. Damage assessment may be based entirely on *a priori reasoning* by the designer, or it may involve the system itself performing a number of checks to determine what damage has occurred. In either case this assessment is similar to *fault identification* and will be guided by the causal relationships derived from the knowledge base which is in turn defined by the system structure.

To fix ideas more precisely about the function of the diagnostician the system structure of an unfailed system may be defined by a state space model as:

$$\dot{X}(t) = A\ X(t) + B\ U(t), \qquad Y(t) = C\ X(t)$$

After the occurrence of a plant failure of type i (sensor failures are not

491

considered here but could be included at a later stage if necessary), the state space representation takes the form:

$$\dot{X}(t) = (A + Q_i \Delta A_i)X(t) + B U(t) + Q_i \Delta B_i U_i(t),$$
$$Y(t) = C X(t)$$

where the matrix Q_i is referred to as the *failure mode* matrix and $U_i(t)$ is a vector consisting of fault inputs and/or unknown disturbances. To demonstrate certain interesting features of the failure mode matrix the concept of plant failure sensitivity is introduced. The *plant failure sensitivity matrix* corresponding to the particular failure mode Q_i is defined as:

$$S = [Q_i \quad AQ_i \quad A^2Q_i \quad A^3Q_i \quad \dots \quad A^{n-1}Q_i]$$

It is the controllability matrix corresponding to the pair (A, Q_i). The significance of this matrix S is that if it is of rank n then all states of the system are sensitive to the particular failure mode Q_i. On the other hand if the rank is less than n it follows that certain states are insensitive to the failure. By appropriately balancing (Moore, 1981) the triple (A, Q_i, C) or transforming the states of the system by a similarity transformation, the transformed states can be ordered according to the failure sensitivity or failure observability. Failure observability is important if the states are to be monitored and is guaranteed if the matrix CS is of full rank.

The ordering of states according to their relative sensitivity or observability may be exploited to design failure sensitive or failure insensitive filters that may be used to monitor the systems. After a failure is detected a hypothesis for the possible failure mode is generated. This in itself may be a multi-step process involving fault identification and confirmation as there may exist a number of failure modes corresponding to each fault. The process of damage assessment is then carried out by comparing the characteristics of the monitored variables to a table of pre-computed characteristics maintained in the form of a library corresponding to each possible failure mode.

The purpose of the comparison is to test whether the hypothesis could explain all the abnormal data gathered from the monitoring system. This is the basis on which the diagnostician functions and it is principally responsible for the selection of the appropriate failure mode matrix Q_i

using a practical "*reasoning*" technique. Once the failure mode is identified the complete plant model parameters can be estimated by parameter identification techniques to complete the identification of the failure. For a finite dimensional plant all possible failure modes can be covered in this way; i.e. the characteristics of all possible failure modes can be predicted. However in practice process plants are basically dimensionally indeterminate even if they are finite dimension and so there may exist a set of unpredictable failures. Any failure mode not identified by the above process is considered to be unpredictable. In such cases the failure mode matrix is assumed to be $Q_1 = I$, where I is the identity matrix. The practical benefit of failure mode determination is that the number of parameters that need to be estimated may be considerably reduced for certain failure modes thus making the parameter identification process more reliable as well as robust. Thus the diagnostic step involves damage assessment and failure identification. Although there are several techniques that play a critical role in the diagnostic step, three of these are extremely important and are:

1) Failure Modes, effects and criticality analysis

The principle here is to consider each mode of failure in turn for every component of a system and to ascertain the effects of the failure on the relevant aspects of the system operation.

2) Fault tree analysis

This is an analysis technique which starts from consideration of system failure effects, referred to as 'top events', and proceeds by determining how these can be caused by individual or lower level failures or events which correspond to a known set of failure modes. It is a strictly top-down approach and multiple failures are considered as a matter of course.

3) Methods for identifying the important failure modes based on statistical experiment design and Taguchi methods.

These methods are based on the analysis of variance and are also the basis for design for variation inherent in parameters, environments and processes. The latter can form the basis for making strategic AI (Artificial Intelligence) type pruning decisions.

Practical diagnostic reasoning can be implemented by one of several methods. However the most common method is based on the 'Generate and Test Paradigm'. The steps involved are:

i) Generation of a hypothesis from a knowledge of the active residuals. Residuals that exceed the threshold of the detector are said to be active.

493

Error residuals that are active over a length of time indicate that certain assessments or actions are appropriate. The hypothesis are generated from decisions supported by observations of active error residuals.

ii) Establish the cause of the fault based on the previously obtained Fault Tree analysis database.

iii) Establish the expected symptoms based on the failure modes and effects database.

iv) Compare the expected symptoms and the actual features of the residuals to test the hypothesis.

v) If the test is successful the reasoning is complete and the hypothesis is deemed to be true. If not one must adopt a strategy to alter the hypothesis suitably and repeat the process.

vi) The alteration of the hypothesis could be based on data obtained from:

1) the current comparisons of the expected symptoms and the observed features or,

2) the knowledge base already generated in the past.

vii) If the test is still not successful and all hypothesis alteration procedures have been exhaustively searched it is necessary to alter the symptoms. This means that the failure modes and effects analysis must be repeated (possibly in real time) in the light of the new data to generate a new set of symptoms. This symptoms analysis is generally quite expensive to perform in real time and usually the system may adopt some form of default reasoning which results in a graceful degradation of performance.

The most important steps in this process are certainly the generation of a suitable hypothesis and the adoption of an appropriate strategy for altering the hypothesis. In practice adopting such a methodology can be simplified by the use of some sophisticated techniques of database restructuring. Techniques developed in the literature to restructure the database permit the use of *decision trees* and other graphical techniques as well as the associated data structures. Using such techniques it is possible to effeciently program the methodology of fault diagnosis for fast real time applications. They also form the basis for the development of *Control Advisory Systems* both for the control of a process plant and cockpit management of modern aircraft.

6. References

Basseville, M. (1988), "Detecting Changes in Signals and Systems: A survey", *Automatica*, <u>24</u>, 3, 508-516.

Chandrasekhar B. and Milne, R. (1985), 'Reasoning about structure, behaviour and function,' In: *SIGART Newslett.*, (Chandrasekhar, B. and Milne, R., Eds.) <u>93</u>, 4-59.

Favier, G. and Senges, P. (1989), "A Knowledge-Based System for CAD in Automatic Control and Signal Processing", *IEEE Control Systems Society Workshop on Computer Aided Control System Design (CACSD), December, 1989*, Tampa, Fl., U.S.A.

Frank, P. M. (1990), "Fault Diagnosis in Dynamic Systems using Analytic and Knowledge based redundancy : A survey of some new results", *Automatica*, <u>26</u>, 3, 459-474.

Gertler, J. J. (1988), "Survey of Model based Failure detection and isolation in Complex Plants", *IEEE Control Systems Magazine*, 12, 3-11.

Grimshaw, J. D. and Anken, C. S. (1991), 'A Distributed Environment for Testing Cooperating Expert Systems', In: AGARD Conf. Proc. on *Machine Intelligence for Aerospace Electronic Systems*, AGARD-CP-499, paper no. 2.

Hayes-Roth, B. (1985), "A Blackboard Architecture for Control", *Artificial Intelligence*, <u>26</u>, 251-321.

Jakob, F. and Suslenschi, P. (1990), "Situation Assessment for Process Control", *IEEE Expert,* 4, 49 - 59.

Laffey, T. J., Cox, P. A., Schmidt J. L., Kao S. M. and Read J. Y. (1988), ' Real Time Knowledge Based Systems', *AI Magazine*, <u>9</u>, 2, 27 - 45.

Milne, R. (1987), 'Strategies for Diagnosis', *IEEE Trans. on Syst. Man Cyber., Special Isuue on Causal and Diagnostic Reasoning*, <u>SMC-17</u>, 3, 333-339.

Montoya, R. J., Howell, W. E., Bundick, W. T., Ostroff, A. J., Hueschen R.M. and Christine M. Belcastro, (Editors) (1982) 'Restructurable Controls,' NASA CP 2277, *Proceedings of a workshop held a NASA Langley Research Centre*, Hampton, Virginia.

Moore, B. C. (1981), "Principal Component Analysis of Linear Systems: Controllability, Observability and Model Reduction," *IEEE Transactions on Automatic Control*, <u>Ac-26</u>, 2, 17-32.

Ng, Keung-Chi and Abramson, Bruce (1990), "Uncertainty Management in expert Systems", *IEEE Expert*, 4, 30-48.

Patton, R. J., Frank, P.M. and Clark, R. N. (1989)(Eds.) '*Fault Diagnosis in Dynamic Systems*', Prentice Hall.

Pau, L.F.(1986), 'Survey of expert systems for fault detection, test generation and maintenance,' in *Expert Systems*, Vol. 3, No. 2, pp 100-111.

Pomeroy, B. D., Spang, H. A. and Dausch, M. E. (1990) "Event-based architecture for diagnosis in control advisory systems", *Artificial Intelligence in Engineering*, 5, 4, 174 - 181.

Pomeroy, B. D. and Irving, R. R.(1990), 'A Blackboard approach in Diagnosis in Pilot's associate", *IEEE Expert*, 8, 39 - 46.

Skinner, J. M. and Luger, G. F.(1991), 'A Synergistic Approach to Reasoning for Autonomous Satellites,' In: AGARD Conf. Proc. on *Machine Intelligence for Aerospace Electronic Systems*, AGARD-CP-499, paper no. 5.

Stifel, J.M., Dittmar, C. J. and Zampi, M. J.(1988), 'Self Repairing Digital Flight Control System Study', *Final Report for Period Jan. 1980-Oct. 1987*,' ARWAL-TR-88-3007, GE Aircraft Control Systems Dept., Binghampton, NY., U.S.

Tzafestas, Spyros G.(1989), 'System Fault diagnosis using the Knowledge based methodology', Chap. 15, In: Patton el at(1989), pp 509 - 594.

Willsky, A. S.(1976), "A survey of design methods for failure detection systems ", *Automatica*, 12, 601.

Expert Governed
Numerical Optimization

R.A. Vingerhoeds

Automatic Control Laboratory, University of Ghent.

Abstract

Numerical optimization techniques can solve many problems in an elegant manner. However, applications on large scale are hampered by convergence problems. Artificial intelligence techniques have complementary characteristics to procedural programming languages. Using expert system technology a good part of these convergence problems can be solved. Using a balanced combination of conventional programming languages and expert system programming languages, a totally new approach for numerical optimization has been created, which has been given the name Expert Governed Numerical OPtimization (EGNOP).

Keywords

Numerical Optimization; Artificial Intelligence; Expert Systems.

Introduction

From extensive studies, the practical use of numerical optimization methods has become clear. In many domains applications can be found, where problems to be solved can be represented in the form of a system description and a cost function to be minimised. Numerical optimization can solve many problems given in this format in an elegant manner. However, applications on large scale of these techniques are hampered by the following problems:

- the technology is not very well known in industry,

- much knowledge on numerical optimization is required for successful applications,

- much initial preparations are needed when numerical optimization programs are not coupled with existing analysis programs,

- a good convergence to a solution can not be guaranteed beforehand,

- much effort may be needed to obtain a practical result.

497

In general, much knowledge is required on the use of the numerical optimization methods involved, and the actual application may be cumbersome. Even more, interpreting and evaluating the results of such optimization runs, requires a thorough knowledge on the problem domain involved. These two issues do pose a barrier to using numerical optimization techniques by people who are no experts in both domains.

The recently introduced EGNOP concept (Expert Governed Numerical OPtimization) [Vingerhoeds, Netten, Ramon and Boullart 1990] can solve a good part of the existing problems. Using a balanced combination of conventional programming languages and expert system programming languages, a totally new approach for numerical optimization has been created. The main goal of the EGNOP concept is to make numerical optimization techniques accessible for a larger group of users, who are not experts in both domains. In this text, Expert Governed Numerical Optimisation will be discussed, as an example of new directions that can be found using expert system technology. Special attention will be given to the combination of expert system programming languages and numerical (procedural) programming languages.

Optimization

Optimization is a design activity, of which applications can be found in a variety of domains. One may think of optimal design of the aerodynamic flow around wings, where a minimum drag is required with a maximal lift. Or one may think of the design of a construction where a maximal strength is desired using a minimum of construction material (as light as possible). In control, one may think of minimum-time problems, minimum fuel consumption problems, etc. In optimization a certain wished behaviour is translated into a criterion to be minimised. This is an elegant manner of solving a given problem.

The central goal in optimization is to find a solution for a problem, in which a given cost function is to be minimised or maximised. The problem definition consists of a system description, a cost function and often a set of constraints. The mathematical description of the system defines the behaviour of the system with respect to its environment in terms of differential equations or other mathematical relations. The cost function describes the criterion to be optimised. This criterion may be a very complex function of the design variables and the input or disturbance signals. The design variables are variables of which the values are to be the result of the optimization. The feasible region for the solution of the problem may be limited by means of equality or inequality constraints. The optimization program performs an iterative process, in which a starting point for the design variables is adapted until the convergence criteria are met. A optimization program determines in each iteration a best search direction, followed by a line minimisation in this direction. Depending on the status of the constraints, the solution is reached or a problem has occurred during execution.

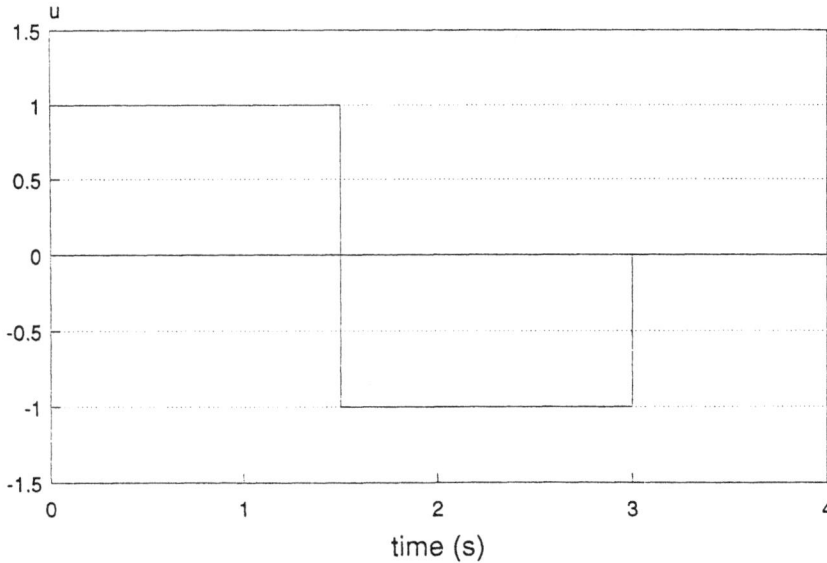

Figure 1: Control sequence nuclear reactor

In the following example, a minimum time problem is presented, where a criterion will be minimised, resulting in the required control sequence.

Example (from [Noldus 1990])
In nuclear power centres somtimes fast starting of the reactor is required. The reactivity is determined by the state of the control bars in the reactor. If u_0 is the maximum speed with which these bars can be displaced. There is a maximum speed:

$$| u | \leq u_0$$

The goal is to bring the reactor in a minimum time from a stationary state n_0 to a new setting n_f, with $n_f > n_0$. In figure 1 the resulting control sequence is shown.

Note that the control sequence is one of the type bang-bang control. It can be shown that for such control problems bang-bang control is optimal. Of course, for implementations in real systems bang-bang control sequences can not always be realised. The steps are very sharp and can usually not be realised in real systems.

Using analytical techniques to find the minimum of a criterium, as was done in the example, solutions can be found for simple problems. For more complex problems however, these techniques are not sufficient. Some problems can not be solved using these analytical techniques. In these situations use can be made of numerical techniques to minimize functions. In this way also bigger, more complex problems can be handled. For many engineering problems, analytical techniques are not sufficient to bring a solution and numerical methods are therefore necessary to solve the problem.

499

In the following example a non-linear function will be minimised.

Example
Given is a set of nonlinear differential equations, differentiated with respect to the dimensionless time τ:

$$\dot{x}_1 = t_f x_2$$
$$\dot{x}_2 = t_f(-x_2 + 0.1x_2^3 + x_3)$$
$$\dot{x}_3 = tf(-x_3 + u)$$

The time is defined in the interval $[0,1]$, with $\tau = t/t_f$. t_f is the end-time of the problem. The system has to be controlled from the point $x_0 = (0,1,0.9)^T$ to the point $x_f = (3,0,0)^T$ in the time t_f. The following cost function has to be minimised:

$$f(\underline{x}) = (x_1(t_f) - 3)^2 + x_2(t_f)^2 + x_3(t_f)^2$$

On the control action, an limit of 1 ($| u | \leq 1$) is imposed. After a numerical n optimisation, the following switch times resulted:

$$t_1 = 1.889$$
$$t_2 = 3.416$$
$$t_3 = 3.851$$

This is also a bang-bang control, but here a smoother transition from one control state to another is implemented. The resulting control sequence is given in figure 2.

The convergence of the iterative process is determined by the chosen algorithm and the difficulty of the problem at hand. It can be influenced by many factors, such as the starting points of the design variables and the operational parameters which tune the optimization. Selecting the best algorithm is a problem in itself. It requires a good understanding of both the mathematical complexity of the physical problem and the applicability of the algorithms. Furthermore, at different stages in the optimization process, different algorithms may give the best performance. For most algorithms, a starting point or an initial design and operational parameters should be provided by the user. Faulty values may affect the convergence drastically and may even result in divergence. Some methods are less sensitive to these numerical values than others, but they may have a less favourable behaviour in the neighbourhood of the optimum.

In general the user has to ensure proper convergence by taking actions like changing the values of the operational parameters, the starting point, restart options, stepwise finetuning of the accuracies and the penalty factors, changing to other algorithms, etc. In conventional

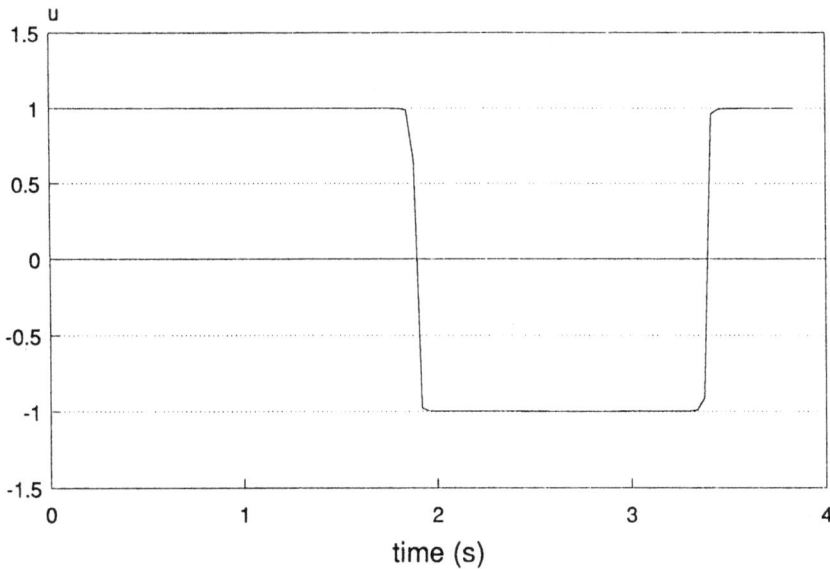

Figure 2: Control sequence nonlinear problem

programming languages many attempts have been made to ensure proper convergence by making use of IF-THEN structures. These can contain convergence tests at several stages of the optimization process as well as plain heuristics. When the decision process gets more complex this leads to a very difficult to read program and still the user remains in charge to monitor the optimization process and to decide what actions are to be taken to ensure the convergence. This requires much skill and experience from the user. The effect is that such a program can only be used efficiently after a long training period. Furthermore, the user can only change parameters etc. after the execution of an optimization run. A modified problem formulation will again require considerable computational efforts.

A second problem that arises in practical use of numerical optimization programs concerns the interpretation of the final results of an optimization run. These can be rather impractical for the end-user or even incorrect. The cause may be found in an incorrect or an incomplete formulation of the problem. Often such conclusions could not have been predicted at the beginning, but they can only be deduced from the results themselves. In the problem reformulation, domain dependent knowledge is used.

From the above it is clear that for efficient use of numerical optimization algorithms, knowledge is required on the numerical optimization algorithms, the problem at hand and on the required solutions. Several runs are usually needed to find a practical result and each run has to be executed in full before the user can improve the problem formulation. The total number of evaluations and the CPU-time will then increase considerably and so is the time spent by the user himself.

501

Constrained / Unconstrained Optimization

A non-linear optimization can be described as follows:

Find an $\underline{x} \in \Re^n$, for which holds that $f(\underline{x})$ has an absolute or relative minimum. For $\underline{x}$ some constraints may exist:

$$g_i(x_1, \ldots, x_n) \geq 0 \qquad i = 1, \ldots, m$$
$$h_j(x_1, \ldots, x_n) = 0 \qquad j = 1, \ldots, p$$

It will be assumed that the functions f, g_i and h_j are continuous and have first and second order derivatives.

The criterion to be minimised is called $f(\underline{x})$ and the vector $\underline{x}$ contains the design variables. These variables are assigned values during the optimization process. The result of an optimization is formed by a set of values for $\underline{x}$, for which f reaches a minimum. Furthermore, some restrictions may be imposed upon the solutions. These restrictions, called constraints, may be either equality constraints or inequality constraints and are represented with the functions g_i and h_j. Clearly constrained optimization asks for a different approach then unconstrained optimization.

Unconstrained Optimization

For unconstrained optimization many algorithms have been developed. No constraint functions g_i or h_j has been defined; there are no restrictions on the solution space.

A necessary condition for a (local) minimum is:

$$f_x(\underline{x}) = 0$$

This gradient has the dimension n. The second order derivatives can be collected in the so-called Hessian-matrix F_{xx}. For a relative minimum, the following condition is necessary: $f_x(\underline{x}) = 0$ and F_{xx} non-negative definite. A sufficient condition for a strict relative minimum is: $f_x(\underline{x}) = 0$ and F_{xx} positive definite.

For convex functions these can be simplified, because they always fulfill the non-negativitiy requirement. A local minimum is than a global minimum.

Constrained Optimization

When functions g_i and / or h_j are defined, imposing restrictions on the solutions space, this solution space B can be defined as:

502

$$B = \{\underline{x} \mid g_i(\underline{x}) \geq 0, i = 1, \ldots, m \mid h_j(\underline{x}) = 0, \; j = 1, \ldots, p\}$$

A solution must be found such that in the optimum point $\underline{x}^*$ the following holds:

$$f(\underline{x}^*) \leq f(\underline{x}) \qquad \forall x \in B$$

In order to take constraints into accound when optimising a cost function several methods have been developed. In general, the techniques consist of introducing some kind of penalty factors for violating the constraints and adding these terms to the cost function, which consequently can be minimised, as if it were an unconstrained optimisation problem.

Loss functions were designed especially for equality constraints. These functions, also referred to as exterior penalty functionstake the constraints squared and sum these to the cost function. Thereby a penalty factor $r(> 0)$ is used, which can be reduced in successive iterations. The resulting object function is:

$$L(\underline{x}) = f(\underline{x}) + \sum_{j=1}^{p} h_j^2(\underline{x})/r$$

Inequality constraints may be handled as well with loss functions, by treating them as equality constraints when they are violated and else disregarding them. When $f(\underline{x})$ and $h_j(\underline{x})$ are convex and $g_i(\underline{x})$ is accidently concave, then the result will still be convex.

Another method of taking constraints into account is formed by the so- called barrier functions or interior penalty functions. These are more suitable for inequality constraints. The following object function results:

$$B(\underline{x}) = f(\underline{x}) - r \; ln \; g_i(\underline{x}) \qquad r > 0$$

These functions both have advantages and disadvantages. Under certain circumstances the optimal point $\underline{x}^*$ that is found does not fulfill the constraints. Sometimes there are restrictions on linear searches, etc. Therefore mixed functions can be used, which try to use the positive sides of the loss functions and the barrier functions. The can start from any point, there are no restrictions on linear searches, the optimal point $\underline{x}^*$ lies in the allowed interval.) and the barrier function ($\underline{x}^*$ lies within B).

A special technique of taking into account the constraints is formed by the so-called Augmented Lagrange functions. These have favourable characteristics. When only equality constraints are taken into account, the following Lagrange function can be defined:

$$L(\underline{x}, \underline{\lambda}) = f(\underline{x}) - \underline{\lambda}^T \underline{h}(\underline{x})$$

This function has to be minimised with respect to $\underline{x}$ and then maximised with respect to $\underline{\lambda}$. Because the defined Lagrange function is not necessarily convex in the neighbourhood of the optimum, a convexification, using an extra term, has to be performed:

503

$$L_a(\underline{x}, \underline{\lambda}) = f(\underline{x}) - \underline{\lambda}^T \underline{h}(\underline{x}) + w \underline{h}^T(\underline{x}) \underline{h}(\underline{x}) \qquad w > 0$$

In the optimal solution pair $(\underline{x}^*, \underline{\lambda}^*)$ the following holds:

$$L_{a_x}(\underline{x}^*, \underline{\lambda}^*) = 0$$
$$L_{a_\lambda}(\underline{x}^*, \underline{\lambda}^*) = 0$$

This yields:

$$f_x(\underline{x}^*) = \underline{h}_x(\underline{x}^*) \underline{\lambda}^*$$
$$\underline{h}(\underline{x}^*) = \underline{0}$$

It can be proven that these equations, also called the Kuhn-Tucker conditions, have to be fulfilled for an optimum in constrained optimisation. As long as w is chosen big enough to keep the Hessian matrix F_{xx} positive definite, $\underline{x}^*$ can be regarded to be the solution of the problem. Inequality constraints, again, can be treated by regarding them to be equality constraints when they are violated and else disregarding them.

Function Minimisation Techniques

In the heart of numerical optimisation, function minimisation techniques are used to minimise the constructed object functions. There are several groups of numerical function minimisation methods, each of them having a specific behaviour regarding CPU-time and requiring a specific amount of information. Here, a general introduction to these methods will be given. Please note that here only a very brief introduction will be given to illustrate the text. For more in depth information about these methods, their characteristic and their backgrounds the reader is referred to the literature (for example [Lootsma 1984a and 1984b]).

Methods of the order 0 0th order methods use only the function values of the object functions, to calculate a next $\underline{x}$ and are also called direct search methods. A well known method in this class is the method of Hooke and Jeeves. Starting from the current point $\underline{x}_i$, a new point is searched, such that the successive function values decrease during the iterations:

$$f(\underline{x}_0) > f(\underline{x}_1) > \ldots > f(\underline{x}_i) > \ldots > f(\underline{x}_n)$$

These methods are very robust, but tend to slow down in the neighbourhood of the optimum.

504

First order methods First order methods have a better convergence than 0th order methods. Next to the function values of the object function, also the gradient $f_x(\underline{x}_i)$ of the object function is used for determining a new point. These methods are also regarded to be direct search methods. A well known example is the gradient method, which searches in the direction of the negative gradient $f_x(\underline{x}_i)$ of the object function and is also often referred to as the 'steepest decent' method. A new point in the iteration process can be found using:

$$\underline{x}_{i+1} = \underline{x}_i - k f_x(\underline{x}_i)$$

The variable k has to be chosen such that the new function value is lower than the previous value and has a very big influence on the convergence. The choice of k can be chosen constant or can be modified during the optimization process. When the difference between two successive $\underline{x}$ becomes smaller than a given criterium, the iteration process stops.

As a variant on the gradient method, the so-called conjugate gradient method can be used, which generally has a better convergence. The successive search vectors are all orthogonal (conjugated) and can be calculated using the gradients of the last two iterations. In order to avoid slow convergence in the neighourhood of the optimum, a reset of the search vector to a normal gradient step every $n + 1$ iterations is advisable.

Second order methods Second order methods use besides the function values of the object function and the gradient, also the second order derivatives. This method belongs to the so-called indirect search methods, because the optimum is found by iterative solving the optimality condition. The Newton-Raphson formule is used to solve $f_x(\underline{x}) = \underline{0}$. Using the method of Newton results in a stationary point, of which the nature has to be determined. The following holds:

$$\underline{x}_{i+1} = \underline{x}_i - k F_{xx}^{-1}(\underline{x}_i) f_x(\underline{x}_i)$$

If the object function is exact quadratic, then will this method lead to the desired point in one step with $k = 1$. When this is not the case and also the local behaviour of the cost function can not good be approximated by a quadratic function in the neighbourhood of the optimum, then is this method not very efficient. The convergence area is limited. Second order methods are quite computer time consuming and a appropriate moment to apply these methods is when the current $\underline{x}$ is in the neighbourhood of the optimum. Far from the optimum, this method is not very efficient.

Quasi-second order methods A disadvantage of second order methods is that they are quite computer time consuming. The calculation of especially the second order derivatives asks a lot of time. The so-called quasi-second order methods need less time, but have a comparable convergence. A very well known method in this class is the method of Broyden. In successive iterations, the inverse of the Hessian matrix F_{xx} is approximated,

505

using a so-called update formula. These approximations of are inexact, but the errors on the estimates decrease with $\mid \underline{x}_k - \underline{x}_{k-1} \mid$. As such this method converges to second order methods. Whenever an update of F_{xx}^{-1} fails and the new approximation turns out not to be positive definite, then a reset of the approximation to a unity matrix can be performed. The successive search vectors are all conjugated.

Line minimisation techniques   Opposed to the methods described earlier, line minimisation techniques search in a straight line for a point with a lower function value. These techniques are very usefull in combination with one of the other function minimisation methods. Once a good search direction has been obtained, using one of the other methods, a line minimisation technique can find the point with the lowest function value in that search direction. This is done by varying the value of in an efficient manner until a minimum has been found. It is important to ensure a good combination of a search algorithm and a line minimisation technique.

A number of line minimisation techniques are used frequently: the unity step, the second order and third order line minimisation techniques and the Golden Section method. In practical implementations of these methods, often a lot of heuristics are used.

Problems with existing programs

In short the problems with existing optimisation programs can be seen in two aspects: convergence ensurance and interpretation / evaluation of results. These problems of hamper frequent use of non-linear optimisation programs. Easier, but less powerfull, methods are often preferred by users, in order to avoid the problems.

Convergence ensurance

The convergence of an optimisation is dependend on the problem at hand. Moreover, at different stages of the optimisation another approach for the optimisation might give a better result. The choice of the 'correct' optimisation method and within the methods a 'good' set of values for the operational parameters is vital and requires a lot of experience from the user.

Optimisation method.   The choice of an optimisation method is of vital importance. The method should 'fit' the problem at hand. Some methods are very sensitive to variations in the start values for the design variables, other not. Some methods have a very good performance far from the optimum, but slow down in the neighbourhood of the optimum. Other methods may have exactly the oposite characteristics. Every method has its own behaviour, needs another amount of information for the determination of a new step and consumes a different an amount of CPU time. At different stages in an optimisation, an other method might give a better performance. It is clear that there exists no unique choice for an optimisation method. This leaves the user with a very complex choice.

Operational parameters. Within an optimisation method, operational parameters tune the performance. The choice for the values for these parameters has a big impact on the convergence. A good choice leads to a good performance, a less fortunate choice can even lead to divergence of the iteration process. Examples of the operational parameters are the desired accuracy and the restart values of the methods. The choice of the values for the operational parameters is vital and requires considerable knowledge on the optimisation methods.

Concluding The user has to choose the method and the values of the operational parameters. This is not an evident task, since it directly includes the problem at hand, the behaviour of the methods and the influence of the operational parameters. Furthermore, during an optimisation the choices might have to be changed for an optimal convergence. Usually, however, the choices can only be made at the end of an optimisation run. In this way, a lot of iterations will have been lost and a lot of time wasted. Choosing the methods and the values of the operational parameters requires a lot of experience and for a good, efficient use of the programs, the user should be an expert on the domain of numerical optimisation.

Interpretation of the results

When an answer has been found, the validity of the answer has to be verified. Sometimes the answers are not very practical for the user. Allthough the answer fulfills the criteria specified in the optimisation, the practical use is not guaranteed. This stems from the fact that many criteria can not be formulated in the form of mathematical equations. A reformulation of the optimisation problem can lead to a new optimisation run, with as goal a better estimate of an optimal, usable answer.

Expert Governed Numerical Optimization

The recently introduced EGNOP concept (Expert Governed Numerical OPtimization) can solve a good part of the existing problems. Using a balanced combination of conventional programming languages and expert system programming languages, a totally new approach for numerical optimization has been created.

As opposed to conventional (procedural) programming languages, expert system programming languages can be used efficiently for implementing reasoning mechanisms. Conventional programming languages are very good for numerical operations, but not very suitable for reasoning mechanisms. In numerical optimisation, using conventional programming languages, many attempts have been made to solve the mentioned problems using many

IF-THEN statements. This usually led to hard to read programs and the result was often not the desired solving of the problem.

Using the best of both programming techniques, a good improvement can be achieved. The reasoning mechanisms can be ported to an expert system structure, tuning the numerical routines, written in procedural languages. The goal for this is to make nonlinear optimisation techniques available for a larger group of users, that are not necessarily experts on the domain of numerical optimisation. Optimisations can be done more efficiently, when the methods are tuned by such an expert system structure. The interaction time between the program and the user is minimised and from the user no special knowledge is expected.

Decisions that the expert system structure can take involve on-line selecting an other optimisation method if the progress of the optimisation is not sufficient. A higher order method or a lower order method may be selected. Also within the methods, an other line minimisation technique may be selected, or other values for the operational parameters may be chosen.

Expert system technology is used to govern numerical optimization. The total layout of the new optimization approach can graphically be represented in the form of a sphere, consisting of two shells and a core. The inner shell governs the numerical opimization in an application independend manner. This entails ensuring a proper convergence, using a user specified method, changing to another method if this other method promises a better convergence and tuning the operational parameters. The core consists of a library of numerical routines. The inner shell and the core replace existing numerical optimization programs, but adding the mentioned new functions (self-tuning, etc.), which could previously not be offered to the user. The outer shell contains a user interface and an evaluation module, which is application dependend. The evaluation module can perform another way of optimizing the solution. Several constraints can be taken into account, that can not be represented mathematically, but can be represented in a logic manner. These types of constraints are called soft-constraints. The outer shell can reformulate the problem, taking into account the results of a previous opimization run. In this way an optimization on two levels can be obtained. Both shells are strictly separated expert systems, with each dedicated tasks.

Algorithms that can be chosen from for the numerical optimizations are collected in a library in the core. As control over the optimization is transfered from the user and the existing IF-THEN structures to the expert system layer, the routines in the core are called from the inner shell.

The inner shell governs the actual optimization process and is to ensure that the convergence is proper. The advantage of using an expert system closely monitoring the actual optimization is that the expert system can react immediately to events that occur during the optimization, where human experts could only change settings at the end of an opti-

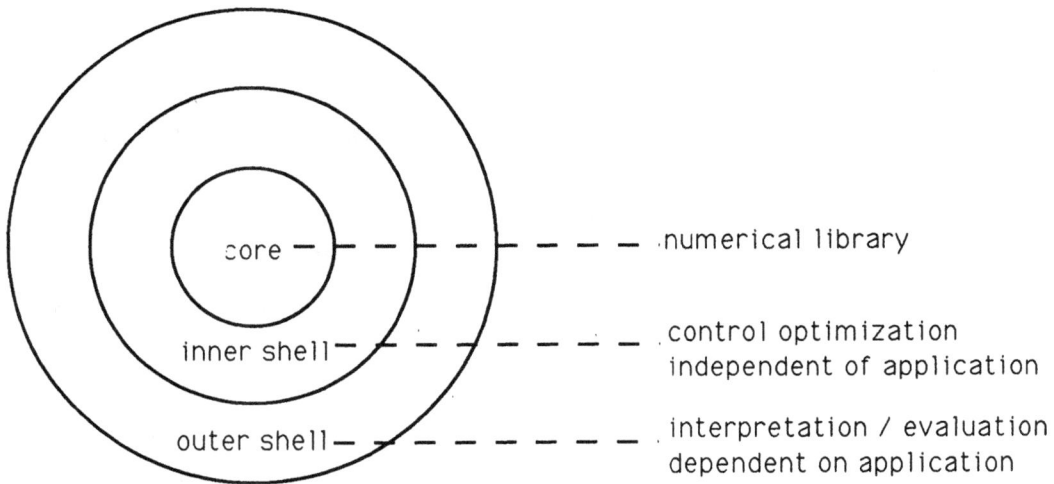

Figure 3: Schematic layout EGNOP

mization run. This leads to a faster interaction with the optimization process and avoids too many unnecessary function evaluations. As the control is stored in the inner shell rather than performed by a human expert, the control over the optimization is also more consistent. The inner shell can for instance change the starting points or the operational parameters. Also it can choose another algorithm or a different line minimisation technique. In this way numerical optimization will become more robust. The inner shell can control the method as it used to be done in the original algorithms, but with additional possibilities through the use of expert system technology.

The outer shell performs the evaluation and interpretation of the results of an optimization run. Based on this evaluation process, it can decide on whether an acceptable solution has been found or that another optimization run is needed to obtain a realistic, practical result. In that decision domain specific knowledge is being used and sometimes conflicting interests have to be dealt with. Within an expert system environment, such a problem can be treated efficiently. Mathematical information as well as heuristic information, based on the experience of experts, can be included in the decision process. Actions for the preparation of a new optimization run can be performed in the outer shell. The user can for example be asked for new values of the starting points or, based on previous optimizations, the system should be able to come up with a choice itself. The outer shell is in communication with the user and informs the user on the current status of the optimization runs and on the decisions that were taken. When required, the user can modify the decisions or suggestions of the system. However, such an interference takes time and is unfavourable for a continous process. For some applications, like on-line process control systems it is required that the system can operate without any interference from users. Therefore, the system must be

509

able to work autonomously.

Concluding, the problems as formulated earlier can be solved in the following manner:

- much knowledge required for using numerical optimization is captured in the inner shell,

- convergence of the optimization run is closely monitored by the inner shell and immediate actions can be taken to improve the convergence,

- from the core, the inner shell can choose the best method for the specific application at any stage in the optimization process,

- the outer shell can be developed to realise a communication between the user, the analysis routines and the numerical optimization programs,

- the outer shell can perform an initial evaluation of the obtained results for the user and may indicate a new optimization run to be performed.

The main advantages of the new concept are:

- almost no interaction between the user and the optimization process is required,

- the performance of the optimization process is increased, because it is closely monitored and changes in the control are made immediately by the system,

- general applicability of the system, because of its highly modular structure including the application independend inner shell and core.

EGNOP combines the numerical power of conventional programming languages and the possibilities of expert system techniques to implement reasoning mechanisms. The result is a very powerfull optimization environment, which tunes itself, switches to other methods, can adjust operational parameters if needed and can perform an evaluation of the obtained results. In this way, the user is relieved from having to be a true expert on numerical optimization, but it also allows for continuous running stand-alone applications, such as controlling aircraft.

Implementation aspects

For using expert system techniques flexibly in numerical optimization a balance has to be found between conventional programming languages and expert system programming languages. In the first approach these languages are C and Fortran 77 for the numerical side of the system and OPS5 (inner shell) and Nexpert Object (outer shell) for the expert system side. In numerical optimization, a solution can be found in an interative manner. Depending on the status of the constraints, at the end of the iteration process a solution

510

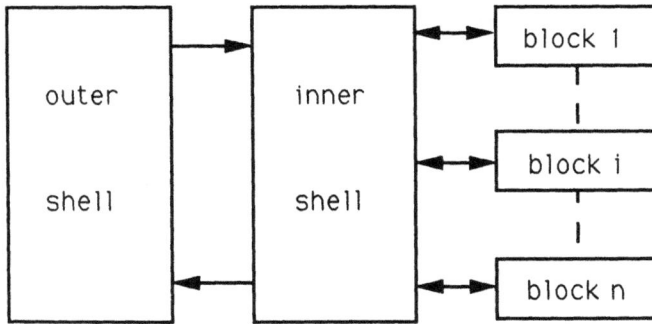

Figure 4: Schematic layout implementation

has been found or a problem has occured during the optimization.

The proposed implementation for expert systems for numerical optimization consists of splitting the original routines in logical and numerical blocks. The numerical blocks are implemented in C and the logical blocks in an expert system programming language. For the first implementation on MicroVAX, OPS5 was chosen as programming language for the inner layer of EGNOP. The result is shown in figure 4. It is more flexibly extendable with other optimisation methods, line minimisation techniques, etc, but it asks more from the programmer to implement it.

The basic structure of COPWAL [de Wit 1983] can be used as starting point for the main structure. COPWAL allows for constrained optimization, using the method of Broyden as function minimization technique and augmented Lagrange functions to take into account the constraints.

In the first implementation of the new approach, three optimisation methods are included; the method of Hooke and Jeeves (modified by de Wit), the conjugate gradient method and the method of Broyden. Two line minimisation techniques are available for the conjugate gradient method and the method of Broyden; the second order line minimisation and the Golden Section method. The COPWAL concept has been modified. If an optimisation method fails, an other will be chosen. Whenever the progress is unsufficient, an other method will be chosen as well. A selection mechanism takes care of these changes in optimisation method. The user does not have to interfere. Also within an optimisation method several changes can be imposed by the system, like selecting an other line minimisation technique, changing operational parameters, etc. The checks on the overall progress of the optimisation is done within the general scheme of the inner layer, as it is the same for all methods. So, at a predetermined amount of function evaluations, a jump will be made to the general scheme to evaluate the progress and take actions if necessary.

511

Other moments when a jump from the optimisation method to the general scheme will be made, is when the maximum amount of evaluations is exceeded (after which the program stops) or when an update failure in the search vector occurs. In this case a reset to a gradient method will be done. But, after twice such an error in the same method, the method will be left in favour of the method of Hooke and Jeeves, in order to get to a better region of the solution space. It has to be remarked however, that such update failures are not always indications of a bad convergence. The studies in [Ramon et al. 1989] showed that sometimes such errors appeared when arriving right on the optimum of the problem. A switch to the method of Hooke and Jeeves will in such cases result in a quick termination of the optimisation process.

The inner shell fulfills together with the core the task now performed by existing optimization programs, added with the self-tuning functionality. The biggest problem in setting up the inner shell can be found in creating the selection mechanism for new methods or line minimisation techniques and the adjustment of operational parameters. The selection mechanisms present in the first implementation are still open for refining. Based on a number of experiments in a variety of domains, the mechanisms will be modified in the near future. The foundations for the mechanisms remain however in tact. The knowledge for these foundations comes from literature and from own experience in the domain. It should be noted that in the literature much emphasis is put on memory allocation in the computer for selecting a certain method. With the progress in computer science over the last 15 years, however, this aspect becomes less and less important. The convergence aspects are therefore the most important issues to take into account.

The main concern in the implementation has been that the system should be as flexible as possible. This flexibility was aimed for with respect to the user as well as to future expansions of the system. The latter is not as trivial as it may seem, especially with respect to the numerical routines. During execution of the numerical routines, checks should be made at several stages, and operational parameters are adjusted. It should also be possible to leave one optimization method for another during execution, when convergence problems occur. This flexibility has been realised by splitting up the original algorithms into blocks for the numerical operations and into blocks which contain heuristics and logical operations as was discussed before. The numerical blocks are placed in the core, while the logical blocks are incorporated in the inner shell. In this way, the numerical blocks can be called individually from the inner shell. Similar operations in different algorithms can make use of the same blocks. For different operations specific blocks can be called. When temporarily another optimization method is preferred, the expert system can call a different block, in stead of an entire different algorithm. This allows changes in algorithms as well as simple changes in operational parameters. It also allows the system to be expanded gradually in later stages by adding new blocks in the core and in the inner shell. In EGNOP, adding a new method to the system, only asks to add some rules that enable the activation of this new method. Existing methods already present in the system do not have to be changed.

A direct consequence of the EGNOP approach can be illustrated for the COPWAL algorithm. Whenever a first attempt with the method of Broyden within COPWAL fails, or did not result in a suitable answer, then the optimization run is terminated. The user has to make a new attempt with different starting points, changed operational parameters, etc. In the current approach however, the inner shell can detect this failure and can respond to it immediately by taking actions as mentioned above or for example choosing an alternative method. When an optimum has been found with the alternative method, a new attempt with Broydens method can be performed. Alternatively, the process can be directed to the phase directly after the method of Broyden. The means of detection and reaction of the expert system immediately offers much extra performance and less interaction with the user. Also within the method of Broyden the inner shell can infere with the ongoing optimization, by means of changing line minimisation methods etc.

Example

As example to illustrate the working of the EGNOP concept, an optimal height problem for an aircraft will be solved. It is requested to minimize the fuel consumtion of the aircraft per unit of distance. The aircraft is in horizontal, symmetric flight, where an equilibrium exists of the forces acting on the aircraft:

$$t_h cos(\alpha + \epsilon) - D = 0 \qquad horizontal\ forces\ equilibrium$$

$$t_h sin(\alpha + \epsilon) + L - mg = 0 \qquad vertical\ forces\ equilibrium$$

In these equations, t_h is the thrust, α is the angle of attack, ϵ is a constant, L is the lift of te aircraft, D is the drag of the aircraft and mg is the weight of the aircraft.

The design variables are:

$$x_1 = \sqrt{M} \qquad Mach\ number$$
$$x_2 = \sqrt{\alpha}/0.3 \qquad angle\ of\ attack$$
$$x_3 = t_h/23.500 \qquad thrust$$
$$x_4 = h/50.000 \qquad height$$

The question is to determine an optimal combination of the Mach number, the angle of attack, thrust setting and the height where the fuel consumption is minimal. This fuel consumption is defined as a weighed quotient of the thrust and the Mach number. In the optimization 2 equality constraints (the force equilibria) and 7 inequality constraints are taken into account. The inequality constraints define the allowable solution space. The system description itself is a highly nonlinear combination of the design variables.

Two optimization runs will be discussed, illustrating the working of EGNOP. Thereby special attention will be given to the selection mechanism for the methods. In the following

513

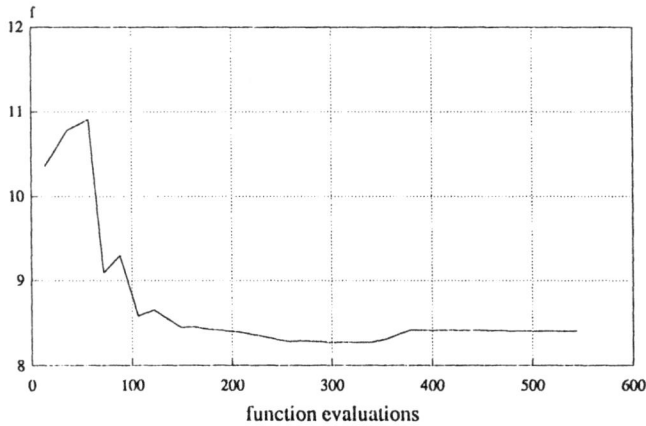

Figure 5: Evolution cost function first experiment

tests the optimization will first be started using the method of Broyden and next with the method of Hooke and Jeeves.

The first experiment starts by selecting the method of Broyden, which is a quasi-second order method. The current problem is well suited for a quasi-Newton approach. It appears that after starting the optimization with the method of Broyden will be used through the entire optimization. The optimum has been found after 544 function evaluations. The first fase of the optimization (loss function) has been finished after 326 function evaluations, initiating the Lagrange optimization phase. In the figures 5 and 6, the changes in the cost function and the object function can be seen. During this experiment, the reset factor for the Hessian matrix F_{xx}^{-1} has not been a major issue in the optimization. The number of evaluations after which a reset was performed remained more or less constant throughout the optimization. The final results are (final cost function f=8.40638):

$$
\begin{aligned}
x_1 &= 0.907070 & M &= 0.822777 \\
x_2 &= 0.390495 & \alpha &= 2.619917 \ degrees \\
x_3 &= 0.167039 & t_h &= 3925.42 \ lb \\
x4 &= 0.209405 & h &= 10470.23 \ m
\end{aligned}
$$

In the second experiment the method of Hooke and Jeeves, a 0th order method, was selected as start choice. After 291 function evaluations, the system decided that the obtained progress was not sufficient and the method of Broyden was selected. The optimal solution was reached after 816 function evaluations, which is clearly less good than the previous result. The interference of the selection mechanism was necessary. Remarkable is that

514

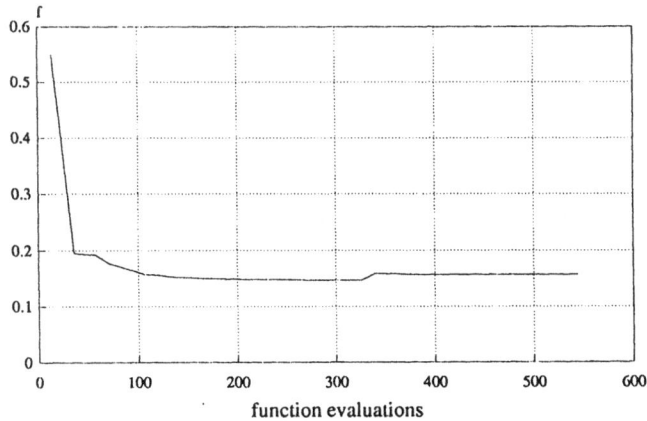

Figure 6: Evolution object function first experiment

the first phase of the optimization (loss function) was already completed after 94 function evaluations. The method of Hooke and Jeeves did not really find a good result in this phase. In figure 8, it can clearly be seen that in the first part of the optimization the cost function changes a bit chaotically and only after the method of Broyden was selected a good convergence was obtained. The result is a cost function of f=8.40707:

$$
\begin{aligned}
x_1 &= 0.907061 & M &= 0.822761 \\
x_2 &= 0.390423 & \alpha &= 2.618947 \; degrees \\
x_3 &= 0.167049 & t_h &= 3925.66 \; lb \\
x4 &= 0.209118 & h &= 10455.91 \; m
\end{aligned}
$$

Based on these and other experiments that were conducted during the development of EGNOP, a number of observations can be made.

- Reset factors, regularly resetting the search vector of the conjugated gradient method and the method of Broyden, can have a positive effect on the convergence. The best moment of the resetting, however, depends on the progress that the method has during the ongoing optimization. The mechanism implemented in EGNOP takes these factors in account and offers a flexible, self-adjusting resetting procedure.

- The selection mechanism for the optimization methods and for the line minimization techniques acts properly. This can be verified from the experiments. At difficult moments in the optimization, both line minimization methods are used after one another to obtain a new workable point in the optimization.

515

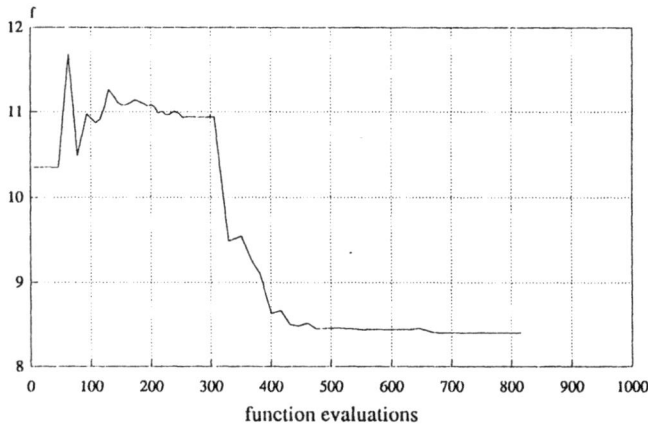

Figure 7: Evolution cost function second experiment

- From the highly modular set-up of the program, a powerfull evaluation environment for new optimization techniques has been created. Using the current construction of expert system techniques and numerical routines, changes can be implemented easily and the effect of those changes can be verified step by step, by using the expert system environment.

Conclusions

The impact of EGNOP as new approach for numerical optimization techniques has been discussed.

The aim of EGNOP is to make numerical optimization accessible for a large group of users. Thereby the user is relieved from the need to be a true expert on numerical methods, because an expert system structure takes over parts of his tasks. For specific applications, an outer shell can be developed for the evaluation phase, which allows also a good use of the system by people who have less knowledge on the problem domain.

EGNOP offers a new view on numerical optimization techniques. On the issue of ensuring proper convergence, good progress has been achieved. The main results can be summarized as follows:

- The optimization methods themselves have been looked at and the in conventional programming languages added decision rules have been taken out. These rules often did not result in the desired effect, which left the user in charge to ensure a proper

516

convergence. Within EGNOP, the function of those rules has been taken over by an expert system structure. Expert System techniques are better suited for implementing reasoning mechanisms. Through this approach, the methods have become more robust.

- At a higher level, a decision structure has been constructed, which allows a choice between several optimization methods, depending on the phase and of the progress of the ongoing optimization.

- As side-effect of this research a flexible 'test-bed' has been created, which allows easier testing of new optimization techniques. Because of the now existing expert system structure, new techniques can be implemented and tested fast and efficiently.

Future research will focus on the following main topics:

- The decision mechanisms for the optimization techniques and line minimisation techniques have to be evaluated, based on extensive testing in several application domains. The rules can be adjusted based on those results. One of the aspects where an refining can bring good new results is the resetting of the search vector in the method of Broyden and the conjugate gradient method.

- Extending the decision structure, however, also slows down the performance of the system. Expert system environments that generate C-code may bring a solution here. This C-code may be optimised itself afterwards. However, existing OPS5 implementations that generate C-code do not support floating point numbers, which for optimization purposes are needed.

- For design activities the solution space does not always consist of a continuous space. Sometimes the answers have to be at discrete values. Discrete numerical optimization is a separate domain. Based on the experiences gained in the development of EGNOP, a good new approach for discrete optimization can be made and is currently being developed.

References

Boullart, L. and R.A. Vingerhoeds (1989). A Rule-Based Programming Language Example. In: *Artificial Intelligence Handbook* (A.E. Nisenfeld and J.R. Davis, eds.), Instrument Society of America, 1989.

Lootsma, F.A. (1984a). Algorithms for unconstrained optimization, lecture notes a85A, Mathematical Department, *Delft University of Technology*.

Lootsma, F.A. (1984b). Algorithms for constrained optimization, lecture notes a85B, Mathematical Department, *Delft University of Technology*.

Noldus, E. (1990). Optimal Control, (in Dutch), lecture notes, Automatic Control Laboratory, *University of Ghent*.

Ramon, H., R.A. Vingerhoeds, B.D. Netten, and E. Noldus (1989). A practical application of a numerical optimization technique in feedmill industry. In: *Proc: Mathematical Programming, a tool for Engineers*, Mons, In: *Engineering Optimization*, 18, 107-120 (1991).

Vingerhoeds, R.A., B.D. Netten, H. Ramon, and L. Boullart (1990). Expert Governed Numerical Optimization, In: *Proc: European Simulation Multiconference*, June 10-13, Nuremberg.

Vingerhoeds, R.A., B.D. Netten, and L. Boullart (1990). On the use of expert systems with numerical optimization, In: *Proc MIM-S2'90*, September 3-7, Brussel.

Vingerhoeds, R.A. (1992). Expert Governed Numerical OPtimisation (EGNOP), a new view on optimisation techniques, (in Dutch), PhD-thesis, *University of Ghent*, submitted.

Wit, C. de (1983). COPWAL, Algol procedure for Constrained Optimization by means of Augemented Lagrange functions, report RC-TWA-83001, *Delft University of Technology*.

Author Index

521

Subject Index

527

528

www.ingramcontent.com/pod-product-compliance
Lightning Source LLC
Chambersburg PA
CBHW060952210326
41598CB00031B/4805